# Ceramic Coatings - Applications in Engineering

# Ceramic Coatings - Applications in Engineering

*Editor*

**Hitesh Dave**

# Ceramic Coatings - Applications in Engineering
Edited by **Hitesh Dave**

ISBN: 978-1-68117-202-6
Library of Congress Control Number: 2016934749

© 2017 by
SCITUS Academics LLC,
www.scitusacademics.com
Box No. 4766, 616 Corporate Way,
Suite 2, Valley Cottage,
NY 10989

**Notice**

Reasonable efforts have been made to publish reliable data and views articulated in the chapters are those of the individual contributors, and not necessarily those of the editors or publishers. Editors or publishers are not responsible for the accuracy of the information in the published chapters or consequences of their use. The publisher believes no responsibility for any damage or grievance to the persons or property arising out of the use of any materials, instructions, methods or thoughts in the book. The editors and the publisher have attempted to trace the copyright holders of all material reproduced in this publication and apologize to copyright holders if permission has not been obtained. If any copyright holder has not been acknowledged, please write to us so we may rectify.

# Preface

Ceramic materials are inorganic, non-metallic materials that are processed and used at high temperatures. They are highly resistant to corrosive compounds. Ceramic materials are harder; more resist to heat and frictions lasting longer than other materials which occurred by machining. These main properties make ceramic materials used unique or together with other materials. To make it usable with other materials, ceramic materials are generally coated on. These coatings may be thick and thin depending on the functional application. There are a wide range of ceramic coatings that can be applied to metal components in order to enhance their functional properties. Ceramic coatings can be categorised in terms of thickness. Thick coatings can be deposited in numerous ways but the most common are thermal spraying and enamelling, such as thermally sprayed alumina or tungsten carbide, or the enamel coatings on white wares. Ceramic coatings are often used as barrier materials to enhance the interaction between moving metal parts, such as in the automotive industry. However, they are also increasingly being employed to augment certain manufacturing processes, and exhibit potential for improving the efficiency of some fabricating methods. Ceramic coatings are sturdy and have a high level of lubricity, but due to oxidation concerns, they are typically used in temperatures under 1,200 degrees (F). However, this allows them to be applied to hot forging dies, which operate at lower temperatures. Ceramic coating increases the operational lifespan for these dies, allowing them to produce a greater number of parts before wearing down. This book entitled Ceramic Coatings - Applications in Engineeringis intended to state the latest advancement in ceramic coatings technology in various industrial fields. The book covers topics associated to the applications of ceramic coating in engineering.

# Table of Contents

# CHAPTER 1
# Coatings for Superalloys

*Mathias C. Galetz[1]*

[1] DECHEMA-Forschungsinstitut, High Temperature Materials, Frankfurt am Main, Germany

## ABSTRACT

High-temperature coatings for superalloys can be divided into three categories: Two of them, diffusion and overlay coatings, are both used to protect a system from oxidation and corrosion. The third type, thermal barrier coatings, protects the substrate from thermal degradation.All types of coatings are designed to avoid direct contact between a hot gas or detrimental components from the environment and the base metal alloy. The aim is to enhance the lifetime of high temperature applications, which is otherwise often limited by corrosion or thermal degradation. In this chapter, the different coatings and their durability are discussed, which ideally matches the aging process of the substrate underneath. Besides the corrosive attack the coatings must also often be able to withstand certain mechanical and thermal mechanical loads as well as erosion. Nevertheless, all coatings have a limited thickness and get consumed in service. Therefore it is essential to understand and choose coatings that show slow degradation mechanisms.

**Keywords:** Diffusion coatings, MCrAlYs, TBC

## INTRODUCTION

Superalloys are operated in industrial environments containing corrosive species such as sulfur, chlorine- or carbon-containing compounds, water vapor, alkali and alkaline-earth metal salts, or ashes such as vanadates [1,2]. At elevated temperatures, such compounds cause a wide range of attack types on most metallic alloys such as oxidation, carburization, sulfidation, hot corrosion, or a combination of different mechanisms [3,4]. But even in air the oxidation resistance of nickel- and cobalt-based superalloys is not sufficient for continuous operation above 1000°C [5]. Oxide scales grow too fast and the subsurface zone of the alloys is changed and loses its mechanical strength [6-8]. Despite these limitations, advances and improvements of industrial technology have led to more efficient processes and more powerful engines with increased operation temperatures, and the alloys used for their construction are pushed toward the applicable limit even as far as mechanical properties are concerned, such as fatigue, tensile strength, and creep resistance [9]. Therefore higher additions of alloying elements for corrosion protection such as Cr, Al, or Si cannot be used as they either lead to embrittlement or lower the melting point and therefore the creep strength at the high target temperatures [10,11]. The only way is to apply high-temperature coatings to face aggressive corrosive high-temperature atmospheres and make processes possible which could not be operated efficiently and reliably without such coatings.

Diffusion- and metal-based overlay coatings are both used to protect a system from oxidation and corrosion [12]. Such coatings are designed to avoid direct contact between the base metal alloy and a hot gas carrying detrimental species by growing a thin, self-healing oxide scale in situ.

Besides oxidation and corrosion protection, the other major factor affecting the overall life of substrates at high temperatures are heat-flux effects. Nickel- and cobalt-based alloys are employed in gas temperatures up to 1400°C [13]. Without proper protection, the temperatures could easily induce softening or even reach the melting point of the alloys. The employed materials are the actual major constraints that determine the maximum gas temperatures in many processes of today's aircraft as well as energy conversion industry. Without cooling and protection by coatings, the efficiency and speed of jet engines would also be limited to a very low level. In these cases, only a combination of two coatings allows operation; a so-called bond coat (the bond coat belongs to the class of diffusion or overlay coatings, which are discussed in detail below) in combination with a ceramic layer on top to form a thermal barrier coating system. The ceramic layer also reduces the attack indirectly by lowering the metal surface temperature and thus decelerates the degradation of the metallic coating and substrate underneath.

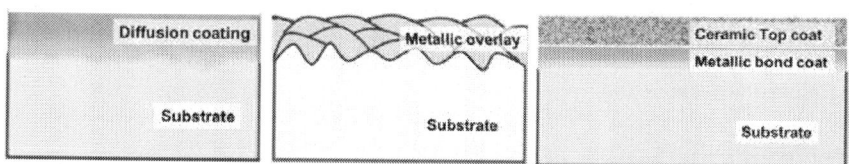

**Figure 1.** Schematic sketches of the general configuration of the three different types of coatings used for high-temperature applications (from left to right): diffusion coatings, metal overlay coatings, and thermal barrier coatings

Only by combining one of the different types of coatings schematically shown in Figure 1 with a load-carrying superalloy underneath, safe operation over a reasonable lifetime can be ensured. In operation the coatings must be able to sustain certain mechanical and thermo-mechanical loads as well as erosion on top of the oxidative or corrosive attack. Ideally, the durability or lifetime of protective coatings for parts such as turbine blades or vanes matches that of the aging processes of the substrate underneath. However, since all coatings are more or less thin films, the mechanical lifetime of the substrate is often longer than the period of coating degradation at high temperature. Therefore, after a certain time in service, the coatings residues are stripped and the coatings are remanufactured several times during the lifetime of the blades [14].

# REQUIREMENTS FOR PROTECTIVE COATINGS AT HIGH TEMPERATURES

Unlike aqueous corrosion conditions at low temperatures, where usually either an anodic protection is applied or even very thin barrier coatings can protect from degradation by separating an electrolyte from the material, high-temperature degradation processes are controlled by transport processes in or through coatings via diffusion. High temperatures are especially demanding because all metals tend to become thermodynamically stable in their oxidized form or tend to react with other gas components such as nitrogen, carbon, or sulfur. For example, even highly corrosion-resistant platinum suffers from significant oxidative degradation above 1000°C in air [15]. Which corrosion products form in an environment depends on the partial pressure or activity of different potential reaction partners in the atmosphere. The idea behind all oxidation- and corrosion-resistant coatings is to create a reservoir of scale-forming elements from which a thermodynamically stable, slow growing, adherent scale can be formed, consisting of corrosion products, usually thermally grown oxide on the surface. An overview of the requirements for coatings for high temperatures is provided in Table 1.

**Table 1.** Requirements for high-temperature coatings

- Slow-growing, thermodynamically stable scale formation and high concentration of the scale-forming elements
- Thermal stability, no detrimental phase changes
- Erosion resistance
- Good adhesion
- Low interdiffusion of the substrate
- Mechanical compatibility: similar modulus and thermal expansion to avoid stresses at the interface
- No crack initiation, not too brittle
- Good processability
- Low price

Even with stable protective oxide scales, the service conditions will usually create flaws such as cracks in the scale. Most important is the self-healing potential or ability for reformation, properties which for example a pure ceramic coating cannot provide. In Figure 2 a pure ceramic coating is compared with a thermally grown oxide scale that builds up in situ. In the first case, if a crack occurs, the atmosphere can attack the metal below, which cannot protect itself. Ceramics also always possess a certain porosity and cannot prevent the gas from reaching the metal below, which means they offer no effective long-term corrosion protection. Ceramic coatings are therefore almost exclusively used to lower substrate temperatures and in combination with one additional coating of the diffusion or metallic overlay type which can form slow-growing oxides. Diffusion or overlay coatings possess a sufficient reservoir of protective elements underneath the scale to allow the reformation of the protective scale many times before the reservoir of the protective elements is depleted.

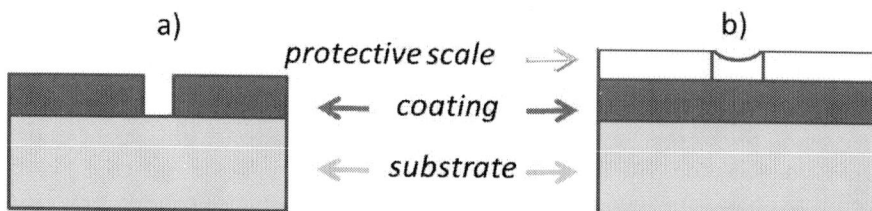

**Figure 2.**Comparison between a ceramic coating without self-healing properties (a) and a reservoir coating that allows the reestablishing of the protective scale in case of local failure (b)

The potentially protective elements, which qualify for scale formation, must form thermodynamically more stable corrosion products than the main elements in the superalloy such as nickel or cobalt. The graph in Figure 3 is a simplified Ellingham-Richardson Diagram that shows the standard free energy of oxide formation as a function of the temperature. Generally, for scale formation on nickel- or cobalt-based alloys, elements with a very high affinity to oxygen such as aluminum, chromium, or silicon are important, but also titanium or tantalum would theoretically qualify. The drawback for example in the case of titanium is that it allows fast oxygen diffusion through its oxide scales, which have high growth rates even at moderate temperatures. Additionally, they often tend to spall off easily.

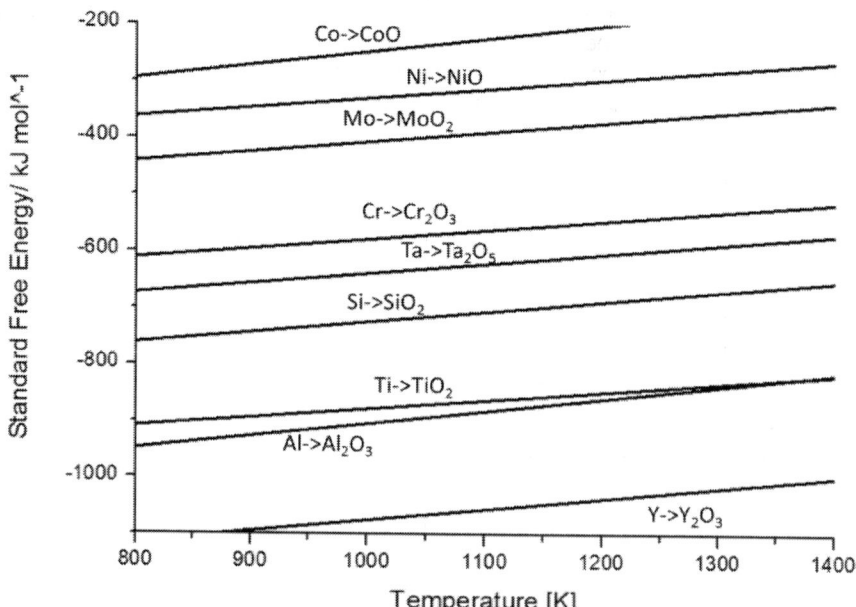

**Figure 3.**Simplified Ellingham-Richardson Diagram

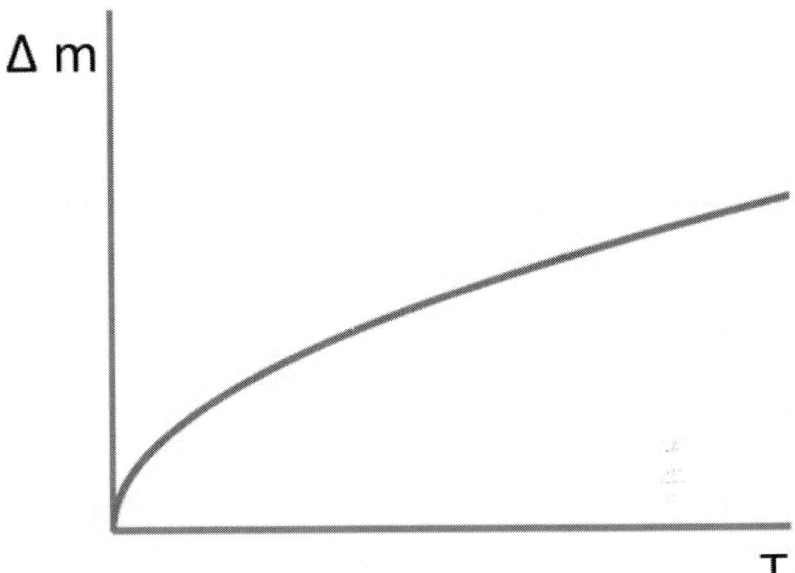

**Figure 4.** Parabolic weight gain of ideal scale growth

So not only the criterion of thermodynamically preferred reaction has to be fulfilled, the scales also have to be slow-growing. Only scales that form a dense and stable crystal lattice effectively limit diffusion and guarantee a slow growth and weight gain as shown in Figure 4. From the curve, the growth rate constant $k_p$ of the thermally growing oxide can be determined according to the function:

$$(\Delta m)^n = k_p t$$

where $\Delta m$ is the weight gain per surface area of the metal [g/cm$^2$], t is the exposure time [s], and n is the rate exponent. If the scale is protective, stable, and adherent, the value of n is usually close to 2 and the weight gain shows a parabolic diffusion-controlled behavior. The diffusion is generated by the anion and cation concentration gradient in the scale. The speed of diffusion is determined by the crystallographic structure of the oxide, the defect structure, and the microstructure, for example grain size and grain boundary distribution. In most cases, the faster diffusing species is a metallic cation, but anion diffusion is also possible [16], for example it is predominant in silica formers in a wide range of temperatures. Figure 5 shows the parabolic rate constants at 1000°C normalized to the behavior of slow-growing alumina for nickel, chromium, and silicon. It can be seen that Cr, Al, and Si offer much slower growth rates than the base metal nickel. Cobalt is not included in the diagram since its oxidation rate is even ten times higher than that of nickel.

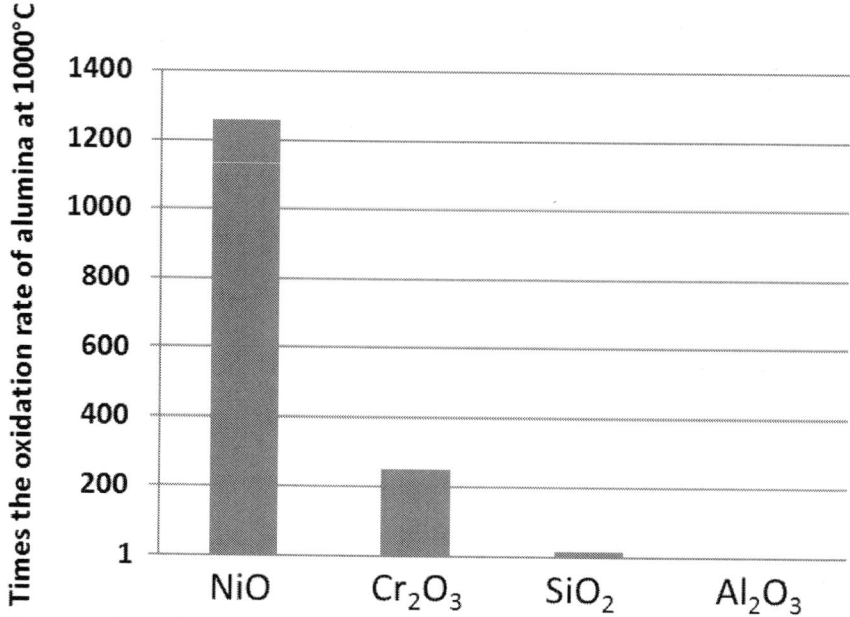

**Figure 5.** Parabolic rate constants of Fe, Co, Ni, Cr, Si, and Al oxide formers at 1000°C; data from [17] as $k_p(g^2cm^{-4}s^{-2})$

Further requirements derive from the mechanical interaction of the scale with the coating underneath. During formation growth stresses occur due to the volume change, when a scale is formed by including elements from the environment such as oxygen. As a first hint on the stresses, the classical approach of Pilling and Bedworth from 1923 can be used, according to which the Pilling-Bedworth Ratio is defined as:

$$PBR_{metal} = \frac{Volume\ of\ oxide}{Volume\ of\ metal}$$

It must be higher than 1 in order to guarantee a continuous scale formation, as ceramic scales are highly prone to failure by tensile stresses. On the other hand, it must not be too high either, because when high compressive stresses occur they also induce failure by cracking and buckling of the scales. In Table 2 the values of different oxides on pure metals are shown.

**Table 2.** Pilling-Bedworth Ratio of different oxides on pure metals [18]

| Oxide | MgO | $Al_2O_3$ | $ZrO_2$ | NiO | $TiO_2$ | CoO | $Cr_2O_3$ | $Ta_2O_5$ | $Nb_2O_5$ |
|-------|-----|-----------|---------|-----|---------|-----|-----------|-----------|-----------|
| PBR | 0.81 | 1.28 | 1.56 | 1.65 | 1.75 | 1.86 | 2.07 | 2.5 | 2.68 |

It should be noted that the PBRs of alloys are not the same as those of pure metals, so for example the PBR of an alumina scale growing on intermetallics from the Ni-Al system is significantly different to the PBR of pure Al (Table 3). So the alloy system on which an oxide grows at high temperature must be considered as well.

**Table 3.**Pilling-Bedworth Ratio of $\alpha$-$Al_2O_3$ on different intermetallic aluminides [19]

| $\alpha$-Al2O3 on | $Ni_3Al$ | NiAl | $NiAl_3$ |
|---|---|---|---|
| PBR | 1.71-1.88 | 1.64-1.78 | 1.48-1.57 |

After formation of the scales, the occurring stresses are dominated by the mismatch in the thermal expansion coefficients between the ceramic scale, the coating, and the alloy beneath.

When the mismatch is too high, for example with silica scales on nickel- or cobalt-based alloys, extensive cracking occurs during cyclic exposure, providing less protection than alumina and chromia, oxides which have a higher thermal expansion coefficient closer to that of the alloys underneath [20,21]. Still, even alumina or chromia scales suffer some cracking in service, but coatings for high temperatures are designed to allow self-healing of the scales.

As already mentioned, such high-temperature coating systems can be classified as diffusion coatings or overlay coatings. Ceramic thermal barrier coatings also fall into the category of overlay coatings, but these will be covered separately.

# DIFFUSION COATINGS

Diffusion coatings are based on the enrichment of protective metallic elements close to the surface by diffusion from a reservoir outside. As for all diffusion processes, the parameters time, temperature, and phase composition are critical. The process conditions have to be optimized for each material because the phase composition and the microstructural evolution of diffusion coatings depend highly on the substrate which is coated. Figure 6 illustrates the general idea of diffusion coatings. A reservoir with a very high concentration of the element that should be enriched is created at the surface either in the gas phase or in a solid/liquid state. Due to the concentration gradient at high temperatures, elements start to diffuse into the substrate, and elements of the alloy have a tendency to diffuse outward. Depending on which process is faster, the inward or outward diffusion process is more pronounced. In both cases these processes eventually achieve the aim of changing the composition close to the surface by enrichment with the elements from the diffusion reservoir.

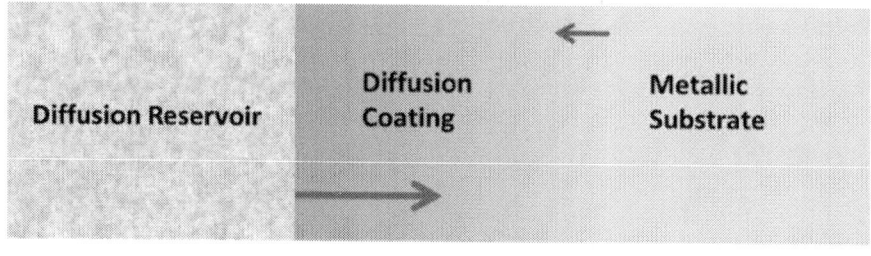

$$c_1 \qquad\qquad\qquad c_o$$

**Figure 6.** Principle of diffusion coatings manufacturing

Several different manufacturing methods have been developed for diffusion coatings. Figure 7 gives an overview of the most important procedures. These include pack cementation, which can be done in-pack or out-of-pack (also called above the pack, which resembles the chemical vapor deposition (CVD) processes). Other processing techniques include slurry-reaction coatings, galvanic and metal foil coatings, or a hot-dipping process in a metal melt. They all have in common that a reservoir of the diffusing metal is first applied to the surface and later diffused via heat treatment.

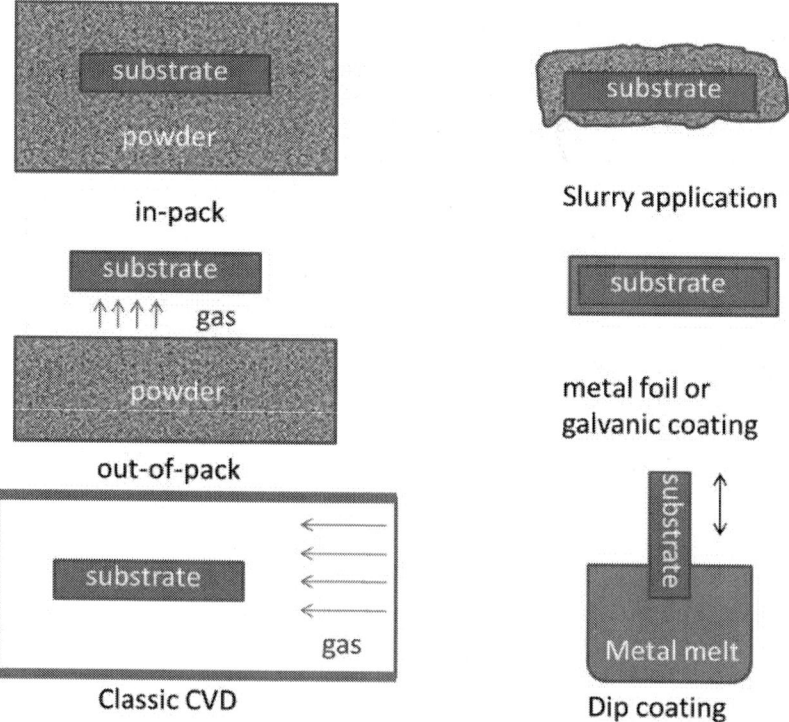

**Figure 7.** Different methods to produce diffusion coatings

## 3.1. Pack Cementation Processes

As early as in 1914 the first "pack cementation" process was reported by Allison and Hawkins [22]. They diffused aluminum into iron and steel, but the process did not receive much attention up to the 1950s. Since then, pack cementation with chromium has been widely used to enhance the corrosion behavior of low-alloyed steels [23]. In the 1970s, the aluminizing process for nickel-based alloys became popular, especially in the aircraft industry. Today more than 80% of stage 1 airfoils are coated with aluminum diffusion coatings [12], most of them by the pack cementation process. It falls into the category of chemical vapor processes and is carried out at high temperatures (600-1200°C) in an inert gas atmosphere, in vacuum, or under reducing conditions. The the powder pack is composed of an inert filler (usually $Al_2O_3$), a halogen-carrying compound as an activator, and a metallic powder of the elements to be enriched. In the 1990s, with a better theoretical understanding of the processes, co-deposition of more than one element at a time became possible and was first developed for Al in combination with Cr [24-28] to improve the corrosion behavior when sulfur is part of the hot gas. Recently, investigations have been focusing on the modification of aluminide scales by co-diffusion with reactive elements such as Y, Ce, or Hf [29,30]. Small amounts of these elements are known to improve scale adhesion [31]. Platinum modification can also be beneficial under certain conditions since it facilitates the formation of stable alumina scales. Pt has to be galvanically applied and interdiffused before aluminum is enriched via pack cementation [32]. Silicon can also be added to the aluminide coating. Similar to chromium, it enhances the resistance against sulfur-carrying gases [33]. However, its application is usually limited to small amounts because silicon bears the risk of forming low-melting phases in nickel- or cobalt-based systems [34].

In industrial applications, the reaction is conducted in tightly sealed containers in which the components are placed. The parts can either be embedded in the reactive powder mixture ("in pack") or suspended "out-of-pack" (also called "above the pack"). In both cases either an inert or a reducing atmosphere is used. The entire container is heated to 100-200 °C to remove any residual moisture and oxygen. In the following step it is heated to the actual process temperature, which depends on the diffusing elements and on the substrates. For superalloys the temperature typically lies in the range of 900-1100°C. The holding time is dependent on the particular system and usually varies between 1 and 10 h. The optimized parameters must be evaluated for each system, usually with the help of thermodynamic calculation programs to investigate the temperature and activity of the different metal halogens which carry diffusion metals to the surface [35,36]. If the activities of the diffusing metal halides at the surface are too high, brittle phases occur frequently; if their activities are too low, too little diffusion takes place [37].

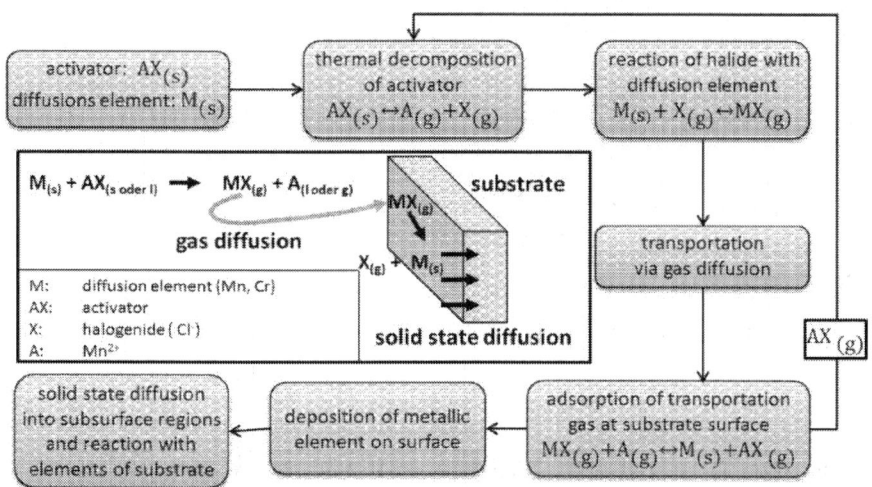

**Figure 8.** Major steps during the pack cementation process [38]

Figure 8 summarizes the major steps of the classical pack cementation process. Metal halides are formed and transported via the gas phase to the surface of the component to be coated, where the halide interacts with the sample surface and dissociates into the metal and the halogen anion. The metal atom diffuses into the substrate surface zone. The formation of the coating resembles that of diffusion couples, and the driving force for the interdiffusion is the activity or concentration gradient between the environment (which contains the diffusing elements such as Al) and the surface of the components [29]. After releasing the metal at the surface, the halogen reacts with new metal atoms from the powder to keep the process continuously alive. In any case, halogens (most common are chlorine and fluorine, but bromine or iodine can also be used) serve as an activator. The contents of activator and metallic powder determine the activity of metallic halogens and therefore the amount of metal which can react at the surface. The activator is added in the form of compounds such as $NH_3Cl$, $AlF_3$, or as HCl gas. The metallic powder, for example Al, can be used as pure element powder but also in an alloy form with nickel or chromium, for example CrAl powder, to lower its activity.

In any case, at the surface several intermetallic phases form according to their thermodynamic stability and local phase composition. Typical intermetallic phases that can form within the substrate metals by aluminization are $CoAl_3$, $Co_2Al_5$, or CoAl in cobalt-based alloys and $NiAl_3$, $Ni_2Al_3$, NiAl, or $Ni_3Al$ in nickel-based alloys. As mentioned earlier, the formation of the phases at the surface is controlled by the activity (the powder composition and activator), the temperature, and the duration of the process. Goward and Boone [39] classified diffusion coatings into low- and high-activity coatings, based on observations on nickel-based alloys. A "high activity" pack structure is usually observed in pack cementation processes at lower temperatures in combination with a high aluminum activity at the surface. In this case, Ni2Al3 is formed by inward diffusion of aluminum as the dominant mechanism and the diffusion of Ni is rather low. Additionally, in aluminum-rich phases the diffusion of aluminum is favored, even in aluminum-rich NiAl. With lower Al content, its mobility decreases as well. After a high-activity process, Ni2Al3 is present at the surface, while closer to the coating–substrate interface aluminides less rich in aluminum can be found. Ni2Al3 coatings usually require a second heat treatment to transform the rather brittle phase into the desired NiAl phase. Instead, if a lower activity is used in the pack in combination with a higher temperature, Ni diffusion is faster and NiAl forms directly. As a result, these coatings grow by outward diffusion and elements of the substrate that have little solubility in the NiAl phase such as refractory metals are enriched in the substrate close to the interface as part of the so-called interdiffusion zone. In Figure 9, a high-activity coating is compared with a low-activity coating on a nickel-based superalloy.

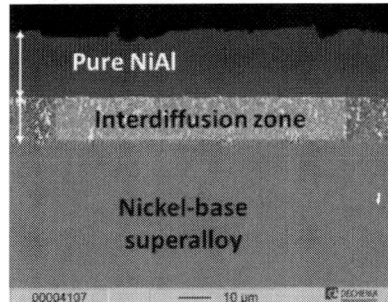 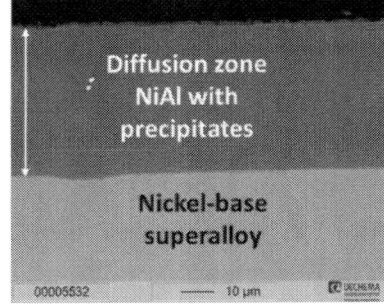

**Figure 9.**Comparison of a high-activity coating (left) with a low-activity coating (right) on a nickel-based superalloy substrate (PWA 1484)

Such coatings can be applied even inside of pipes. In this case, the tube is filled with the powder mixture and then sealed at the ends. The tube represents both the container and the part to be coated. This is an industrially established technique for huge components such as 15-m-long furnace tubes [40]. After the coating process, in case of "in pack," sometimes inert particles from the ceramic filler remain on the surface and must be removed. If the parts have small holes or channels, these can even be blocked by residues of the inert filler. To avoid such drawbacks, turbine blades and vanes are usually coated by using the "out of pack" technique. In all cases the huge advantage is that, since it is a gas phase process, complete coverage of the surface can be ensured.

## 1.2. Slurry Coatings

Compared with the pack cementation process, slurry coatings were developed much later in the 1970s and 1980s [39,41]. Although the processing and formation mechanisms are totally different, they offer similar microstructure and features as coatings that were applied by pack cementation. In Figure 10, the diffusion layers obtained with these two different coating methods are shown for a nickel-based alloy (CM247). The coating on the left was produced by the pack cementation described in detail above. The other coating was achieved via the slurry aluminizing route. After 1000 h of exposure in air at 1050°C, diffusion layers of both coatings look very similar and are still protective.

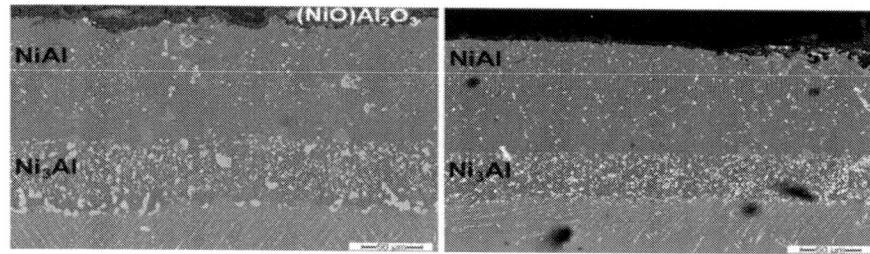

**Figure 10.**Diffusion coatings on a CM 247 nickel-based alloy after exposure in air at 1000°C for 1050 h. The left coating was applied by the slurry process (slurry composition: 40% Al, polyvinyl alcohol, water), the coating on the right via pack cementation (powder mixture: 5% Al, 1% NH4Cl, rest Al2O3, 1000°C, 1 h in Ar/5% H2)

Thus, the slurry route represents an interesting alternative for many systems or components, especially since slurry systems can be applied on components via common immersion, painting, or spraying methods, which is a big advantage for large parts. The slurries usually contain a metal powder, sometimes an activator, and either an organic, water-based binder or chromium-phosphate acidic binder [42,43]. Water-based activator-free systems are preferred nowadays, as the use of chromium-phosphates or halogen activators is hazardous and toxic. When the coating is applied, it also has to be heat-treated. In a first step, organic binders are burnt-out in the temperature range of 300-450°C and subsequently the metal powder reacts and diffuses into the substrate at higher temperatures between 600 and 1100°C [44]. In contrast to the pack cementation process no gas phase is present. Instead, above the melting point metallic melt from the metal powder wets the substrate and reacts via a combustion synthesis (exothermic formation of intermetallics) mechanism, · which is extremely fast [44]. In the beginning of the process aluminum-rich phases are formed that usually require further heat-treatment so that the layers can be converted into the desired NiAl or CoAl phases (similar as for pack cementation low-activity coatings). Modification of coatings applied by this slurry technique with other elements such as chromium and silicon is also possible by alloying with aluminum particles used for the slurry or by mixing two metallic powders. Most recently, even low-activity coatings have been manufactured by the slurry technique in one step, as described in detail elsewhere [43].

Figure 7 shows an overview of application methods for diffusion coatings, including aluminum foil, galvanic or dip coatings, whose formation mechanisms can be compared to slurry coatings where a certain amount of aluminum is directly deposited on the surface and heat-treated. However, since it is much more difficult to apply homogeneous layers, such techniques are hardly used compared with the slurry technique.

# 4.    OVERLAY COATINGS

## 4.1. Mcraly-Coatings

In contrast to diffusion coatings, no elements of the substrate are incorporated in overlay coatings. Therefore, such coatings offer the possibility to apply totally different compositions compared to the base materials, and their properties can be perfectly optimized to fulfill the requirements listed in Table 1, such as being corrosion-resistant as well as mechanically and thermally compatible with the substrate. Although this flexibility allows a wide range of compositions, almost all systems for superalloys are based on a general MCrAlY composition and contain usually more than four elements with M = Co or Ni or a mixture of them plus aluminum, chromium, and an element from the group of the reactive elements such as Yttrium (MCrAlY is the common term for such types of coatings and Y stands for Yttrium) [12]. One requirement is an Al content of 10-12 wt%, which is less than in the diffusion coatings. However, a higher chromium content favors alumina formation and makes such systems reliable alumina formers. The still rather high amount of aluminum in MCrAlY coatings for superalloys forms intermetallic phases, but in this case only the β-NiAl or CoAl phase is present in the coatings. In contrast to the diffusion coatings, such phases are surrounded by a metallic nickel- or cobalt-based gamma solid solution matrix. The resulting two-phase microstructure β+γ (see Figure 11, left) increases the ductility of the coating over purely intermetallic coatings and thereby gives higher thermal fatigue resistance.

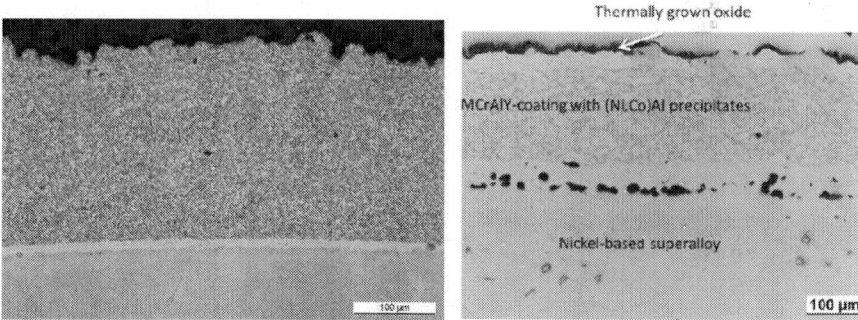

**Figure 11.** NiCoCrAlY coating on nickel-based superalloy; on the left a light microscopy picture after manufacturing after application [45], on the right a MCrAlY oxidized for 500 h at 1050°C in air is shown. Close to the surface the β-precipitates are dissolved and at the interface interdiffusion and interdiffusionpores (Kirkendall-pores) can be seen

One other advantage over diffusion coatings is the significant amount of Cr in the MCrAlY coatings, which enhances the corrosion and oxidation resistance. Regarding only oxidation resistance, NiCrAlY shows the best protection, but when sulfur is present in the gas, Co-based MCrAlY systems have a higher resistance. Additionally, NiCrAlY shows a phase transition above 1000°C.

Therefore, usually well-balanced NiCoCrAlY or CoNiCrAlY compositions are used in industrial applications, also depending on the type of superalloy. Besides the major elements, these systems are enhanced by the addition (<1%) of reactive elements (RE) such as yttrium, cerium, lanthanum, hafnium, or zirconium. The addition of rhenium was also shown to be beneficial [46,47].

When employed at high temperatures, the coatings degrade mainly by Al depletion due to both oxide formation and interdiffusion with the substrate. The beginning of the β dissolution is already visible close to the surface and at the interface to the superalloy after 500 h at 1050°C as shown in Figure 11, right. Similar to diffusion coatings, this aluminum depletion alters the microstructure, as visible close to the surface and at the interface to the substrate. When the β-phase is totally dissolved, the coating quickly loses its protective properties. Compared to diffusion coatings, the composition and performance is optimized, but the manufacturing of overlay coatings is much more expensive. Overlay coatings are produced by one of the following processes: Physical vapor deposition (PVD) process, thermal spraying process, or overlay welding (also called cladding). For MCrAlY overlay coatings, mainly thermal spray processes or rarely PVD processes are used.

Thermal spraying processes are a group of surface coating processes in which a spray material is partly or fully melted inside of a spray gun (electrical arc discharge is usually used as the source of energy) and accelerated toward the surface of the component to be coated in the form of micrometer-size particles. The resulting coatings are formed by the accumulation of numerous sprayed particles. Adhesion occurs primarily due to a mechanical interlocking with the surface roughness of the substrate, while metallurgical processes hardly take place and a subsequent heat treatment is required. When MCrAlY powders are sprayed in the molten state, aluminum and chromium are likely to oxidize "in-flight." Therefore, only the following three thermal spray methods are commonly used to coat superalloys with MCrAlYs: Low pressure plasma (LPPS), vacuum plasma (VPS), and high velocity oxyfuel (HVOF) spraying. Atmospheric plasma spraying (APS) is only used for the application of zirconia thermal barrier coatings for turbine blades, as discussed later. For plasma spray processes an anode and a cathode are incorporated into a spray gun producing an electric arc to ionize the operating gas. The dissociation and ionization produces heat that allows melting and spraying even refractory elements or ceramics. Several gases can be used, but typically one of the following is chosen: argon, hydrogen, helium, or their mixtures. When the powder is injected in the plasma jet, it is melted by the high temperature of the plasma torch and also propelled toward the substrate. Because of the required low pressure or inert atmosphere for MCrAlYs, the effort and cost for this process is high. High-speed processes such a high-velocity oxy-fuel spraying (HVOF) are much more efficient and have earned high industrial acceptance. By increasing the particle velocity during spraying, unwanted oxidation reactions can also be minimized without the requirement of operating in vacuum [48]. Furthermore, due to the high deformation during the solidification process in the HVOF process, very low porosities can be achieved (<1%). Another option is Physical Vapor Deposition (PVD) processes [12,49]. PVD processes are conducted in high vacuum to avoid reactions between the metal atoms and the process gas. In order to transform the coating material into the gas phase, generally two methods are employed: heating until high vapor pressures occur; or bombardment with high energetic electrons or ions to release atoms from the target. For the components of high temperature coatings, which usually have high evaporation temperatures, the second method is preferred. Various modifications of this method exist, but the electron beam heating is the most well-established one. Concentrated electron beam rays induce a high energy density on the surface of the target so that locally extremely high temperatures occur and even refractory alloys or ceramic material can be evaporated. In order to apply alloys such as MCrAlY coatings, it is important to note that the partial pressures of the different elements are very different. The composition of the melt bath or of the master alloy has to be adapted with respect to the difference in the vapor pressures during the coating process. It can be calculated, but it must be considered that they change even during the process. Especially in the beginning of the coating process, the parameters are far from equilibrium and therefore cannot produce the desired composition. For metallic alloys, different elements have to be added during the process to regulate and adapt the master alloys in a way that the manufactured coating always has the right composition. Secondly, the achieved microstructure is highly textured, with elongated grains oriented perpendicular to the surface of the substrate. This microstructure and process is of highest technical relevance for two-component ceramic yttrium oxide-zirconium oxide layers, serving as thermal barrier top coats for superalloys.

## 4.2. Thermal Barrier Coatings

Thermal barrier coatings (TBC) are used to reduce the heat flow into the metal below in order to reduce oxidation rates and to protect it from thermal softening and accelerated creep. The main application of these systems is in stationary gas and aircraft turbines. Modern TBC systems decrease the temperature load in the superalloys beneath by 200°C, or vice versa allow an increase in operating temperature in that range to improve the turbine efficiency. TBCs consist of a combination of a bond coat which can be either a MCrAlY overlay coating or a diffusion layer with a ceramic top coat. A schematic TBC system is given in Figure 12. It shows the temperature gradient and the typical configuration on a superalloy used, e.g., as a turbine blade.

All classical TBC systems, which have been developed and improved throughout the last 30 years, rely on a partially stabilized zirconia as ceramic top coat, which offers a very good combination of necessary properties [46]. Zirconia was chosen because of its low thermal conductivity in combination with a rather high thermal expansion coefficient closer to metals than that of most other ceramics [50]. In addition, it is compatible with the thermally grown alumina oxides (TGO) scales on top of the bond coat. One challenge is that ZrO is allotropic and shows three phase modifications (Figure 13).

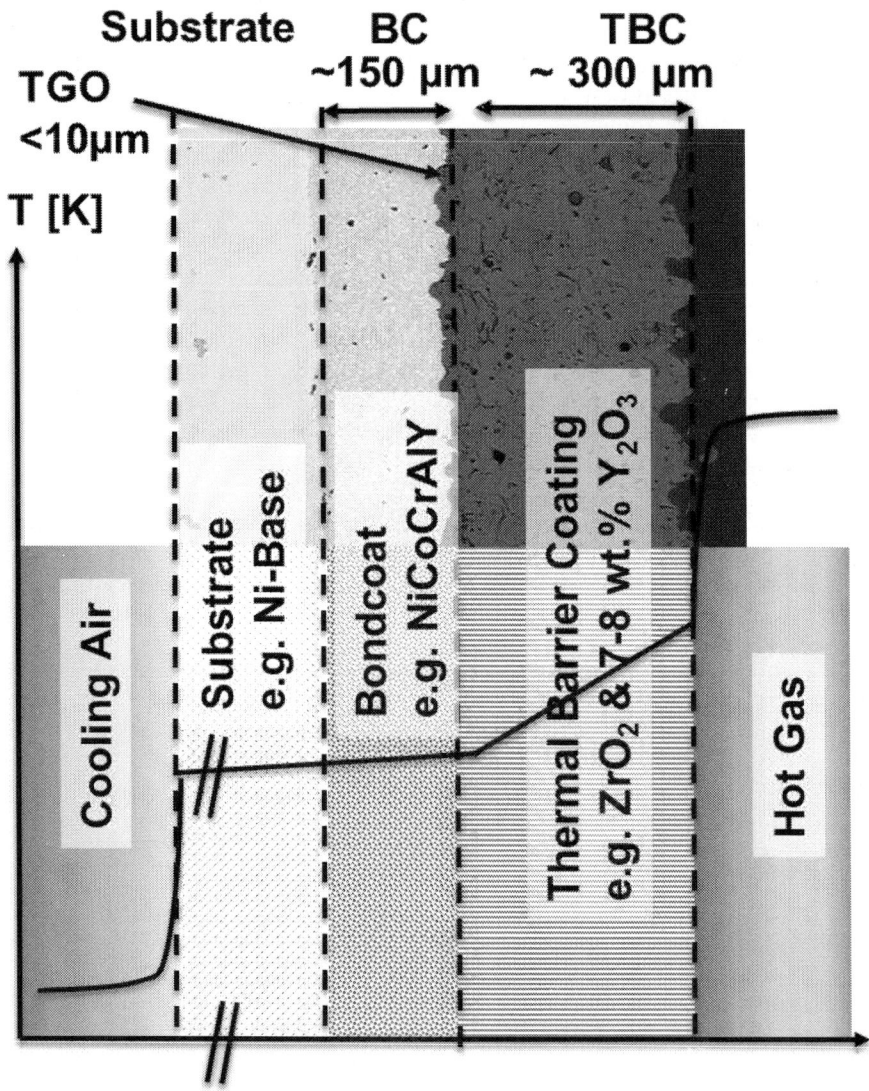

**Figure 12.**Different layers of a thermal barrier system and the typical temperature gradient

Especially the phase transformation from tetragonal to monoclinic at about 1170°C that goes along with a volume change of about 3-9% would induce detrimental cracks. This phase transformation can be suppressed by adding oxides of Yttrium, Cer, Magnesium, Niobium, or Calcium [51]. In Figure 13, the impact of yttrium addition on the phase formation is shown. In commercial technical systems, about 8 % YO1.5 is added to receive metastable partially stabilized tetragonal zirconia and suppress detrimental phase transformations. Another modification that is technically used is MgO-stabilized zirconia, which requires about 20-25 wt.% MgO. This zirconia slowly destabilizes above about

1000°C and can only be used for components of diesel engines and not in turbines. Any phase destabilization or an increased sintering behavior at higher temperatures determines the upper temperature limit of the application of zirconia, because it destroys the necessary porous structure and induces cracks. Such limitations trigger the investigation of other ceramics for even higher-temperature applications than the systems today allow. Especially several ziconates with a low thermal conductivity are looked at, of which the most promising candidate is gadolinium zirconate. Such coatings are developed as a two-layer ceramic system with classical zirconia under the novel ceramics [52]. Due to the fact that ceramics are prone to tensile strains, the microstructure has to be designed carefully in order to allow at least a certain strain tolerance. The ceramic layers today are either manufactured via atmospheric plasma spraying (APS) or electron beam physical vapor deposition (EB-PVD). In both cases, a certain strain tolerance is achieved by adjusting the microstructure [53].

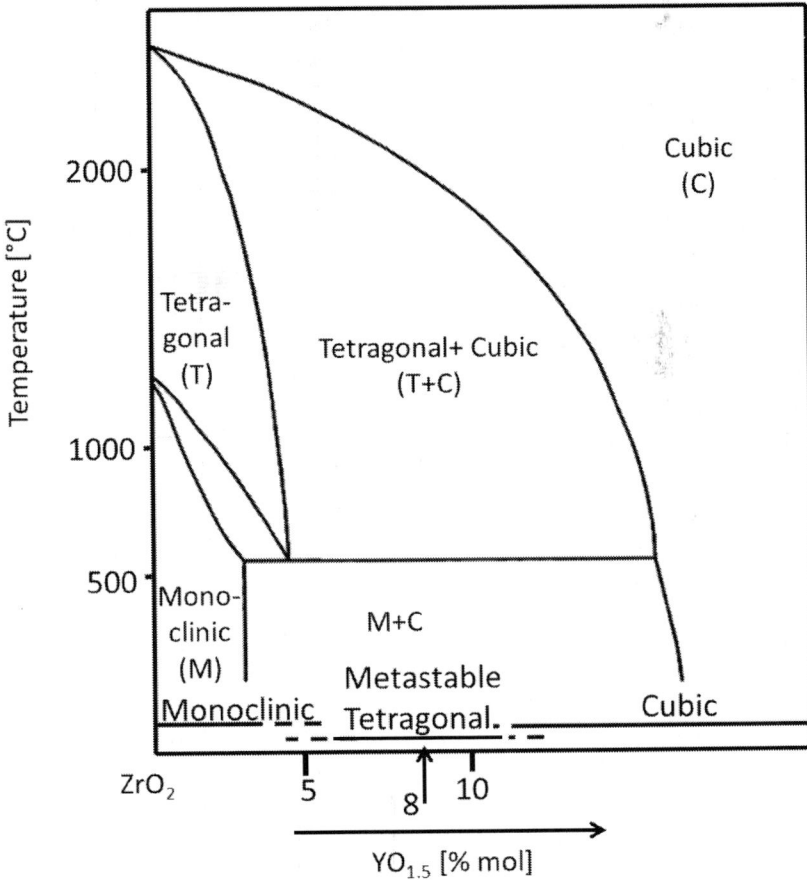

**Figure 13.**Phase diagram of ZrO2-Y2O3 showing the allotropy of zirconia and the technically used range in which the tetragonal lattice modification remains stable

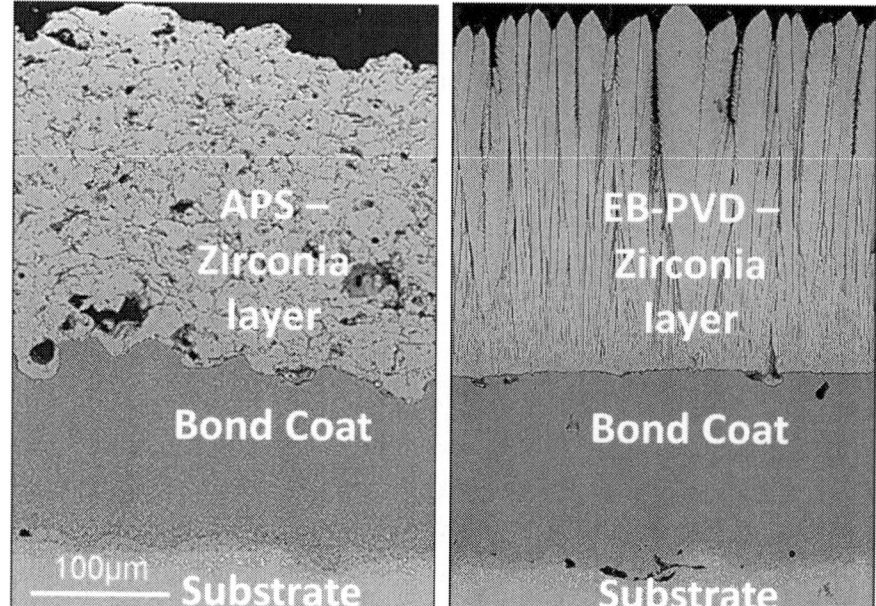

**Figure 14.** Comparison of TBC systems with APS-sprayed zirconia and EB-PVD zirconia top layers showing the different morphology of the ceramic coating on top of a sprayed MCrAlY bond coat on the left and a pack cementated bond coat on the right

In Figure 14, on the left the APS-sprayed microstructure is schematically shown. It possesses a pancake-like structure. The strain tolerance is created by a heavily intertwined network of fine cracks. This network separates the ceramic coating into "segmented flakes" providing a higher tolerance upon strain application through the temperature gradient or difference in thermal expansion to the material below. EB-PVD (Figure 13 on the right) coatings always exhibit the typical columnar structure. This method has the big advantage that, during the EB-PVD process, the crystals of the scales grow epitactically and perpendicular to the surface. The ceramic columns are separated from each other by small gaps, providing an exceptional tolerance to deleterious tensile strains parallel to the surface. These coatings are preferentially used for aircraft turbine blades, while for stationary turbine blades or aircraft nozzle parts often APS coatings are employed, which can be produced at lower cost. One drawback of EB-PVD zirconia top coats is that they are sensitive to calcium-magnesia-alumina silicates (CMAS) attack [54]. CMAS can derive from the use of turbines in sandy desert areas or when volcanic ash enters a turbine. The CMAS melt in the heat chamber and are deposited on the turbine blades. When they penetrate into the TBC, they cause failure and spallation. Other factors with an influence on the lifetime are hot corrosion mechanisms by sulfur, calcium, or vanadium deposits on the coatings that can reduce the lifetime [55], often by reacting with the stabilizers (e.g., yttrium), thereby destroying the resistance against phase transformation of the zirconia. When the ceramic scale remains intact, in service an alumina scale grows slowly under the zirconia ceramic. This thermally grown oxide scale (TGO) must not exceed about 3 μm in thickness to guarantee mechanical integrity and avoid spallation [56].

## 5. OUTLOOK

By using diffusion coatings, overlay coatings, and ceramic top coats for high temperatures, the otherwise often very short lifetime of unprotected materials can be enhanced and makes processes and applications possible for which otherwise no material is available. At the same time, high-temperature processes have been constantly changing due to changes in fuel or operating conditions, and the thermal operation limit of superalloys has continuously increased over the years. In the future, even systems such as molybdenum-based alloys might become interesting with a thermal application potential well above the nickel- and cobalt-based superalloys of today [9]. Such developments induce also a strong driving force to further develop high-temperature coating systems in order to keep pace with such increased operating conditions, and to allow efficient and reliable operation of metallic high-temperature materials.

## REFERENCES

1.  R.C. Reed. The Superalloys: Fundamentals and Applications, Cambridge University Press; 2006, 283-335
2.  R. Subramanian, A. Burns, W. Stamm. Advanced Multi-Functional Coatings for Land-Based Industrial Gas Turbines. Proceedings of ASME Turbo Expo 2008: Power for Land, Sea and Air. ASME. 2008, 1-10.
3.  D.J. Young. High Temperature Oxidation and Corrosion of Metals, Elsevier; 2008, 139-455.
4.  P. Kofstad. High Temperature Corrosion, Essex UK: Elsevier Applied Science; 1988, 139-181.
5.  N. Vialas. Substrate effect on the high temperature oxidation behavior of a Pt-modified aluminide coating. Part II: long-term cyclic-oxidation tests at 1050-|C, Oxidat Metals. 68 (2007) 223-242.
6.  S. Dryepondt, D. Monceau, F. Crabos, E. Andrieu. Static and dynamic aspects of coupling between creep behavior and oxidation on MC2 single crystal superalloy at 1150 -|C, Acta Materialia. 53 (2005) 4199-4209.
7.  V. Seetharaman, A.D. Cetel. Superalloys, TMS, 2004, pp. 207-214.
8.  Srivastava, S. Gopagoni, A. Needleman, V. Seetharaman, A. Staroselsky, R. Banerjee. Effect of specimen thickness on the creep response of a Ni-based single-crystal superalloy, Acta Materialia. 60 (2012) 5697-5711.
9.  M. Heilmaier, M. Krüger, H. Saage, J. Rösler, D. Mukherji, U. Glatzel, R. Völkl, R. Hüttner, G. Eggeler, C. Somsen, T. Depka, H.J. Christ, B. Gorr, S. Burk, Metallic materials for structural applications beyond nickel-based superalloys, J Minerals Metals Mater Soc. 61 (2009) 61-67.

10. H. Harada. Superalloys: Present and Future. Current Topics on High Temperature Materials, JSPS Report of the 123rd Committee on Heat Resisting Materials and Alloys. 48 (2007) 357-363.

11. C.T. Sims. A History of Superalloy Metallurgy for Superalloy Metallurgists. In: M. Gell, C.S. Kortovich, R.H. Bricknell, W.B. Kent, J.F. Radavich (eds.). Proceedings of Superalloys 1984. Warrendale, Pennsylvania, USA: The Metallurgical Society of AIME; 1984.

12. G.W. Goward. Progress in coatings for gas turbine airfoils, Surf Coatings Technol. 108-109 (1998) 73-79.

13. R.C. Clarke, M. Oechsner, N.P. Padture. Thermal-barrier coatings for more efficient gas-turbine engines, MRS Bull. 37 (2012) 891-898.

14. S. Poupard, J.-F. Martinez, F. Pedraza, Soft chemical stripping of aluminide coatings and oxide products on Ni superalloys, Surf Coatings Technol. 202 (2008) 3100-3108.

15. H. Jehn. High temperature behaviour of platinum group metals in oxidizing atmospheres, J Less Common Metals. 100 (1984) 321-339.

16. C.E. Birchenall. Diffusion in Oxides: Assessment of Existing Data and Experimental Problems. Proceedings of the Symposium "Mass Transport in Oxides," October 1967. 1967, p. 119-127.

17. N. Birks, G.H. Meier, F.S. Pettit. High Temperature Corrosion. In: J.K. Tien, T. Caulfield (eds.). Superalloys, Supercomposites and Superceramics. Academic Press; 1989, p. 439-489.

18. A.S. Khanna. Introduction to High Temperature Oxidation and Corrosion, ASM International; 2002, 203-206.

19. C. Xu, W. Gao. Pilling-Bedworth ratio for oxidation of alloys, Mater Res Innov. 3 (2000) 231-235.

20. M. Schütze. Mechanical aspects of high-temperature oxidation, Corr Sci. 35 (1993) 955-963.

21. M. Schütze. Mechanical properties of oxide scales, Oxidation Metals. 44 (1995) 29-61.

22. Allison H.B.C, L.A. Hawkins. Gen Elect Rev. 17 (1914) 947-951.

23. F.C. Kelley. Chromium Impregnation, Metals Handbook, American Society for Metals, Cleveland, Ohio; (1948) 705-706.

24. P.A. Choquet, E.R. Naylor, R.A. Rapp. Simultaneous chromizing and aluminizing of iron-base alloys, Mater Sci Engin A. 121 (1989) 413-418.

25. F.D. Geib, R.A. Rapp. Simultaneous chromizing-aluminizing coating of low-alloy steels by a halide-activated, pack-cementation process, Oxidation Metals. 40 (1993) 213-228.

26. W. Da Costa, B. Gleeson, D.J. Young. Codeposited chromium-aluminide coatings II. Kinetics and morphology of coating growth, J Electrochem Soc. 141 (1994) 2690-2697.

27. W. Da Costa, B. Gleeson, D.J. Young. Codeposited chromium-aluminide coatings I. Definition of the codeposition regimes, J Electrochem Soc. 141 (1994) 1464-1471.

28. B. Gleeson, W.H. Cheung, W. Da Costa, D.J. Young. Co-deposited chromium-aluminide coatings and their high-temperature corrosion resistance. Proceedings of the 8th Asian-Pacific Corrosion Control Conference, Bangkok. 2000, p. 43-48.

29. R. Bianco, R.A. Rapp. Pack cementation aluminide coatings on superalloys: codeposition of Cr and reactive elements, J Electrochem Soc. 140 (1993) 1181-1190.

30. Z.D. Xiang, P.K. Datta. Formation of Hf- and W-modified aluminide coatings on nickel-base superalloys by the pack cementation process, Mater Sci Engin A. 363 (2003) 185-192.

31. B.A. Pint, J.A. Haynes, T.M. Besmann. Effect of Hf and Y alloy additions on aluminide coating performance, Surf Coatings Technol. 204 (2010) 3287-3293.

32. M.C. Galetz, X. Montero, H. Murakami. Novel processing in inert atmosphere and in air to manufacture high-activity slurry aluminide coatings modified by Pt and Pt/Ir, Mater Corrosion. 63 (2012) 921-928.

33. H.W. Grünling, R. Bauer. The role of silicon in corrosion-resistant high temperature coatings, Thin Solid Films. 95 (1982) 3-20.

34. E.Y. Lee, D.M. Chartier, R.R. Biederman, R.D. Sisson. Modelling the microstructural evolution and degradation of M-Cr-Al-Y coatings during high temperature oxidation, Surf Coatings Technol. 32 (1987) 19-39.

35. V.A. Ravi, P.A. Choquet, R.A. Rapp. Chromizing-aluminizing coating of Ni- and Fe- base alloys by the pack cementation technique. In: T. Grobstein, J.K. Doychak (eds.). Oxidation of High-Temperature Intermetallics. Minerals, Metals Materi Soc.; 1989, p. 127-145.

36. R. Bianco, R.A. Rapp, N.S. Jacobson. Volatile species in halide-activated diffusion coating packs, Oxidat Metals. 38 (1992) 33-43.

37. Naji, M.C. Galetz, M. Schütze. Improvements in the thermodynamic and kinetic considerations on the coating design for diffusion coatings formed via pack cementation, Mater Corrosion (2015).

38. D. Fähsing, M.C. Galetz, M. Schütze. Eindiffundieren metallischer Elemente. In: G. Spur, H.-W. Zoch (eds.). Handbuch Wärmebehandeln und Beschichten. Carl Hanser Verlag GmbH & Co; 2015, in press.

39. G.W. Goward, D.H. Boone. Mechanisms of formation of diffusion aluminide coatings on nickel-base superalloys, Oxidation Metals. 3 (1971) 475-494.

40. B. Ganser, K.A. Wynns, A. Kurlekar. Operational experience with diffusion coatings on steam cracker tubes, Mater Corrosion. 50 (1999) 700-705.

41. B.G. McMordie. Oxidation resistance of slurry aluminides on high temperature titanium alloys, Surf Coatings Technol. 49 (1991) 18-23.

42. A.J. Rasmussen, A. Agüero, M. Gutierrez, M.J.L. Östergaard. Microstructures of thin and thick slurry aluminide coatings on Inconel 690, Surf Coatings Technol 202 (2008) 1479-1485.

43. X. Montero, M.C. Galetz, M. Schütze. Low-activity aluminide coatings for superalloys using a slurry process free of halide activators and chromates,Surf Coatings Technol. 222 (2013) 9-14.

44. M.C. Galetz, X. Montero, M. Mollard, M. Günthner, F. Pedraza, M. Schütze. The role of combustion synthesis in the formation of slurry aluminization, Intermetallics. 44 (2014) 8-17.

45. W. Stamm. Siemens AG, Personal communication 2015

46. Robert. Current status of thermal barrier coatings – An overview, Surf Coatings Technol 30 (1987) 1-11.

47. N. Czech, F. Schmitz, W. Stamm. Improvement of MCrAlY coatings by addition of rhenium, Surf Coatings Technol. 68/69 (1994) 17-21.

48. D. Toma, W. Brandl, U. Köster. The characteristics of alumina scales formed on HVOF-sprayed MCrAlY Coatings, Oxidation Metals. 53 (2000) 125-137.

49. R.D. Jackson, M.P. Taylor, H.E. Evans, X.-H. Li. Oxidation study of an EB-PVD MCrAlY thermal barrier coating system, Oxidation Metals. 76 (2011) 259-271.

50. X.Q. Cao, R. Vassen, D. Stoever. Ceramic materials for thermal barrier coatings, J Eur Cera Soc. 24 (2004) 1-10.

51. R.L. Jones. Some aspects of the hot corrosion of thermal barrier coatings, J Therm Spray Technol. 6 (1997) 77-84.

52. Y. Zhang, J. Malzbender, D.E. Mack, M.O. Jarligo, X. Cao, Q. Li, R. Vaßen, D. Stöver. Mechanical properties of zirconia composite ceramics, Cera Int. 39 (2013) 7595-7603.

53. L. Singheiser, R. Steinbrech, W.J. Quadakkers, D. Clemens, B. Siebert. Thermal barrier coatings for gas turbine applications – Failure mechanisms and life prediction. In: D. Coutsouradis (ed.). Materials for Advanced Power Engineering: Proceedings of the 6th Liege Conference, Part III. Kluwer Academic Publishers; 1998, p. 977-996.

54. N. Chellah, M.-H. Vidal-Sétif, C. Petitjean, P.-J. Panteix, C. Rapin, and M. Vilasi. 8th International Symposium on High-Temperature Corrosion and Protection of Materials, 2012.

55. B. Bordenet, W. König, G. Witz, and H.-P. Bossmann. 8th International Symposium on High-Temperature Corrosion and Protection of Materials, 2012.

56. H.E. Evans. Oxidation failure of TBC systems: An assessment of mechanisms, Surf Coatings Technol. 206 (2011) 1512-1521.

# CHAPTER 2

# Development of Atmospheric Plasma Sprayed Dielectric Ceramic Coatings for High Efficiency Tubular Ozone Generators

*Rainer Gadow[1], Christian Friedrich[1], Andreas Killinger[1], Miriam Floristán[1, 2]*

[1]Institute for Manufacturing Technologies of Ceramic Components and Composites (IMTCCC), Universität Stuttgart, Stuttgart, Germany

[2]New Materials Technologies - TTI GmbH,Stuttgart, Germany

## ABSTRACT

Oxidative degradation of hazardous materials by ozone treatment like in sterilization of water, dump waste, pulp bleach and chemical processing, is superior to the traditional chlorine chemistry with respect to by-products and environmental protection. For an efficient and cost effective production of ozone for applications in drinking water and wastewater purification, a new concept of tubular composite material components has been developed. A borosilicate glass tube was coated with a layer system consisting of an intermetallic electrode and a dielectric oxide ceramic surface layer. Thermo-mechanical and dielectric properties are investigated with respect to the use of different thermal spray powders as well as the use of a high and a low energetic atmospheric spray gun. The materials and ozone production system of thermal sprayed ozonizer tubes are described and analyzed.

**Keywords:** Ozone Production, Water Treatment, Thermal Spraying, Dielectric Strength, Permittivity

## INTRODUCTION

E Ozone is an environmentally benign chemical compound having the ability to replace commonly used chlorine compounds in many processes. It is one of the strongest oxidants surpassed only by fluorine in its oxidising power. Ozone is used as a germicide and bactericide for the treatment of potable and waste water as well as for bleaching processes and in further chemical oxidation steps. The use of the ozone as a primary oxidant before chlorination usually will satisfy most of the oxidant demand of the water being treated, thus lowering the subsequent demand for chlorine and minimizing the disinfection by products of chlorination, such as trihalomethanes (THMs) [1].

Ozonation has been used for tackling various industrial wastewaters [2-4]. However, its industrial applications have been handicapped by low ozone transfer efficiency and relatively high production costs [5]. The aim of the research project is the development of a novel powerful ozonizer tube, which cuts down the production costs and thus causes ozone to be an economically competitive alternative to traditionally used chlorine compounds [6-9].

In general, large quantities of ozone are generated by dielectric barrier discharges (DBD) [10-13]. Only these non-equilibrium gas discharges are suitable for an effective ozone generation. Since ozone is not chemically stable and decays during transportation, it is produced in a set of discharge tubes on site where it is needed in a tailored capacity.

The tube coatings are manufactured via Atmospheric Plasma Spraying (APS), a cost effective and very flexible manufacturing process. Various electrically insulating ceramic coatings ($Al_2O_3$, $Al_2O_3$-$TiO_2$, $ZrO_2$ with different stabilisers) are suitable as dielectric barrier material. In this paper, their phase composition as well as their thermomechanical and dielectric properties is studied and the suitability of the most promising coatings in the application is verified.

## 1.1. Principle of Ozone Generation

An ozonizer consists of two electrodes, which are separated by a gap and a dielectric (see **Figure 1**). Applying an alternating high voltage at the electrodes, leads to silent discharges within the gap. This results in partial ionization of the oxygen containing gas stream flowing through the discharge gap (air or pure oxygen) [8,9,14, 15].

The capacitors of most of the commercially available ozonizers are cylindrically shaped instead of plane-parallel. The dielectric prevents the controlled silent discharge to change into a glow or arc discharge that would damage or even destroy the ozonizer due to high currents. A silent discharge evenly distributed across the surface of the electrodes results in an optimum yield of ozone. If $e_r$ of the dielectric is increased, higher field intensities within the discharge gap are possible, since the dielectric serves as a "protective resistor". This leads to a higher efficiency of the ozonizer.

Ozone is generated within the discharge gap due to several chemical reactions of excited or dissociated oxygen species. Other gas species, which interact in form of three-body reactions, have also an influence on the efficiency of the process. Simultaneously, part of the generated ozone is lost again due to several decomposition processes [7,16,17].

The efficiency P of the ozonizer is in general proportional to the capacity C of the individual capacitor arrangement and the frequency f of the applied alternating voltage: $P = k\,f\,C$.

Thus, the efficiency of the ozonizer can be increased by operating the system either at a higher frequency or by increasing the capacity. Commercially available ozonizers are in general not equipped with a frequency converter and operate at a standard frequency range of 50-60 Hz.

## 1.2. Improved Ozonizer Tubes

Standard ozonizer tubes consist of a borosilicate glass tube with a thermally sprayed metallic coating applied to the inner surface of the tube. A special Atmospheric Plasma Spraying based coating technology was developed by EUROFLAMM in Bremen, Germany [8]. In this arrangement the tube itself serves as a dielectric. The metalized inner tube surface is electrically connected via metal brushes (see **Figure 2**). The borosilicate glass is restricted to a permittivity of $e_r \gg 4.6$. One possible way to improve the efficiency of the ozonizer arrangement is to introduce a dielectric material with a higher value for $e_r$. To keep the glass tube geometry, a straightforward way is to multicoat the glass tube with a metal layer serving as the inner electrode and a dielectric oxide [8]. The glass tube serves as a support, see **Figure 3**.

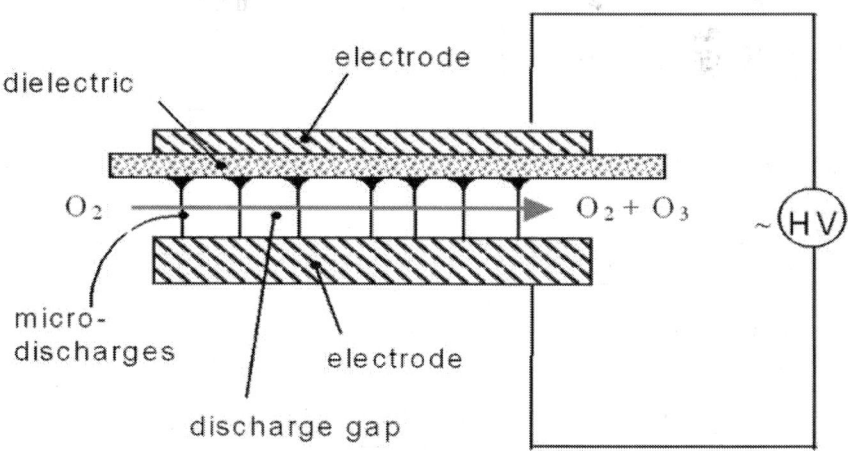

**Figure 1.** Schematic drawing of a capacitor arrangement for ozone production.

**Figure 2.** Commercial ozonizer with tube array.

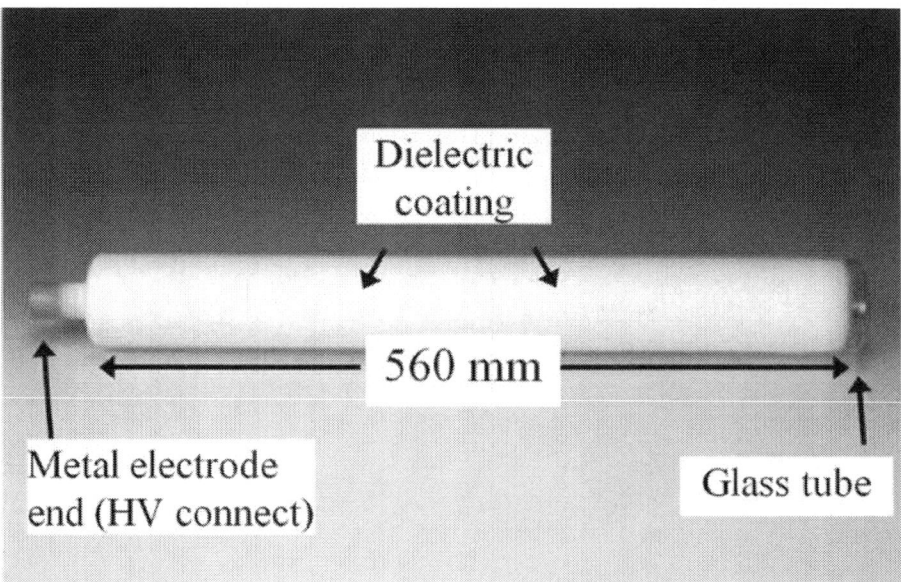

**Figure 3.** Single glass tube with metal dielectric coating applied.

In the present study, results of metal dielectric multilayer structures based on aluminium alloys and several ceramic materials respectively are presented.

# 2. METHODS

## 2.1. Principle of Atmospheric Plasma Spraying

Thermal spraying are a group of processes in which finely divided metallic and non-metallic materials are deposited in a molten or semi-molten state on a prepared substrate [18,19]. Atmospheric plasma spraying is a thermal spraying process in which the energy source to melt the spray material is an electric arc, which generates a plasma jet at atmospheric pressure. Due to the high temperatures achieved in the plasma heat source, over 8000 K [20], this is a very flexible coating technique able to process almost any material.

By means of a high voltage electrical discharge, an arc is created between a water-cooled copper cylindrical anode and a thoriated tungsten cathode situated inside the anode, see **Figure 4**(a). A gas mixture is injected in the torch and flows around the cathode and through the anode nozzle. The interaction of the electric arc with the gas mixture makes the gas atoms dissociate and ionize, leading to the formation of the plasma jet or plasma plume [18].

The powder is transported by a carrier gas and injected in the plasma. The particles are totally or partially melted by the plasma jet and propelled towards the substrate surface. Upon impacting at the surface, the particles deform and rapidly solidify forming splats. The coating is built up particle by particle. The coating structure is lamellar and inhomogeneous presenting oxidized particles, pores and unmelted or partially melted particles, see **Figure 4**(b) [21].

## 2.2. Selected Thermal Spray Powders and Substrate Material

The powders used in this work were commercially available powders as well as custom specified. All the materials used are chemically inert against ozone and oxygen radicals as well. The powders were characterised prior to spraying. Particle size distribution analysis was made with a Malvern Mastersizer by laseropthical method, particle morphology and phase composition were determined by SEM analysis and XRD respectively, see **Table 1**.

### 2.2.1. Metallic Electrode
In general, the metal electrode should present high electrical conductivity, high bonding strength to the borosilicate glass and oxidation resistance. With this aim, two materials were chosen; pure Al and Al/Si alloy. The alloy low thermal expansion coefficient, closer to the one of the borosilicate glass substrate, makes of this material an interesting option for the metallic electrode. However, the Si concentration in the alloy has to be over 50% in order to identify a significant change in the electrical conductivity of the material.

### 2.2.2. Dielectric Material

The thermal spray powders were selected attending to their appropriate thermophysical properties, reduced costs, coating process stability and commercial availability. Several oxide ceramics based on alumina, zirconia and titania alloys have been investigated. These materials showed high permittivity and are electrical insulators. Sintered $TiO_2$ presents a permittivity ten times higher than that of $Al_2O_3$. However, during thermal spraying operations, $TiO_2$ often undergoes a high reduction and rutile is then turned into non-stoichimetric $TiO_{2-x}$, changing therefore its physical properties. This modification can be considered as an n-type semiconductor with properties dependent on the extent of oxygen loss [22-24].

Several α-alumina powders and two types of stabilised zirconia powders with moderately high permittivities have been investigated and will be compared.

### 2.2.3. Substrate Material

The substrate material is a chemically inert borosilicate glass, which is commonly used as an ozonizer tube material. It is characterised by a very low thermal expansion coefficient of $3.3 \ 10^{-6} \ K^{-1}$. This value is quite low compared with the CTE of most of the oxide ceramics that are applied on it. The thermophysical and dielectric properties of substrate and coatings are shown in **Table 2**.

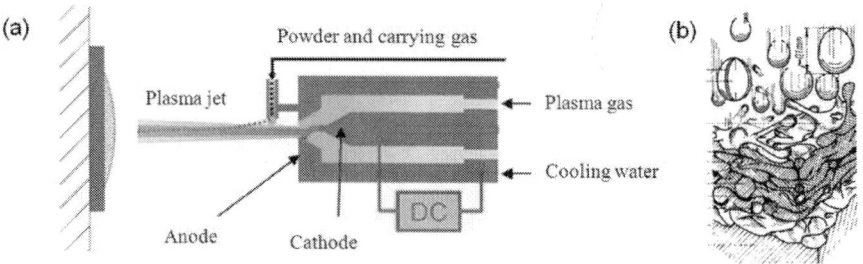

**Figure 4.** (a)Principle of atmospheric plasma spraying, (b)schematic draw of a plasma sprayed coating formation [21].

**Table 1.** Properties of the investigated powders.

| ID | Powder material (supplier) | Phase analysis (XRD) | Grain size distribution ($d_{10}$, $d_{50}$, $d_{90}$) | Grain morphology | Manufacturing process |
|---|---|---|---|---|---|
| 1 | Al (Sulzer Metco) | pure Al | 10.2, 27.4, 56.6 | irregular, tubular, rounded | fused, sprayed |
| 2 | Al/Si 88/12 (Eckert) | Al-Si alloyed | 11.2, 25.1, 49.2 | tubular, rounded | fused, sprayed |
| 3 | $Al_2O_3$ (H.C. Stark) | $\alpha$-$Al_2O_3$ | 16.4, 29.2, 49.4 | blocky, sharp-edged | fused, crushed |
| 4 | $Al_2O_3$ (Sulzer Metco) | $\alpha$-$Al_2O_3$ | 8.3, 16.0, 26.9 | blocky, sharp-edged | fused, crushed |
| 5 | $Al_2O_3$ (Hochrhein) | $\alpha$-$Al_2O_3$ | 8.7, 17.3, 30.1 | blocky, sharp-edged | fused, crushed |
| 6 | $Al_2O_3$ (A.A) | $\alpha$-$Al_2O_3$ | 12.5, 16.96, 22.4 | spherical, phased | n.d. |
| 7 | $Al_2O_3$/$TiO_2$ 97/3 (Hochrhein) | $\alpha$-$Al_2O_3$ $TiO_2$ (rutile) | 7.7, 15.7, 27.5 | blocky, sharp-edged | alloyed, fused, crushed |
| 8 | $Al_2O_3$/$TiO_2$ 94/6 (Hochrhein) | $\alpha$-$Al_2O_3$ $TiO_2$ (rutile) | 7.7, 32.0 | blocky, sharp-edged | alloyed, fused, crushed |
| 9 | $Al_2O_3$/$TiO_2$ 87/13 (Hochrhein) | $\alpha$-$Al_2O_3$ $TiO_2$ (rutile) | 7.7, 15.7, 27.6 | blocky, sharp-edged | alloyed, fused, crushed |
| 10 | $Al_2O_3$/$TiO_2$ 67/33 (H.C. Starck) | $\alpha$-$Al_2O_3$ $TiO_2$ (rutile) | 7.4, 15.5, 27.9 | blocky, sharp-edged | fused, crushed, blended |
| 11 | Mullite (Hochrhein) | $3Al_2O_3 \cdot 2SiO_2$ | 14.6, 29.1, 47.8 | blocky, sharp-edged | fused, crushed |
| 12 | $TiO_2$ (H.C.Starck) | $TiO_2$ (rutile) | 8.2, 14.6, 24.5 | blocky, sharp-edged | fused, crushed |
| 13 | $ZrO_2$ (Hochrhein) | monocline | 7.8, 17.7, 33.1 | blocky, sharp-edged | fused, crushed |
| 14 | $ZrO_2$/$Y_2O_3$ 93/7 (H.C. Starck) | monocline cubic | 9.5, 18.3, 31.6 | blocky, sharp-edged | alloyed, fused, crushed |
| 15 | $ZrO_2$/CaO 95/5 (H.C. Starck) | monocline cubic | 9, 17.6, 29.5 | blocky, sharp-edged | alloyed, fused, crushed |

**Table 2.** Thermophysical and dielectric properties (ground electrode: Rz50 µm) of the thermally sprayed coatings.

| coating ID | Composition Powder supplier Torch model | Universal hardness [N/mm²] | $E/(1-v^a)$ [GPa] | Porosity [%] | Ra/Rz [µm] | Phase-composition | Ratio of the two strongest intensity | Dielectric strength² $E_d$ [kV/mm] | Permittivity $\varepsilon_r$ [1] | Loss angle tan δ [$10^{-4}$] |
|---|---|---|---|---|---|---|---|---|---|---|
| - | Duran Schott - | 3252 | n.d. | 0 | 0 /0 | glass | | 20 | 4.5 | 68 ± 4 |
| A | $Al_2O_3$ Hochrhein PTG | | 132 ± 4 | 9 | 6.5 / 36.3 | $\gamma$-$Al_2O_3$ $\alpha$-$Al_2O_3$ | 1.9 | 6 ± 1.2 | 9 | n.d. |
| B | $Al_2O_3$ Hochrhein F4 | 7831 | 197 ± 4 | 5 | 3.9 / 26.4 | $\gamma$-$Al_2O_3$ $\alpha$-$Al_2O_3$ | 18 | 13 ± 1.2 | 7.6 | 52 ± 6 |
| C | $Al_2O_3$ AA F4 | 6314 | 147 ± 4 | 7 | 2.5 / 17.5 | $\gamma$-$Al_2O_3$ $\alpha$-$Al_2O_3$ | 17.3 | 16 ± 1.3 | 7 | n.d. |
| D | $TiO_2$ H.C. Starck F4 | 6043 | 165.2± 3 | | 4.9 /35.5 | rutile anatase | 3.25 | cd | cd | cd |
| E | $Al_2O_3$ / $TiO_2$ 97/3 Hochrhein F4 | 7424 | 179 ± 3 | 6.3 | 4.3 / 28.5 | n.d. | n.d. | 9.5 ±1.2 | 12.3 | 255 ± 3 |
| F | $Al_2O_3$/$TiO_2$ 94/6 Hochrhein F4 | 8264 | 203 ± 4 | n.d. | 5.0 / 33.1 | n.d. | n.d. | 9.4 ± 1.2 | n.d. | 264 ± 14 |
| G | $Al_2O_3$/$TiO_2$ 87/13 Hochrhein F4 | 7544 | 184 ± 3 | n.d. | 6.0 / 38.1 | n.d. | n.d. | cond. | cond. | cond. |
| H | $ZrO_2$ Hochrhein F4 | 5080 | 144 ± 3 | 3.5 | 3.9 / 26.5 | monoclinic[1] | - | cond. above 3 kV | 28 | 2713 ± 41 |
| I | $ZrO_2$/CaO 95/5 Hochrhein F4 | 5845 | 154 ± 3 | 1.81 | 5.0 / 30.2 | tetragonal[1] cubic | n.d. | 9.7 ± 1.0 | 15.4 | 200 ± 70 |
| J | $ZrO_2$/CaO 70/30 Hochrhein F4 | 5459 | 127 ± 2 | 1.1 | 5.0 / 30.2 | tetragonal[1] cubic | n.d. | 9.9 ± 1.0 | 17.7 | 54 ± 10 |

cond.: conducting [1]expected; [2]not an ideal capacitor n.d.: not determined; [c]electrode roughness $R_q \approx 50$ µm, thickness of dielectric d ≈ 400µm

## 2.3. Experimental setup

### 2.3.1. Coating Deposition and Analytical Characterisation

All coatings have been deposited onto planar glass samples (100 × 100 × 2.5 mm³) via Atmospheric Plasma Spraying. Two different torches have been used. The first is a 80 kW Metco F4 torch and the second is a 15 kW PTG torch especially optimised to manufacture very fine grain sized powders. The torches were controlled by a fully programmable GTV MF-P-1200 unit. A procedure for optimising powder melting, deposition rate and sufficient adhesion to the substrate was carried out for each of the chosen powders. For $TiO_2$, additional experiments have been carried out with a 4P and a TopGun®HVOF gun.

The coating characterisation and microstructure analysis was carried out by SEM and bright field optical microscopy [25,26]. Porosity has been determined from digital image processing of the respective cross section micrographs. Micro hardness of the thermal sprayed ceramic coatings was determined in the cross section in a Fisherscope H 100-V. The Young's modulus was calculated from the elastic deformation energy during the dynamic hardness measurement. Another important aspect of this study was to evaluate the dielectric properties of the oxide ceramic coatings. Therefore the electrical permittivity $\varepsilon_r$ and the loss angle tanδ were determined on planar samples via complex impedance spectroscopy and the dielectric strength $E_d$ was measured using a 40 kV AC-analyzer, according to DIN VDE 0303 Part 21.

### 2.3.2. Ozone Production

Ozone production was carried out in a three tube ozone generator (Ozonia, LN103), which was modified to ensure convenient handling of the glass tubes. The ozonizer is equipped with sensors to record the oxygen gas flow, ozonizer current and voltage as well as the ozone concentration in the process gas, absolute pressure in the ozonizer, temperatures of the in and outlet gas and cooling water. Data recording was performed via personal computer. The parameters used for the ozone production evaluation are given below and were the same in all experiments carried out:

| | |
|---|---|
| Process gas | technical oxygen |
| Process gas flow | $v_g = 0.19$ m³/h |
| Process gas inlet temperature | $T_{g,i} = 20\text{-}25$ °C |
| Vessel pressure | $p_{abs} = 1600$ mbar |
| Cooling water flow | $v_w = 20$ l/h |
| Cooling water inlet temperature | $T_{w,i} = 20\text{-}25$°C |

## 3. RESULTS AND DISCUSSION

According to the aim of this study, the investigated coating systems were analysed focusing on the application in ozonizers. Therefore a metallic electrode, an oxide ceramic and the dielectric properties of the metal-ceram ic-composite were analysed. Finally, the appropriate metal-ceramic-composites were tested in an ozonizer.

## 3.1. Deposition of the Metallic Electrode

The two metallic powders, Al and Al/Si alloy were deposited with low oxidation, and good adhesion to the substrate was obtained even though no shoot peening was performed. This indicates that the bonding mechanism is mainly of chemical nature, in comparison with the typically mechanical bonding mechanisms of thermal sprayed coatings. A higher deposition rate and adhesion was obtained in the Al/Si coatings. This could be explained attending to the lower CTE of the alloy compared with the pure Al, which may be closer to the value of the borosilicate glass. This can lead to reduced residual stresses and improved wettability behaviour due to the increased chemical compatibility between Al/Si and borosilicate glass.

Indeed, adhesion failure induced by residual stresses does not occur in the glass metal interface but within the glass substrate in a depth of about 20 microns. This behaviour indicates a strong physico-chemical bonding between the alloy and the borosilicate. However, further investigations have to be done to characterise the exact bonding mechanism.

During plasma spraying a defined heat transfer to the substrate or already deposited coating takes place, which is dependent on the spraying parameters, such as cooling rates, plasma gases, coating speed, etc. In some cases, the substrate to be coated is preheated prior to spraying in order to improve the coating adhesion. **Figure 5** shows the cross section of two Al/Si 88/12 coatings processed with the same parameters except for the substrate pre heating, which was applied only on one of them. Both coatings have good contact to the glass. It can be seen, that although Al/Si 88/12 is an eutectic material, two phases are formed in the coating; a Si-phase and an Al-dendrites. The formation, distribution and growth of the Si-phase is strongly dependent of the temperature curves during the spraying process. Due to the different microstructure, different conductance values are achieved. However all coatings have a sufficient high electrical conductance for the application.

## 3.2. Oxide Ceramic Coatings

All oxide coatings have been deposited onto the glass-metal multilayer substrate. The thermophysical properties of all sprayed oxide coatings are listed in **Table 2**.

### 3.2.1. Al$_2$O$_3$ Coating

Pure $\alpha$-Al$_2$O$_3$ of several suppliers have been plasma sprayed (see **Figure 6**(a) and **Table 2**). During plasmaspray process, some $\alpha$-Al$_2$O$_3$ is transformed into $\gamma$-Al$_2$O$_3$, and the main phase found in the thermally sprayed coating is the cubic $\gamma$-Al$_2$O$_3$. This effect has been extensively discussed in the literature and was explained by McPherson attending to nucleation kinetics [27]. Due to its lower interfacial energy between crystal and liquid, $\gamma$-Al$_2$O$_3$ is more easily nucleated from the melt than $\alpha$-Al$_2$O$_3$, being therefore the metastable phase, if cooled rapidly enough, the one retained to ambient temperature and therefore present in the coating [27]. The cooling rate of APS processes, which can be in the range of $10^6$ K/s [28], strongly determines the phase composition of the coating. Generally, the higher the cooling rate, the more the $\gamma$ phase [29,30]. The presence of $\alpha$-Al$_2$O$_3$ on the coating indicates the uncompleted melting of some of the particles during coating [31] and depends thus on the process parameters [32]. Usually the stable $\alpha$-Al$_2$O$_3$ is, due to its superior chemical and mechanical properties, desired in the coatings. The change of the $\gamma$-phase to the stable $\alpha$ - phase induces a significant volume change ($\gamma$ to $\alpha$: volume change of ~15%) which can result in micro crack formation and related problems.

The intensity ratios of the α and γ XRD peaks of the various plasma sprayed alumina coatings are given in **Table 2**. All ratios having values about 17 or 18 except the with PTG torch sprayed alumina "A" that has a value of 1.9. This corresponds to a very high amount of α-alumina and is probably due to a high concentration of unmolten particles in the coating resulting of the coarse milled powder. Comparing the XRD-ratios of coating "B", which was sprayed using the same powder but the higher energetic F4-torch, confirms this assumption; the PTG torch does not seem to be able to fully melt the reasonable high fraction of alumina powder particles with grain sizes above 20µm and therefore, the cross section shows high inhomogenities in the microstructure, unmolten particles, high porosity and as follows, the coating has a low dielectric strength. Further experiments show that with the use of finer grinded powders, homogeneous very fine structured coatings can be achieved with the low energetic PTG torch.

Especially for the use in dielectric applications, a low porosity is desired to achieve high dielectric strength of the dielectric layer. Lowest porosities have been obtained with the spherolytic alumina powder, coating "C" **Table 2**. The used powder has a very narrow grain size distribution of − 22 + 15 µm, so that the plasma is able to fully melt nearly all powder particles. Moreover, a very stable spraying process can be obtained due to the high flowability of the spherical shaped powder grains and a very high coating quality was achieved.

### 3.2.2. TiO₂ Coatings

Stoichiometric TiO$_2$ powder was deposited by atmospheric plasma spraying. The reduction of this oxide could not be avoided during the plasma spraying process and the material loses its electrical insulating character (coating "D", **Table 2**).

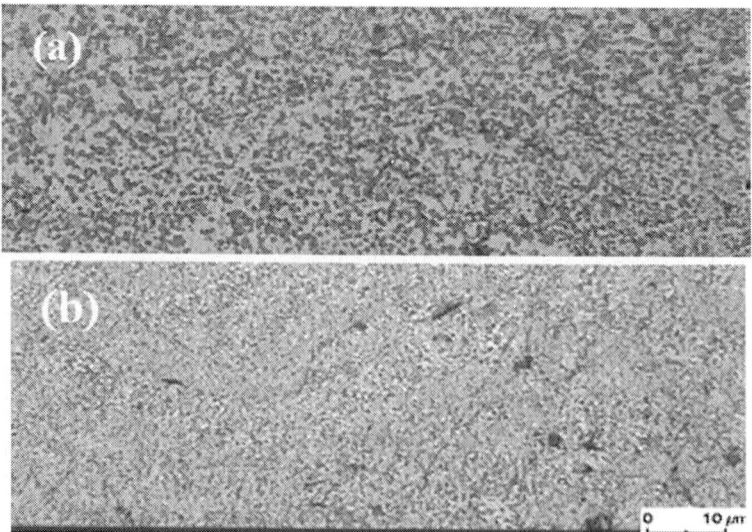

**Figure 5.** Cross section of an Al/Si 88/12 coating with (a) and without (b) preheating the substrate.

Several experiments have been done to deposit $TiO_2$ by flame spraying with the "4P" Metco and the TopGun$^{O}$ HVOF gun in an oxygen enriched flame (small $\lambda$) and simultaneous oxygen cooling gas jet to force the deposition of oxygen stoichiometric $TiO_2$ coatings. The results were the same as in the APS experiments; $TiO_2$ could only be deposited in oxygen understoichiometric modification.

### 3.2.3. $Al_2O_3/TiO_2$ Coatings

Several $Al_2O_3/TiO_2$ oxide powders with different content of $TiO_2$ were plasma sprayed (coatings "E" to "G", **Table 2**). The higher amount of $TiO_2$ in the powders leads to higher coating permittivity values but it gives place also to a decrease in the coating insulating properties, since substoichiometric phases appear in the system. The plasma sprayed coatings exhibit separate phases of $Al_2O_3$ and understoichiometric $TiO_2$ as can be seen in micrographs. No forming of tialite ($Al_2TiO_5$) has been observed, as could be expected from some studies presented in the literature [33].

Coating "F" (**Table 2**) $Al_2O_3/TiO_2$ 94/6 shows the highest permittivity within the alumina based coatings, but the dielectric behaviour is not that of an ideal capacitor, the permittivity strongly depends on the testing frequency. For concentrations above 6 wt% of $TiO_2$, no electrical insulating APS coating can be achieved (coating "G", **Table 2**).

### 3.2.4. $ZrO_2$ Coatings

Due to its high melting temperature, pure zirconia, as well as partially and fully calcia stabilised zirconia, were deposited with the high power F4 torch (coatings "H", "I", "J", **Table 2**). The deposition efficiency of the coating process was about 70%. In **Figure 6**(b), a typical cross section of a zirconia coating is shown, there is an excellent bonding between glass-Al/Si and Al/Si-zirconia. The ceramic shows a low porosity and high homogeneity. The zirconia based coatings have reached the highest permittivity values and a dielectric strength of about 10 kV/mm.

Like $TiO_2$, pure $ZrO_2$ undergoes a structure transformation and becomes conducting during the spraying process. Electrical conductivity starts at voltages above 3 kV. This could be due to vacancies in the crystal lattice comparable to the $TiO_2$-system or due to impurities. Subsequently and although this material presents permittivity values two or three times higher than all the other analysed materials, it is not appropriate for the use in ozonizer tubes.

On the other hand the calcia stabilised zirconia does not show conducting or semi-conducting properties (coatings "I", "J", **Table 2**). The coatings are electrical insulators even at high voltages. The fully stabilised zirconia (coating "J", **Table 2**) has a lower Young's modulus, a higher permittivity and a smaller loss angle than the partially stabilised $ZrO_2$. Therefore it is expected, that this coating presents lower residual stress, higher capacity and lower Ohmic loss. In comparison with all the investigated coatings, it is the most promising one for the application of study.

## 3.3. Dielectric Properties

The measured dielectric strengths and permittivities of the analysed sprayed coatings are listed in **Table 2**. As discussed previously, the use of titania and pure zirconia coatings as dielectrics is not appropriate due to their increased electrical conductivity after being thermally sprayed.

The dielectric strength is the voltage necessary to induce an electric breakdown in an electrical insulating material of 1 mm thickness. During electric breakdown, the material is locally molten and a pinhole is formed. As will be discussed, the dielectric strength of thermally sprayed coatings is also determined by several mechanical properties like the coating roughness of the metallic interface, as well as the porosity and the thickness of the ceramic top layer.

### 3.3.1. Influence of the Metal Interface Roughness

After spraying the metal interface, the surface roughness of the sprayed metal electrode was modified by polishing or by using a finer grain size distribution of the spray powder (−20 + 35 μm). The ceramic layer thickness was about 300 μm. The data shows clearly that the breakthrough voltage depends strongly on the interface roughness (**Figure 7**(a)). The voltage breakthrough is due to strong local electric field inhomogenities caused by discrete peaks sticking out of the metallic interface. Smoothing the surface circumvent strong field inhomogenities and as a result, higher voltages are necessary to create an electrical breakthrough.

Because ozonizer tubes need a high dielectric strength, a minimised electrode surface roughness is desired. On the other hand, the adhesion of the thermal sprayed ceramic layer to the metal electrode is mainly of mechanical nature and increases with higher roughness. Experiments have pointed out, that the electrode roughness should not be under $R_z = 20$ μm in order to prevent delamination of the ceramic layer.

### 3.3.2. Influence of the Oxide Layer Thickness

The maximum reachable coating thickness is limited due to an increase of internal stresses that finally can lead to delamination of the coating or even to destruction of the glass substrate. This is caused by the thermophysical incompatibilities of the ceramic, metal and glass materials, such as their very different coefficients of thermal expansion: $\alpha_{glass} = 3.3 \times 10^{-6}/K$, $\alpha_{Al} = 14 \times 10^{-6}/K$, $\alpha_{ceramic} = 8 \times 10^{-6}/K$. The maximum reachable coating thicknesses are in the range from 400 μm to 600 μm for planar substrates.

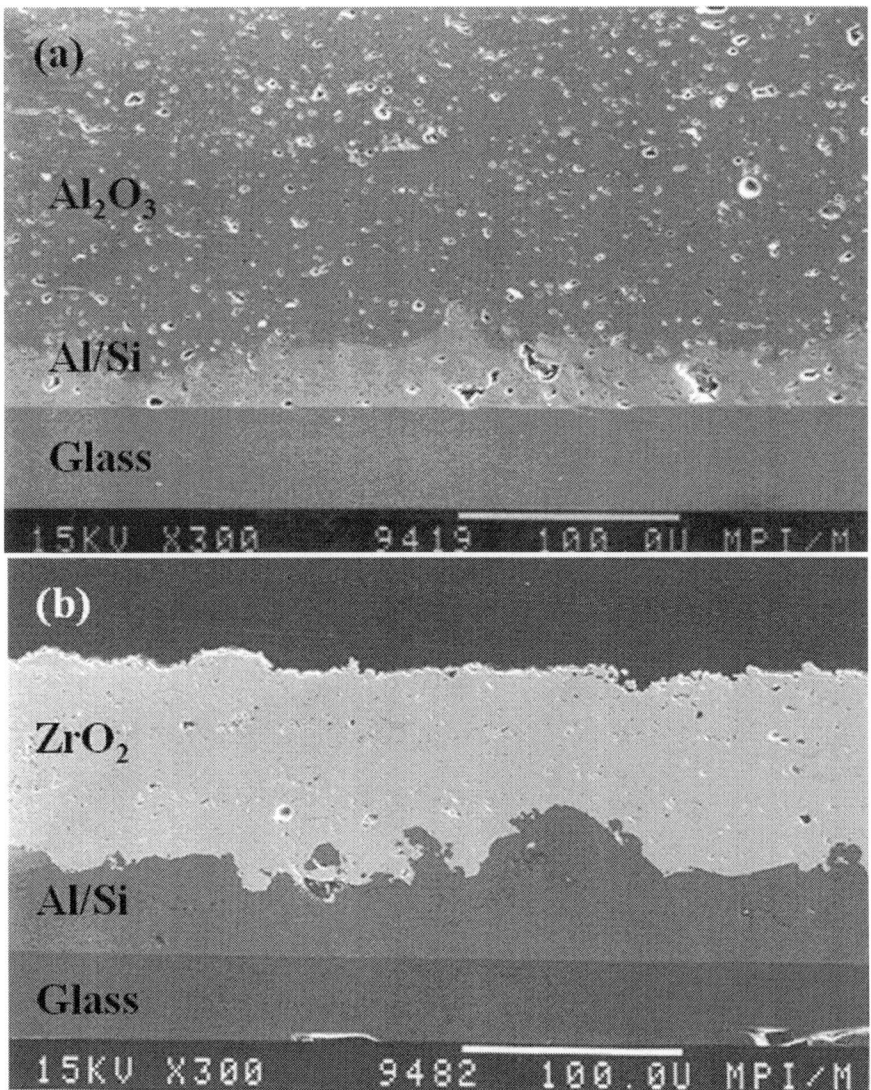

**Figure 6.** (a)Cross section SEM of APS Al2O3 coating and (b)of APS zirconia coating.

**Figure 7**(b) illustrates the breakdown voltage as a function of the coating thickness of the oxide alumina layer. The dielectric strength is more or less constant until thicknesses of around 400 μm and tends to decrease for higher values. A possible explanation for this behaviour is the increase of micro-cracks induced by internal stresses. This effect has already been observed in literature [30,34]. However, this is one of the main problems faced when dealing with thermal sprayed coatings.

## 3.4. Upscaling and Ozone Production

In order to be able to use ozone tubes in the ozone generator, it is necessary that the breakdown voltage of coated tubes reaches at least 13 kV. As the experiments show, thermal sprayed dielectric coatings with a thickness of 1000 μm and above can sustain such high voltages. The deposition of 1000 - 1300 μm thick dielectric coatings on metalized glass tubes caused several problems. As discussed in the last section, the high internal stresses in the coating system can lead to complete destruction of the glass tubes (**Figure 8**).

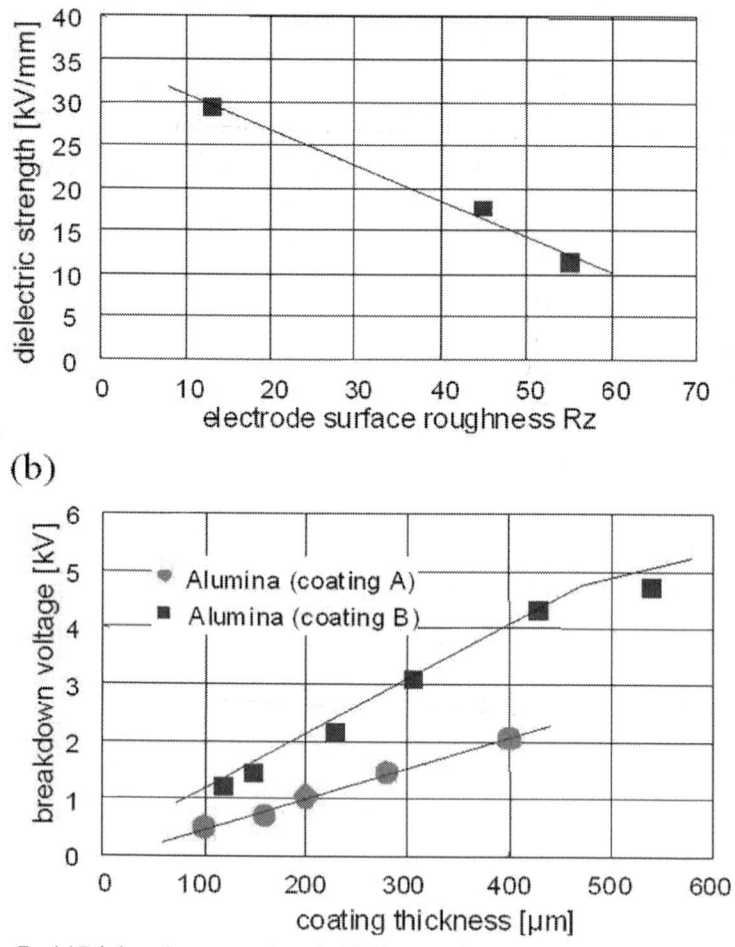

**Figure 7.** (a)Dielectric strength of Al/Si metal alloy and Al2O3 multilayer coatings (coating B, 300 μm) as a function of surface roughness, and (b)breakdown voltage in dependency of the coating thickness for two types of thermal sprayed Al2O3 ceramic coatings.

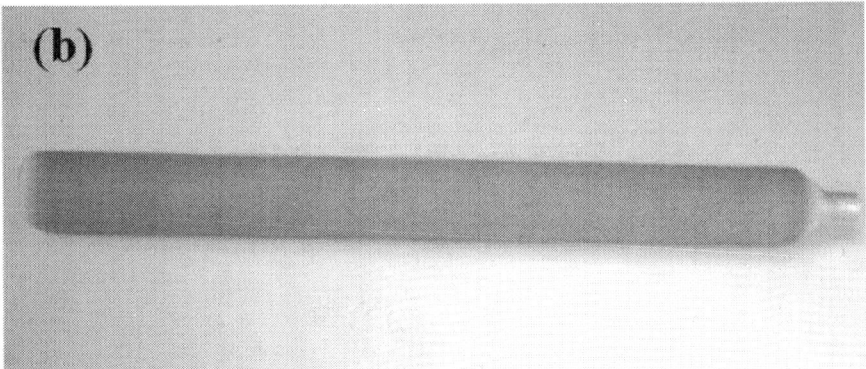

**Figure 8.** (a)Destructed and (b)non-destructed coated glass tube (Al/Si and Al2O3/TiO2) due to different residual stress situation.

The internal stress of multicoated metal-ceramic borosilicate glass was determined by 3D-deformation, drill hole measurement and numerical stress analysis (FEM). Because of these experiments, it was concluded that a successful coating process of glass tubes implies an optimisation of the plasma parameters, the use of sophisticated simultaneous cooling techniques and a stress optimised geometry of the glass tubes.

The glass tubes were coated with 50 μm Al/Si 88/12 and approximately 1000 μm ceramic coating. Presently two different dielectrics were tested, $Al_2O_3/TiO_2$ 97/3 (coating "E") and $ZrO_2/CaO$ 70/30 (coating "J").

**Figure 9**(a) shows the current-voltage curves of the ozonizer with glass and plasma sprayed tubes. The characteristic curves of the two novel tubes are similar, but differ from the behaviour of the traditional glass tubes. All curves show two clearly distinguishable regions with different slopes (named region I and region II in **Figure 9**(a)). In region I no dielectric barrier discharges occur. The capacitor is only charged and discharged, and therefore the current is very small and nearly independent of the voltage. In region II the electrical field in the gap overcomes the Paschen´sche field, DBD appear and ozone is produced.

The lower slopes of the curves for the coated tubes in region II indicate that the Ohmic resistance, and therefore the Ohmic loss, has a lower value. As can be observed in **Figure 9**(a) the voltage necessary for the formation of DBD, and therefore the beginning of ozone production, is approximately 3 kV lower in the case of coated tubes. It should be noted that the maximum efficiency value for ozone production is also reached at lower voltage values.

In **Figure 9**(b) the ozonizer production efficiency of traditional and novel ozonizer tubes are compared. With the novel ozonizer tubes, an increase of the ozone production efficiency up to 30% is observed (15% with $Al_2O_3/TiO_2$ 94/6 and 30% with $ZrO_2/CaO$ 70/30). The decrease of the ozone production efficiency at higher power supply of the ozonizer is due to the partial decomposition of the higher ozone concentration in the used gas.

The ozone production efficiency is correlated to the permittivity. Nevertheless, the often made assumption that the ozone production efficiency is proportional to the capacity of the ozonizer tube is not correct, otherwise the ozone production would increase to at least 200% to 300% with the use of the novel tubes.

There is strong evidence that the ozone production efficiency is a complex function of a series of parameters, i.e. the electron work function of the ceramic material, the surface roughness of the electrode and the dielectric, the gap spacing, the power supply characteristics etc.

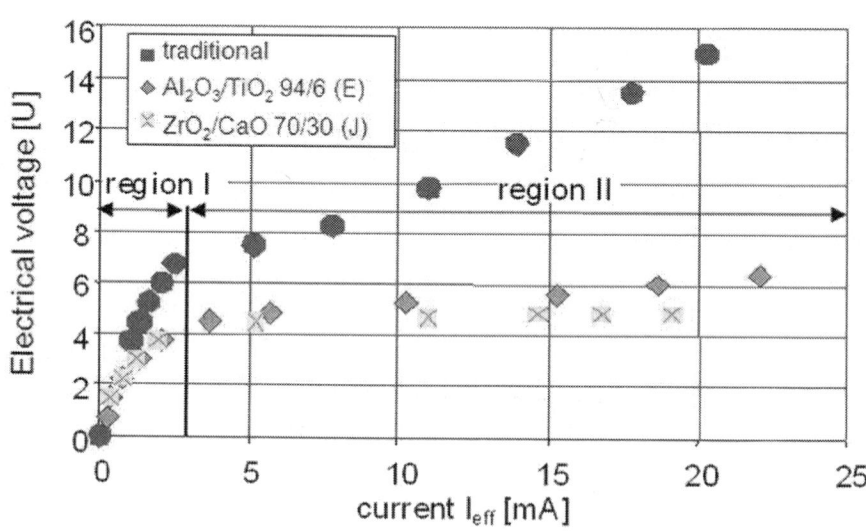

(b)

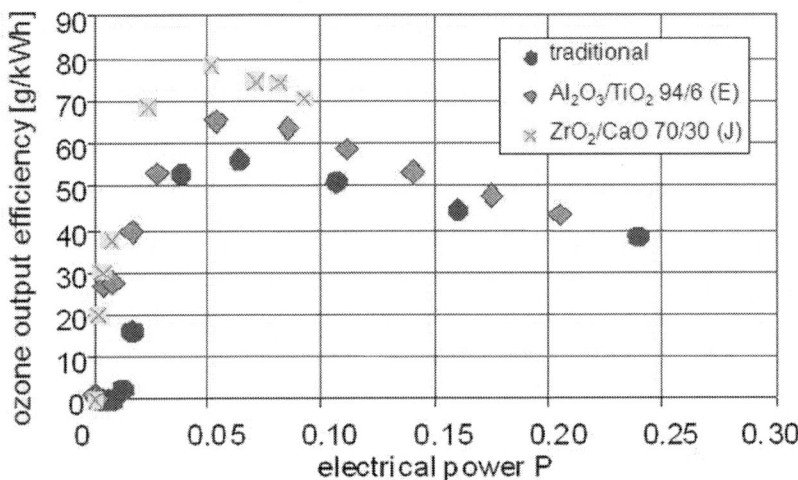

**Figure 9.** (a)Current-voltage curve for traditional and two novel plasma sprayed ozonizer tubes, and (b)ozone production efficiency versus electrical power for traditional and two novel plasma sprayed ozonizer tubes.

## 4. CONCLUSIONS

In this work, the deposition of thermally sprayed metal-ceramic bilayer coatings on borosilicate glass tubes to be used in the production of ozone has been investigated. The coatings consist of an Al/Si intermetallic interlayer representing one of the electrodes, and an oxide ceramic top coating on the glass acting as a dielectric.

Coating experiments on borosilicate glass of the metal layer and different oxide ceramics like $Al_2O_3$, $ZrO_2$ and $TiO_2$ have been carried out. Improved permittivities, with values three times higher than in traditional ozonizer tubes, were obtained. The coating microstructure in terms of porosity, thickness and micro-cracking of the oxide layer and surface roughness of the metallic electrode greatly influence the dielectric strength of the composite. The dielectric breakthrough of the system was successfully increased by the deposition of residual stress optimised 1000 µm ceramic coating.

First results using the new type of thermally coated ozonizer tubes in a laboratory ozonizer already performed an approximately 30% higher efficiency compared to conventional glass tubes.

## REFERENCES

1.  M. A. T. Alsheyaba and A. H. Muñoz, "Optimisation of Ozone Production for Water and Wastewater Treatment," Desalination, Vol. 217, No. 1, 2007, pp. 1-7.

2.  F. L. Evans, "Ozone Technology: Current Status," In: F. L. Evans, Ed., Ozone in Water and Wastewater Treatment, Ann Arbor: Ann Arbor Science Publishers Inc., MI, USA, 1972, pp. 1-13.

3.  R. G. Rice and M. E. Browning, "Ozone Treatment of Industrial Wastewater," Noyes Data Corporation, Park Ridge, New York, USA, 1981.

4.  S. H. Lin and K. L. Yeh, "Looking to Treat Wastewater? Try Ozone," Chemical Engineering, Vol. 100, No. 5, 1993, pp. 112-116.

5.  S. H. Lin and C. H. Wang, "Industrial Wastewater Treatment in a New Gas-Induced Ozone Reactor," Journal of Hazardous Materials, Vol. 98, No. 1-3, 2003, pp. 295-309.

6.  F. Gaia and A. Menth, "Neue Hochleistungs-Ozonerzeu ger und ihr Anwendungspotential," Proceedings 5. OzonWeltkongreß, Berlin, 1981, pp. 325-339.

7.  U. Kogelschatz and B. Eliasson, "Die Renaissance Der Stillen Elektrischen Entladung," Physikalische Blätter, Vol. 52, No. 4, 1996, pp. 360-362.

8.  R. Gadow and G. Riege, "Ozoniser and Method of Manufacturing it," European patent, EP 0 817 756 B1, 1996.

9.  M. Hirth, "Ozonizer," European patent, EP 0 202 501 B1, 1986.

10. M. Labrenz, "Elektrische Gasentladung zur Ozonherstellung," Ph. D. dissertation, University Aaachen, 1983.

11. P. Braumann, "Über die Erzeugung von Gasentladungen zur Herstellung von Ozon," Ph. D. dissertation, University Aachen, 1981.

12. J. Lemmerich, "Die Entdeckung des Ozons und die ersten 100 Jahre Ozonforschung," Sigma Verlag, Berlin, 1990.

13. U. Kogelschatz, "Advanced Ozone Generation," In: S. Stucki, Ed., Process Technologies for Water Treatment, Plenum Press, New York, 1988, pp. 87-120.

14. U. Kogelschatz, "Apparatus for producing Ozone," Europeran patent, EP 0 019 307 A1, 1980.

15. M. Fischer and H. Lang, " Vorrichtung zur Erzeugung von Ozon," German patent, DE 41 07 072 A1, 1991.

16. B. Eliasson and U. Kogelschatz, "Modelling and Applications of Silent Discharge Plasmas," IEEE Transactions on Plasma Science, Vol. 19, No. 2, 1991, pp. 309-323.

17. B. Elvers and S. Hawkins, "Ullmann´s Encyclopedia of Industrial Chemistry A18," VCH Verlag, Weinheim, 1991.

18. L. Pawlowski, "The Science and Engineering of Thermal Spray Coatings," 2nd Edition, John Wiley & Sons Ltd, Chichester, 2008.

19. P. Fauchais, "Understanding Plasma Spraying," Journal of Physics D: Applied Physics, Vol. 37, No. 9, April 2004, pp. 86-108.

20. P. Fauchais, "Understanding Plasma Spraying," Journal of Physics D: Applied Physics, Vol. 37, 2004, pp. 86-108.

21. R. B. Heimann, "Plasma Spray Coating," 2nd Edition, WileyVCH Verlag GmbH & Co. KGaA, Weinheim, 2008.

22. Ohmori, K.C. Park, M. Inuzuka, Y. Arata, K. Inoue, and N. Iwamoto, "Electrical Conductivity of Plasma-Sprayed Titanium Oxide (rutile) Coatings," Thin Solid Films, Vol. 201, No. 1, 1991, pp. 1-8.

23. P. Kofstad, Nonstoichiometry, "Diffusion and Electrical Conductivity in Binary Metal Oxides," Wiley-Interscience, New York, 1972.

24. T. Bak, J. Nowotny and C. C. Sorrel, "Electrical Properties of Metal Oxides at Elevated Temperatures," Key Engineering Materials, Vol. 125-126, No. 1, 1997, pp. 1-80.

25. H. Du, J. H. Shin and S. W. Lee, "Study on Porosity of Plasma-Sprayed Coatings by Digital Image Analysis Method," Journal of Thermal Spray Technologies, Vol. 14, No. 4, 2005, pp. 453-461.

26. S. D. Glancy, "Preserving the Microstructure of Thermal Spray Coatings," Advanced Materials & Processes, Vol. 7, No. 1, 1995, pp. 37-40.

27. R. McPherson, "Formation of Metastable Phases in Flame and Plasma-Preapared Alumina," Journal of Material Sciences, Vol. 8, No. 6, 1973, pp. 851-858.

28. L. Zhao, K. Seemann, A. Fischer and E. Lugscheider, "Study on Atmospheric Plasma Spraying of $Al_2O_3$ Using On-Line Particle Monitoring," Surface and Coating Technologies, Vol. 168, No. 2-3, 2003, pp. 186-190.

29. P. Chráska, J. Dubsky, K. Neufuss and J. Píacka, "Alumina-Base Plasma-Sprayed Materials Part I: Phase Stability of Alumina and Alumina-Chromia," Journal of Thermal Spray Technology, Vol. 6, No. 3, 1997, pp. 320- 325.

30. L. Pawlowsky, "The Relationship Between Structure and Dielectric Properties in Plasma-Sprayed Alumina Coatings," Surface and Coating Technologies, Vol. 35, No. 3-4, 1988, pp. 285-298.

31. S. Yilmaz, "An Evaluation of Plasma-Sprayed Coatings Based on $Al_2O_3$ and $Al_2O_3$ 13 wt% $Tio_2$ with Bond Coat on Pure Titanium Substrate," Ceramics International, Vol. 35, No. 5, 2009, pp. 2017-2022.

32. R. McPherson, "On the Formation of Thermally Sprayed Alumina Coatings," Journal of Material Sciences, Vol. 15, No. 12, 1980, pp. 3141-3149.

33. R. Yilmaz, A. O. Kurt, A. Demir and Z. Tatli, "Effects of $Tio_2$ on the Mechanical Properties of the $Al_2O_3$-$Tio_2$ Plasma Sprayed Coating," Journal of European Ceramic Society, Vol. 27, No. 2-3, 2007, pp. 1319-1323.

34. L. Golonka and L. Pawlowski, "Ceramic on Metal Substrates Produced by Plasma Spraying for Thick Film Technology," Electrocomponent Science and Technology, Vol. 10, No. 2-3, 1983, pp. 143-150.

CHAPTER 3

# Porous Ceramics

*Naboneeta Sarkar and Ik Jin Kim*

[1] Department of Materials Science and Engineering, Institute of Processing and Application of Inorganic Materials (PAIM), Hanseo University, # , Haemi-myun, Daegok-ri, Seosan-city, Chungnam, South Korea

## ABSTRACT

The unique chemical composition and microstructure of porous ceramics enable the ceramic products used in a number of applications such as filtration of molten metals and hot corrosive gases, high-temperature thermal insulation, support for catalytic reactions, filtration of diesel engine exhaust gases, etc. These applications take advantage of special characteristics of porous ceramics such as low thermal mass, low thermal conductivity, controlled permeability, high surface area, low density, and high specific strength. In this chapter, we emphasize on direct foaming method, a simple and versatile approach that allows fabrication of porous ceramics with tailored microstructure along with distinctive properties. Foam stability is achieved upon controlled addition of amphiphiles to the colloidal suspension, which induce in situ hydrophobization, allowing the wet foam to resist coarsening upon drying and sintering.

**Keywords:** Porous ceramics, direct foaming, wet foam stability, Laplace pressure, adsorption free energy, microstructure

## 1. INTRODUCTION

Porous ceramics are widely used in various versatile applications, such as liquid gas filters, catalysis supports, gas distributors, insulators, preforms for metal-impregnated ceramic metal composites, and implantable bone scaffolds [1, 2]. Unlike in metallic or polymeric products, pores have been traditionally avoided in ceramic components because of their inherently brittle nature [3, 4]. However, porous ceramics have become increasingly essential, especially for use in environments involving high temperatures, extensive wear and corrosive media [5, 6]. Porous ceramics are advantageous in such application areas due to their high melting point, tailored electronic properties, and high corrosion and wear resistance, which combine favorably with the features gained by the introduction of voids into the solid material [7-10]. These features include low thermal conductivity, controlled permeability, high surface area, low density, high specific strength, and low dielectric constant. These properties can be tailored for each specific application by controlling the composition and microstructure of the porous ceramic. Changes in open and closed porosity, pores' size distribution, and pores' morphology can greatly affect a material's properties. These microstructural features are highly influenced by the processing route used to produce the porous material [11-15].

Foaming melts by gas injection creates gas bubbles in the liquid by the admixing of gas-releasing blowing agents into the molten metal, or by causing the precipitation of gas which had been previously dissolved in the liquid [16, 17]. The stabilization of such foams can be achieved by surfactants, which form dense monolayers on foam films. The surfactant films can reduce surface tension, increase surface viscosity, and create electrostatic forces to prevent foam from collapsing. The stabilization and destabilization mechanisms of coated bubbles exposed to surfactants to produce metallic foams are discussed elsewhere [18]. Colombo et al. [19] discussed different novel processing methods for cellular ceramics, including the burning out of fugitive pore formers. Established methods of producing porous ceramics employ the burning out of templates. The impregnating of a polymeric template increases struts throughout the material and thus increased the strength of the resulting ceramic foams [20]. The porosity of ceramics produced in this way depends on the template's type, content, and grain size. This limits the maximum useable content of such additives, as too high contents substantially weaken the material. Increased porosity can also be achieved by introducing high-porosity granules—both natural (e.g., diatomite, tripolite, and swelled perlite) and synthesized (e.g., by the crushing of briquettes prepared by foaming) [21]. Chemical formations of gas bubbles within a ceramic mixture can also increase porosity. These include chemical reactions in the ceramic suspension or the decomposition of gas-forming additives. The kinetics of alumina slip swelling for the production of lightweight corundum materials have been investigated [22]. Another method is the impregnation of a polymer cellular matrix with a ceramic suspension and subsequent squeezing out, drying, and thermal treatment to remove the organic components [23]. The addition or embedding of ceramic fibers into the mixture, followed by molding with binders and the subsequent thermal treatment of the molded products, can also yield porous materials [24].

The introduction of air into a colloidal suspension is widely used during processing of highly porous foam ceramics [25, 26]. Uniform, finely cellular foam can be produced by mixing into the ceramic suspension frothing agents that stabilize the resultant three-phase foam. Such cellular structures are preserved under subsequent drying and firing [27]. Much work has sought to develop processing parameters for such syntheses.

Less defective components, as compared with dry processing, have recently been shown to result from the wet processing of powders. It allows better control of the interactions between the powder and the particles and increases the homogeneity of particles' packing in the wet stage, leading to fewer and smaller defects in the final microstructure. This can be achieved either by consolidating the dispersion medium or by flocculating or coagulating the particles in the liquid medium. Such wet methods have recently been developed to incorporate gaseous phases into ceramic suspensions consisting of ceramic powder, solvent, dispersants, surfactants, and gelling agents. The process has been called direct foaming by the hydrophobization of particles' surfaces; the incorporation of the gaseous phase can result from mechanical frothing, injection of a gas stream, gas-releasing chemical reactions, or solvent evaporation [28]. Its simplicity, low cost, and versatility has made it popular for the manufacture of porous ceramics. Fig. 1 outlines common methods of preparing porous ceramics and their corresponding products' degrees of porosity. The fabrication methods of microporous ceramics currently available can be classified as replica techniques, methods that employ sacrificial templates and direct foaming [29]. Ceramics' microstructures and properties depend on their fabrication method. Therefore, consideration of the methods' cost, simplicity, and versatility is important. Stabilization of the introduced species' surfaces is required to overcome coalescence, Ostwald ripening, and phase separation and can be achieved using lower-energy molecules for droplet formation. These provide steric and electrostatic barriers against coalescence [30]. Early twentieth century works by Ramsden and Pickering showed that solid particles adsorbed at liquid-liquid interfaces can stabilize the resulting Pickering emulsions, through the introduced surface active molecules lowering the system's free energy by reducing the liquid-liquid interfacial area [31].

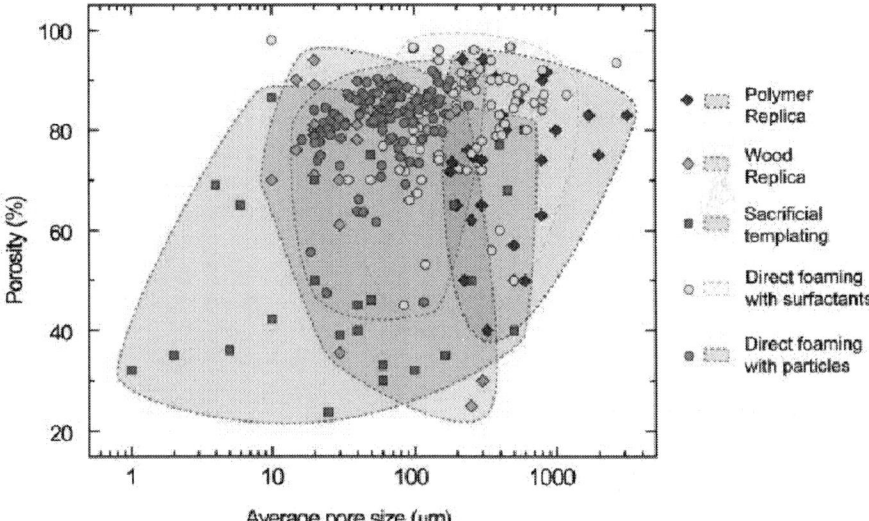

**Figure 1. Typical porosity and average pores sizes achieved via replica, sacrificial templating, and direct foaming routes.[2]**

This chapter explores the stabilization of wet foams by colloidal amphiphilic particles and the development of fabrication techniques of solid macroporous ceramics with tailored microstructures. Each method is discussed and assessed with regard to the versatility and ease of fabrication and its influence on the microstructure and mechanical strength of the resulting macroporous ceramics. Given the importance of ceramics' foam microstructures, the effects of foam precursor suspensions—bubble size, distribution, contact angle, and surface tension—on the resultant porous ceramics' mechanical and physical properties are assessed here. Control of these parameters can allow the tailoring of the microstructures of porous ceramics produced by direct foaming.

## 2. PROCESSING ROUTES TO POROUS CERAMICS

The processes for manufacturing porous ceramics can be classified into following four categories, which is schematically depicted in Fig. 2.

    i.     Replica techniques
    ii.    Sacrificial template
    iii.   Direct foaming

    In this chapter, we reviewed the main processing techniques that can be used for the fabrication of porous ceramics with tailored microstructure. Replica techniques, sacrificial template, and direct foaming techniques is described here and compared in terms of microstructures and mechanical properties that could be achieved. These simple yet versatile approaches give rise to porous ceramics with unique microstructural features that control the properties and functions of the ceramic materials.

### 2.1. Replica Techniques

Replica techniques involve the impregnation of a cellular structure with a ceramic suspension or precursor solution to produce a macroporous ceramic exhibiting a similar morphology to the original porous material (Fig. 2(a)). This is followed by the removal of excess slurry, pyrolysis of the polymeric substrate, and sintering to solidify the foam [6, 13]. Therefore, the ceramic foam replicates the original organic polymer structure. Difficulties of slurry impregnation limit the realization of small cells. The struts contain central holes, which result from the burning out of the polyurethane template. Microcracks and pores also result. Replication generates large amounts of $CO_2$ during firing due to the decomposition of the organic compounds [10, 12]. Suitable biogenic porous structures have been used as templates to form cellular ceramics with particular microstructures that could also be produced by other methods. Those processes for the fabrication of bulk ceramics structures are discussed here.

    This technique, reported in the 1960s, is the first method deliberately used for the production of macroporous ceramics [32]. First, polymeric sponges were used as templates to prepare ceramic cellular structures with various pore sizes, porosities and chemical compositions. In the polymer replica approach, a highly porous polymeric sponge is initially soaked in a ceramic suspension until its internal pores are filled. Binders and plasticizers are also added to the initial suspension to provide ceramic coatings sufficiently strong to prevent the struts from cracking during pyrolysis. This process is explored fully elsewhere [11, 13].

    The resulting ceramic is formed after removal of the polymeric template. The ceramic coating is finally densified by sintering at 1000-1700°C depending on the material. Porous ceramics obtained via sponge replication can reach total open porosity levels of 40%-95% and are characterized by a reticulated structure of highly interconnected pores of between 200 μm and 3 mm. The disadvantages of this technique lie in the formation of the struts of the reticulated structure during pyrolysis of the polymeric template, which significantly weakens the mechanical strength of the resulting porous ceramic [21]. The technique also requires several steps, which lengthen its duration and increases its cost.

## 2.2. Sacrificial Templates

A dispersed sacrificial phase can be homogeneously dispersed throughout a biphasic composite with a continuous matrix of ceramic particles or ceramic precursors. It is ultimately extracted to generate pores within the microstructure (Fig. 2(b)). This method is analogously opposite to replication and results in a negative replica of the original sacrificial template, as opposed to the positive morphology obtained from replication. The method of the sacrificial material's extraction from the consolidated composite depends primarily on the type of pore former employed [33]. A wide variety of sacrificial materials can be used as pore formers, including natural and synthetic organics, salts, liquids, metals, and ceramics. This technique is flexible and can employ various chemical compositions. Various oxides have been used to fabricate porous ceramics using starch particles as sacrificial templates [9, 10]. Nonoxide porous ceramics have also been produced using pre-ceramic polymers and various template materials [34, 35]. Since this method produces a ceramic to the negative of the original template, the removal of the sacrificial phase does not lead to flaws in the struts as can occur using positive replicas. The microstructures obtained by this technique reflect directly the pattern of the sacrificial phase and higher mechanical strengths are generally achievable than by using positive replicas [36, 37].

## 2.3. DIRECT FOAMING

Direct foaming produces porous materials by the incorporation of air into a suspension or liquid medium. The foam structure is then set by high-temperature sintering to obtain crack-free, high-strength porous ceramics. The suspensions are stabilized *in situ* through the hydrophobization of the suspended particles by short chain amphiphilic molecules. The coated, hydrophobic particles irreversibly adsorb to the air-water interface, thus stabilizing it (Figs. 2(c) and 3) [38]. These wet foams can remain stable for several days and show no bubble coarsening, drainage, or creaming. The short-chain amphiphiles modify *in situ* the wetting behavior of the particles' surfaces, as in a Pickering emulsion. Ultrastable wet foams can be produced by direct foaming using particles instead of surfactants as foams stabilizers [16,19, 25]. Porous ceramics' properties are also highly influenced by their chemical compositions and microstructures, with porosity, pore morphology, and size distribution being tailored by different compositions, different physical structures of the starting materials, and the use of different amphiphiles [30-36]. This review focuses on this process.

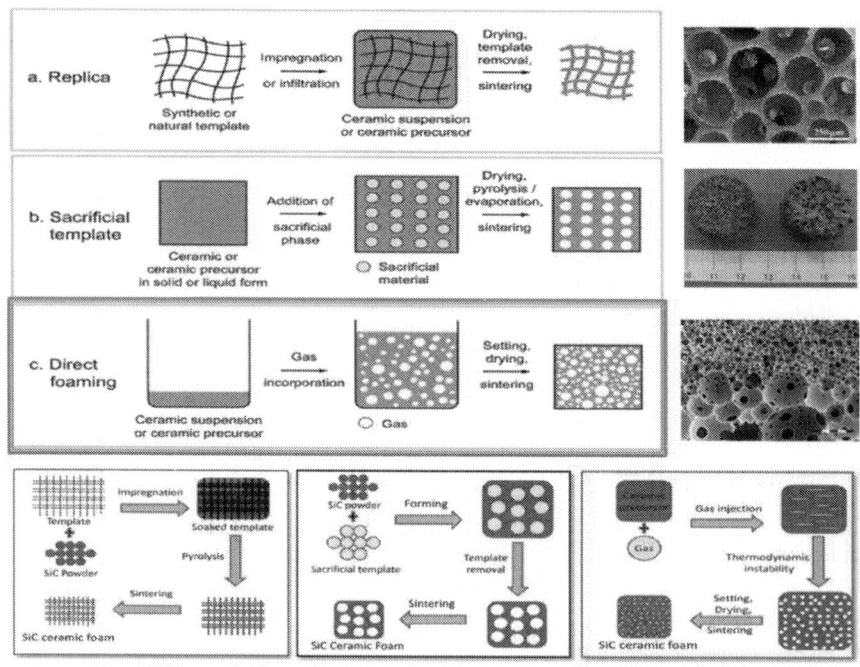

Preparation of SiC ceramic foam by (a) Replica technique, (b) sacrificial Template Method and (c) direct foaming

**Figure 2.** Currently available methods of forming porous ceramics.[2]

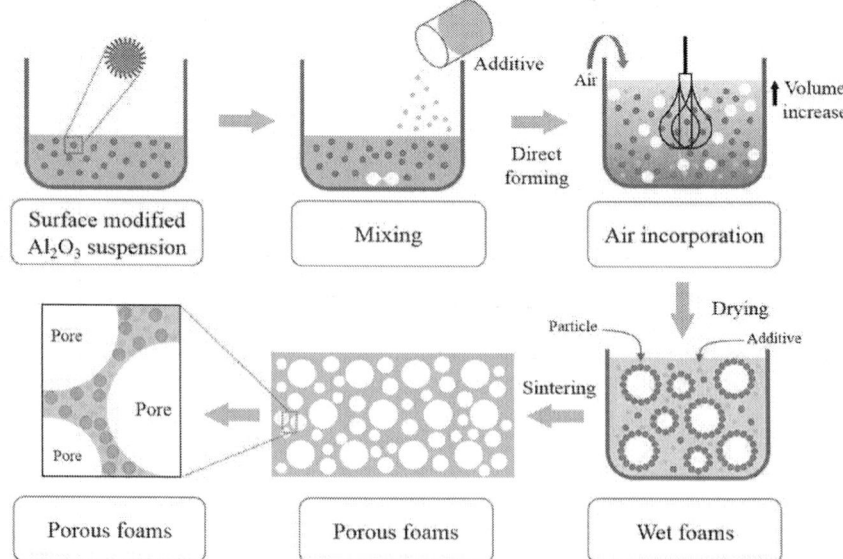

**Figure 3.** *In situ* hydrophobization of particles and solid foam formation by direct foaming.[4]

# 3. PROCESS TO STABILIZATION

## 3.1. Zeta Potential and *In Situ* Hydrophobization

Colloids are suspensions or liquid foams that are generally thermodynamically unstable. The instability arises due to their high gas-liquid interfacial areas, which raise the free energy of the system. To achieve a stable system, free energy must be minimized. The electrokinetic properties of a colloidal system can be described using the zeta potential (Figure 4(a)). Higher charges on the particles' surfaces stabilize a colloidal suspension by preventing the particles from coming into contact and coalescing. Colloids with high zeta potential (negative or positive) are electrically stabilized while colloids with low zeta potentials tend to coagulate or flocculate as shown in Table 1. A suspension's pH affects its charge distribution, and hence its zeta potential. The isoelectric point (IEP) is the pH at which a colloid's zeta potential is zero; it can be used to derive information about the pH ranges in which a colloid is stable. A suspension's pH can be modified to allow dissociated surfactant to adsorb electrostatically as counter ions onto oppositely charged alumina hydroxyl surface groups [39]. The suspension's inorganic particles can be stabilized *in situ* by the particles' hydrophobization with different colloidal particles containing predominantly $-OH_2^+$, $-OH$, and $-O^-$ surface groups. Surfaces with predominantly $-OH_2^+$ and $-OH$ groups can be achieved on inorganic alumina particles at pH 4.5 and pH 9.5, respectively. This could be derived from the zeta potential data for bare alumina particles (Fig. 4(b)), which confirm that the surface exhibits mainly $-OH_2^+$ (positive net charge) and $-OH$ (neutral net charge) groups under those conditions [26, 31, 20]. Amphiphiles of short chain carboxylic acids and gallates are expected to adsorb well onto alumina particles. Propyl gallate has been used to modify the surfaces of particles by ligand exchange reactions [15]. The surface hydroxyl groups ($-OH$ or $-OH_2^+$) were replaced by one or more of the molecule's hydroxyl groups ($-OH$ or $-O^-$). Therefore, the adsorption of gallate molecules does not necessarily require oppositely charged surfaces and amphiphiles and can be used at pH values at which the surface groups and the molecules exhibit the same charge polarity. Hydrophobizing adsorption can change the wettability of particles at the interface of two immiscible phases, and the system is stabilized by the neutral forces between the particles and the amphiphilic coatings. Therefore, the choice of amphiphile depends upon the IEP and the zero net charge of the oxide. Surface hydrophobization can be accomplished by choosing amphiphiles with functional groups that react with the surface hydroxyl groups. Pyrogallol groups can efficiently adsorb on oxide surfaces via ligand exchange reactions [14, 16] and thus can be used with a short hydrocarbon tail to modify the surfaces of particles with intermediate IEPs. The selection of amphiphiles with suitable head groups and tail lengths allows the surface hydrophobization of particles of various compositions.

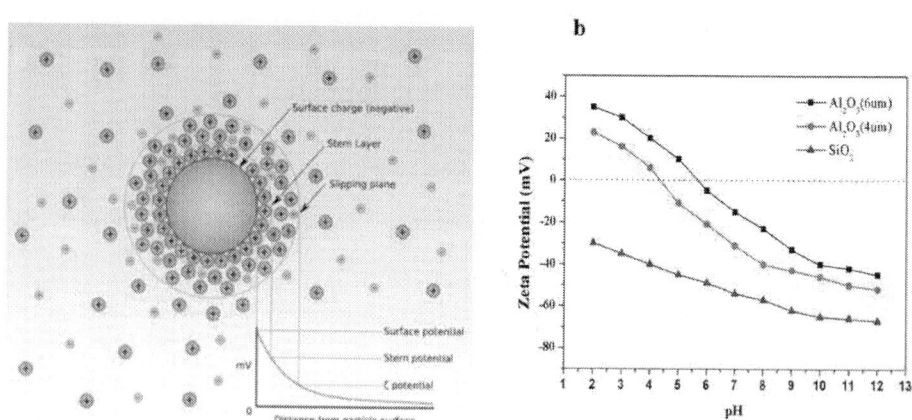

**Figure 4.**(a) The distribution of charges in a colloidal suspension; higher charges at the particles' surfaces can stabilize the system. (b) Zeta potential of raw $Al_2O_3$ and $SiO_2$ colloidal particles.

**Table 1.**Zeta potential as key indicator of the stability of colloidal dispersions

| Zeta potential [mV] | Stability behaviour of the colloid |
|---|---|
| From 0 to ±5 | Rapid coagulation or flocculation |
| From ±10 to ±30 | Incipient instability |
| From ±30 to ±40 | Moderate stability |
| From ±40 to ±60 | Good stability |
| More than ±61 | Excellent stability |

## 3.2. Destabilizing Suspension

Colloidal dispersions can be thermodynamically unstable, with long-term kinetic stability determining their self-life. The main destabilization mechanisms are drainage (creaming and sedimentation), coalescence, and flocculation (Fig. 5). Creaming and sedimentation are caused by gravity: lighter particles float and heavier particles settle. They are reversible in that mechanical agitation (homogenization or simple shaking) will redisperse the suspension. Coalescence and flocculation are not reversible and so affect a suspension's stability. Flocculation is the clustering of colloidal particles via attractive van der Waals forces. It can be overcome or prevented by higher-energy ultrasonification or by generating particles with repulsive interactions [40]. Coalescence is the greatest destabilizing mechanism. It involves smaller particles collapsing into each other, forming larger particles with different properties. Many dispersion techniques have been developed to prevent coalescence [41].

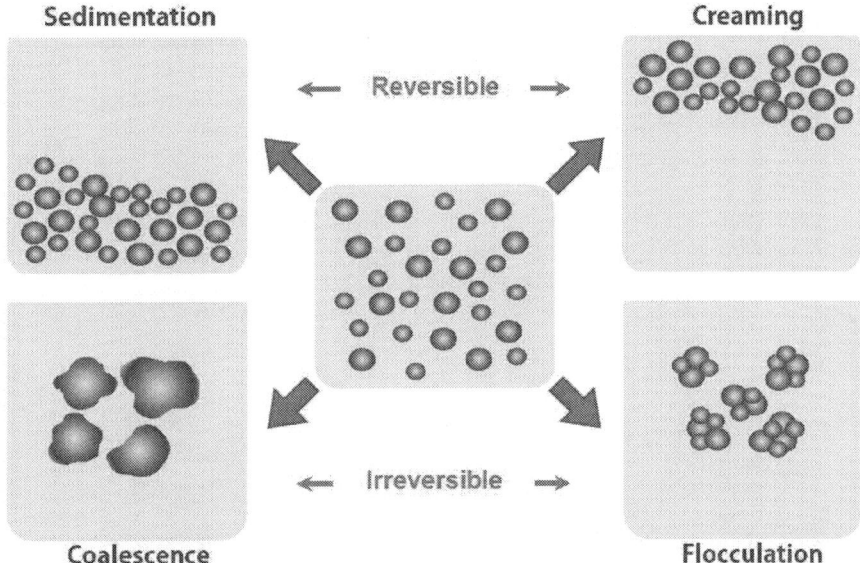

Sedimentation                                          Creaming

← Reversible →

← Irreversible →

Coalescence                                            Flocculation

Figure 5. The destabilization of colloidal suspensions.

## 3.3. Suspension Stability

The foams require the adsorption of particles on the surfaces of air bubbles upon their formation. Alumina particles can be hydrophobized by modification with short-chain carboxylic acids: the carboxylate groups adsorb to the alumina's surface [42], leaving the hydrophobic tail in contact with the aqueous solution. This has been shown to stabilize the dispersion [43]. The hydrophobicity imparted by the first layer of depronated amphiphiles adsorbed onto the surface leads to an energetically unfavorable exposure of hydrophobic species to the aqueous phase. This favors the adsorption of additional molecules from the aqueous phase onto the particles' surfaces to decrease the system's free energy, which determines the stability of a suspension or wet foam. Particles attached to foam and mists' gas-liquid interfaces lower the overall free energy by replacing part of the interfacial area rather than reducing the interfacial tension, as in the case of surfactants [5]. The energy of the attachment, i.e., the Gibbs free energy (G), gained by the adsorption of a particle of radius $r$ at the interface can be calculated using simple geometrical arguments that lead to the following equation (Fig. 6).

$$G = \pi r^2 \Upsilon_{LG} \left(1 - \cos\theta\right) \text{ for } \theta < 90°,$$

where $\theta$ is the contact angle and $_{LG}$ is the gas-liquid interfacial tension. While the maximum energy gain can only be achieved at $\theta = 90°$, contact angles as low as 20° can yield attachment energies in the order of $10^3$ kT in systems of 100 nm particles [2]. The high energy associated with the adsorption of particles at interfaces contrasts to low adsorption energies of surfactants and leads to foams stabilized by particles being more stable than those stabilized with surfactants. It also leads to steric layers which strongly hinder bubbles' shrinkage and expansion, minimizing Ostwald ripening for very long periods of time [46].

The particle systems described in Fig. 6 had adsorption achieved by ligand exchange, whereby a surface hydroxyl group is exchanged for another group. This occurred because of the favorable change in the surface charge by the removal of $(OH_2^+)$, a better leaving group, and replacement with (-OH) [44, 45].

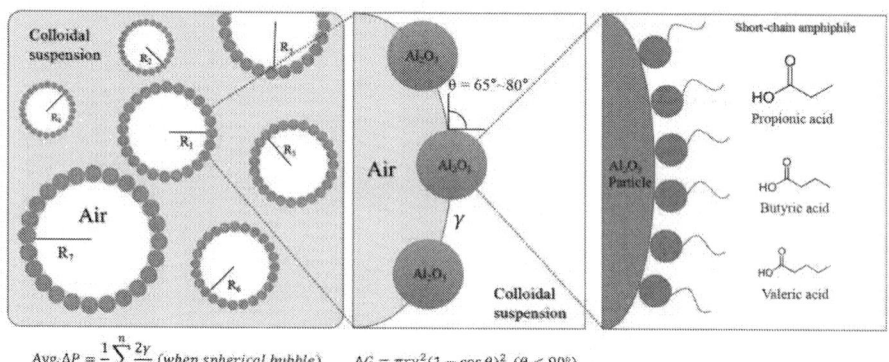

$$Avg. \Delta P = \frac{1}{n}\sum_{i=1}^{n}\frac{2\gamma}{R_n} \ (when\ spherical\ bubble) \qquad \Delta G = \pi r \gamma^2 (1 - \cos\theta)^2 \ (\theta < 90°)$$

**Figure 6.** Foams produced through the adsorption of colloidal particles at the gas-liquid interface.

## 3.4. Contact Angle and Surface Tension

After the stabilizing effects of zeta potential and pH, contact angle and surface tension are important determinants of colloidal systems' properties. Once a suspension is stabilized, the degree of hydrophobization is the main property which affects the production of foam. Given their thermodynamic instability, foams are often kinetically stabilized through the adsorption of surface active molecules or colloidal particles at the gas-liquid interfaces [46, 47]. The adsorbed molecules and particles stabilize the system by inhibiting the coalescence and Ostwald ripening of droplets and bubbles. Adsorption at the fluid interfaces occurs when particles are not completely wetted by any of the fluids, thus exhibiting a finite equilibrium contact angle at the triple phase boundary.

The equilibrium contact angle ($\theta$) is determined by the balancing of the interfacial tensions (Equation 1). A decrease in surface tension upon increasing the initial amphiphile concentration can be observed. However, above a critical amphiphile concentration, surface tension decreased sharply. Above this critical amphiphile concentration, the particles are sufficiently hydrophobic at the air-water interface and decrease surface tension more greatly than that expected from free amphiphiles alone [48]. This significant reduction in surface tension upon particle adsorption was caused by a decrease of the total area of the highly energetic air-water interface. Similar surface tension effects have been observed in systems employing various amphiphiles [18]

Controlling particles' contact angles at the interface is important as it determines their wettability (Fig. 7). Tailoring particles' contact angles via modification of chemical composition enables the creation of foams with a variety of functionalities [19]. Contact angle depends on surface chemistry, roughness, impurities, particle size, and fluid phase composition. Theoretical and experimental work has shown that stabilization is achieved when contact angles are of an intermediate range of 20-86° for oil-in-water foams and of 94-160° for water in oil foams [49]. Contact angle can also be tailored by changing the particles' surface chemistry or adjusting the composition of the fluids. Metallic and ceramic particles can achieve any contact angle ($0 < \theta < 180°$) by reacting or adsorbing hydrophobic molecules on their surfaces [28, 29]. The use of short amphiphiles to tailor particles' wettability is a general and versatile approach for the surface modification of a wide range of ceramic and metallic materials [20].

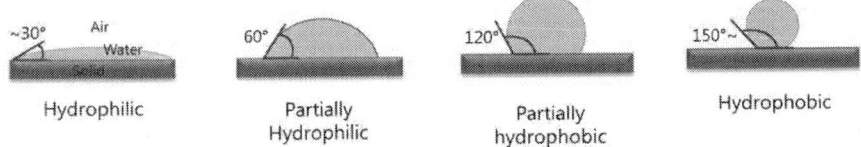

**Figure 7.** The wettability of particles in immiscible phases.[2]

## 3.5. Wet Foam Stability

Liquid foams are thermodynamically unstable due to their high gas-liquid interfacial area. Several physical processes can occur to decrease the overall free energy and destabilize the foam [36]. Drainage occurs through gravity; light gas bubbles rise forming a denser foam layer, while the heavier liquid phase is concentrated below. Coalescence takes place when the thin films formed after drainage is not stable enough to keep adjacent cells apart. Their collapse results in the joining of neighboring bubbles. The stability of the thin films is therefore described in terms of attractive and repulsive interactions between the bubbles. van der Waals forces drive the bubbles closer. They can be overcome by electrostatic forces, steric repulsions force, or ligand exchange reactions. Surfactant or particles adsorbed at the air-water interface can also reduce van der Walls forces [22]. Ostwald ripening or disproportionation is another destabilizing effect that is more difficult to overcome. It occurs due to differences in the Laplace pressures between bubbles of different sizes. Laplace pressure inside a gas bubble arises from the curvature of the air-water interface. The Laplace pressure ($N/m^2$) is the pressure difference between the inner and the outer side of a bubble or droplet. For spherical bubble of radius $R$ and gas-liquid interfacial energy $\gamma$, the Laplace pressure $\Delta P$ is given by $2\gamma/R$. The pressure and force generated for the stabilization can be also calculated through the measurement of bubbles at the intersection. It can be calculated by the equation given below.

$$\Delta P = \gamma \left( \frac{1}{R_1} + \frac{1}{R_2} \right) = \frac{2\gamma}{R}\left(\text{spherical bubble}\right)$$

The difference in the Laplace pressure between bubbles of distinct sizes (R) leads to bubble disproportionation and Ostwald ripening because of the steady diffusion of gas molecules from smaller to larger bubbles over time. This process can be slowed by using surfactants or particles adsorbed at the interface, which decrease the interfacial energy. Wet foam's stability is also related to the degree of hydrophobicity achieved from the surfactant, which replaces part of the highly energetic interface area and lowers the free energy of the system, leading to an apparent reduction in the surface tension of the suspension [49]. Stability also depends on surface charge screening, the electrical diffuse layer around a particle's surface not sufficiently thick to overcome the attractive van der Waals forces between particles. Overcoming the van der Waals attractions requires a stable hydrophobizing mechanism (examined above). Therefore, experiments were conducted as per reported theoretical explanations [18].

These actions' combined effects may collapse the foam within minutes after air incorporation. Foams' life times have been increased from several hours to days and months by the adsorption of the short chain amphiphilic molecules [50], while only a few minutes or hours' stabilization results from the use of long-chain surfactants or proteins at the air-water interface [35]. Unlike other particle-stabilized foams [2], these foams percolate throughout the whole liquid phase and exhibit no drainage over days and months [49] due to the high concentration of modified particles in the initial suspension, which allows for the stabilization of very large total air-water interfacial areas.

# 4. RESULTS AND DISCUSSIONS

## 4.1. CONTACT ANGLE AND SURFACE TENSION

The attachment of particles at gas liquid interfaces occurs when particles are not completely wetted or, in other words, are partially hydrophobic. This enables the production of high-volume stable foam, which produces porous ceramics after drying and sintering. Partially hydrophobic particles remain predominantly in the liquid phase and exhibit a contact angle <90°. Therefore, controlling contact angles of the particles at the interface is important since the angles modify the wettability of the particles by changing their hydrophobicity, as shown in (Figs. 8-10). Generally, lower contact angles improve the wettability. Different contact angles can be achieved by imparting different hydrophobic molecules commonly known as surfactants.

It is shown from Fig. 8 that the average contact angle of the $d_{50} \sim 40$ nm $Al_2O_3$ suspension decreased from 84° to 67° with the increased $SiO_2$ content (1.0 mole ratios in the $Al_2O_3$ suspension). Also, the increasing $SiO_2$ content produced lower adsorption free energy due to the higher interparticle attraction, increasing the viscosity. The suspensions with mol ratios of $SiO_2$ between 0.25 and 0.5 in the suspension show higher levels of attachment energy, resulting in highly stable foam in the sintered porous ceramics. Also, contact angle of around 70°-75° for the nanoparticle suspension leads to better wet foam stability and can give surface tensions of 21-33 mNm$^{-1}$. The required partial hydrophobization of the particles occurs at this point, which leads to porous ceramics with higher porosity.

Fig. 9 shows the effect of suspension added for the mullite phase on the contact angle and surface tension of the aqueous suspensions. From this graph, we can see the suspension exhibits contact angles of 46°-55°, which enables high wet foam stability, as that indicates partial hydrophobization of particles has taken place. We observed that for all the evaluated samples, the surface tension of suspensions decreases, upon increasing the vol% of suspension added for the mullite phase. This can be explained by an increase in surface hydrophobicity of the particles with increasing particle concentration.

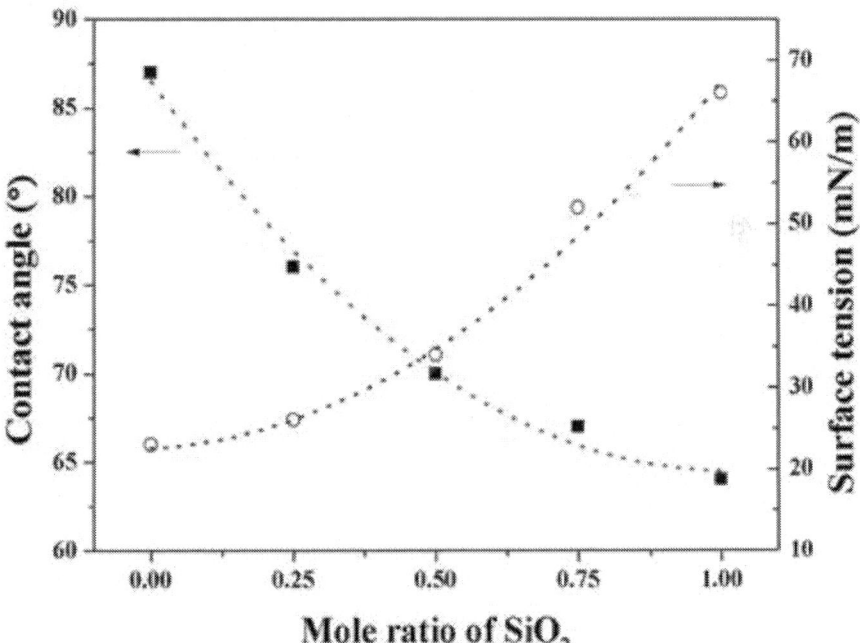

**Figure 8.**Contact angle and surface tension of colloidal suspension with respect to different mole ratio of $SiO_2$.

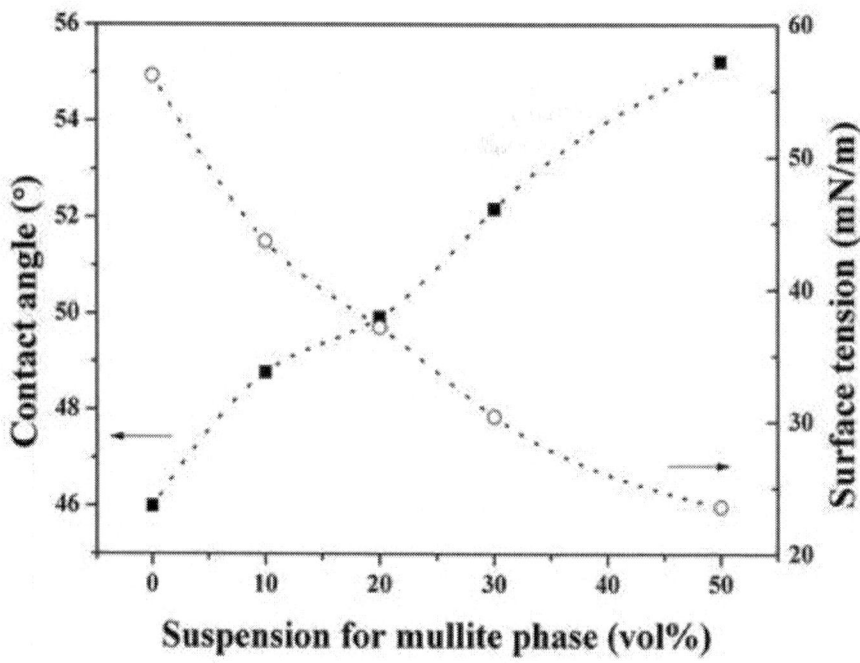

**Figure 9.**Contact angle and surface tension of $Al_2O_3$-$TiO_2$equimolar suspension, with respect to different vol% of 3:2 mole ratio of $Al_2O_3$-$SiO_2$ suspension added for the mullite phase[32]

In Fig. 10, the hydrophobization achieved via amphiphile adsorption was confirmed by contact angle measurements of the aqueous suspensions. As we can see, the 0.05-mol/L concentration of propionic acid was not sufficient enough to impart particle hydrophobicity, which results in unstable foams. From this graph, we can conclude that a contact angle of 65°-72° produces required particle hydrophobicity which enables high wet foam stability.

In Fig. 11, the degree of particle hydrophobization achieved by imparting different concentration of amphiphile was investigated with the help of surface tension measurements. The surface tension of suspensions containing 30 vol% particles and different concentration of amphiphiles is shown in Fig. 11. A decrease in surface tension upon increasing the amphiphile concentrations is observed for all the evaluated suspensions. The reduction in surface tension results from the adsorption of free amphiphile molecules to the air-water interface. The middle and short chain amphiphiles, i.e., butyric acid and valeric acid, respectively, impart relatively low surface energy, which enables sufficient hydrophobicity on the particle surface than the shortest chain amphiphile, i.e., propionic acid, does.

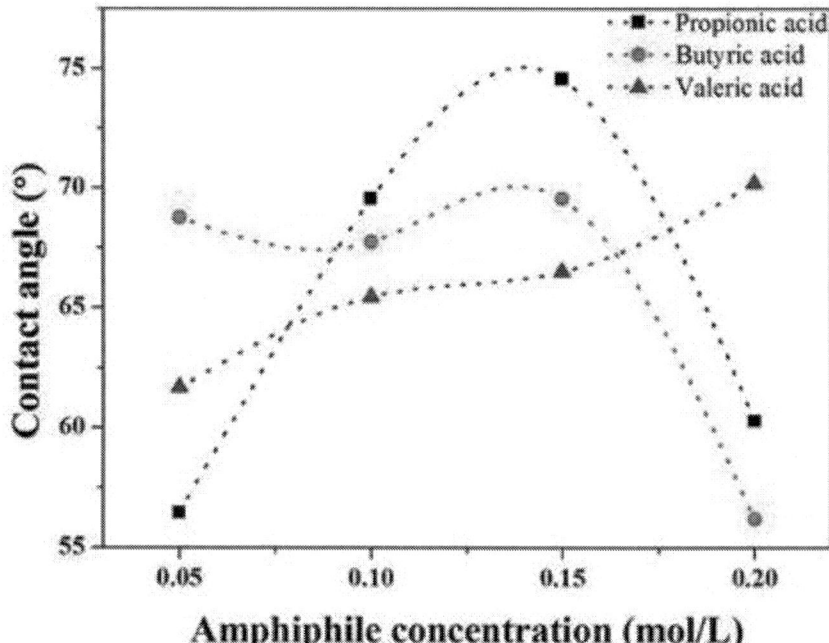

**Figure 10.** Contact angle of suspension with respect to different concentrations of amphiphiles.[4]

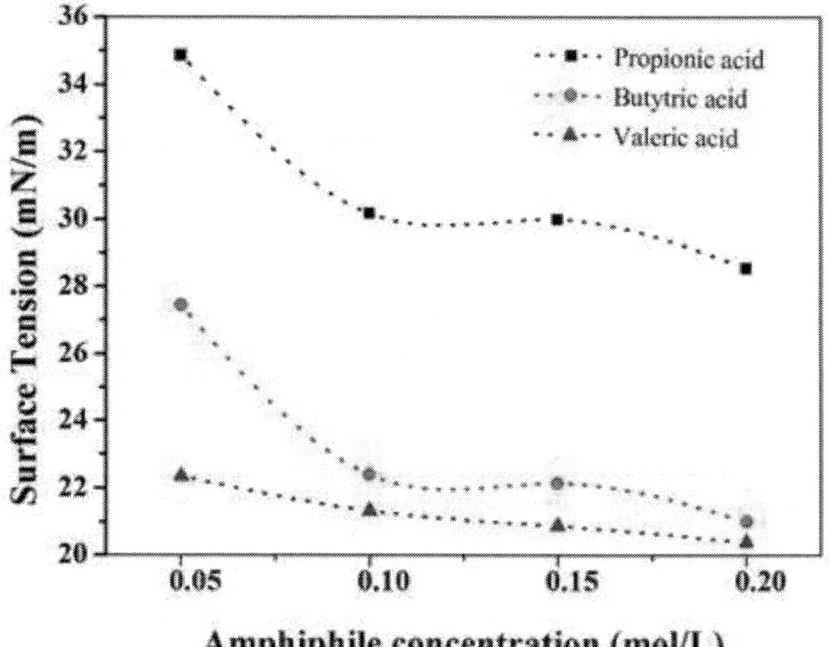

**Figure 11.** Surface tension of suspension with respect to different concentration of amphiphiles.[4]

## 4.2. AIR CONTENT AND WET FOAM STABILITY

The total porosity of directly foamed ceramics is proportional to the amount of air incorporated into the suspension or liquid medium during the foaming process. The pore size, on the other hand, is determined by the stability of the wet foam. High-volume foams are formed upon mechanical frothing which strongly indicates the stabilization of air bubbles due to the attachment of particles to the air-water interface.

In Fig. 12, a relationship between the air content and the different concentration of amphiphiles has been plotted. It can be seen that for all three amphiphiles, the air content gradually increases until it achieves the highest value. This is because the particles were not sufficiently hydrophobized below this concentration (i.e., 0.15 m/mol for propionic acid and 0.10 m/mol for valeric acid). All the investigated suspension reports highest air content (i.e., 69% in case of propionic acid and 58-60% in case of butyric acid and valeric acid) upon achieving sufficient hydrophobization. The decrease in air content at high amphiphile concentration is due to increase of suspension viscosity which resists the air incorporation to the suspension.

In Fig. 13, the influence of the amphiphile concentration on the wet foam stability is displayed. At very low amphiphile concentrations, no stable foams are obtained since the alumina particles are not sufficiently hydrophobized to stabilize the air-water interface of freshly formed air bubbles. Using 0.10 mol/L of amphiphile results in rather unstable foam with wet foam stability of about 72-77%. At a certain point between 0.15 and 0.2 mol/L of amphiphile concentration, wet foams with highest stability are obtained. Propionic acid having the shortest hydrophobic chain requires more concentration (0.20 mol/L) to result sufficient hydrophobization. However, the middle and long chain amphiphiles, i.e., butyric acid and valeric acid, respectively, produce effective hydrophobicity at around 0.15 mol/L. More concentrations of them increase the suspension viscosity results from increasing hydrophobicity of the particles, which prohibits the suspension to be foamed by mechanical stirring.

Fig. 14 establishes the air contents and foam stability of $Al_2O_3$-$TiO_2$ equimolar suspension, with respect to different vol% of 3:2 mole ratio of $Al_2O_3$-$SiO_2$ suspension added for the mullite phase. High-volume foams with air content up to 83% form upon mechanical frothing, which strongly indicates the stabilization of air bubbles, due to the attachment of particles to the air-water interface. We measured the foam stability and observed that on the addition of 10 vol% suspension for the mullite phase, the foam stability suddenly decreased. This is probably due to the high viscosity of the suspension, due to higher particle concentration. However, 20, 30, and 50 vol.% of addition enhanced the foam stability, which might be explained by the optimum surface hydrophobicity being achieved, due to the increased particle concentration.

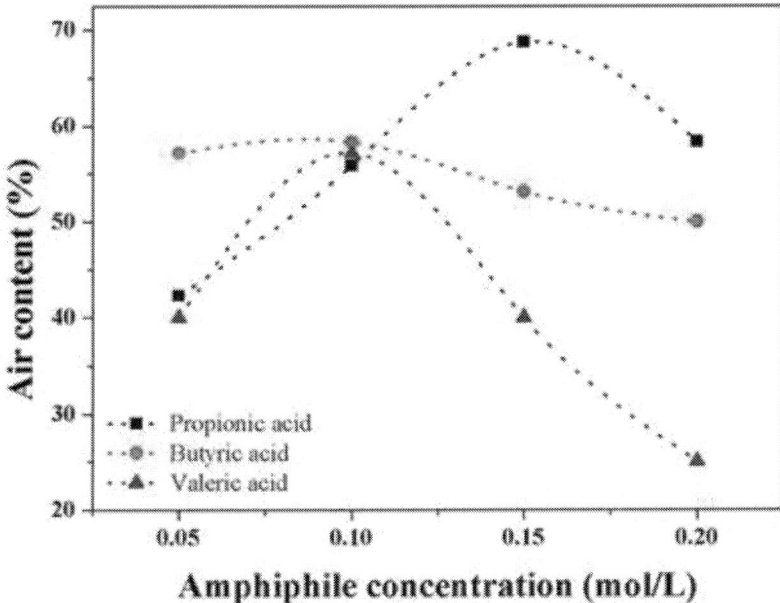

**Figure 12.** Air content of suspension with respect to different concentration of amphiphiles.[4]

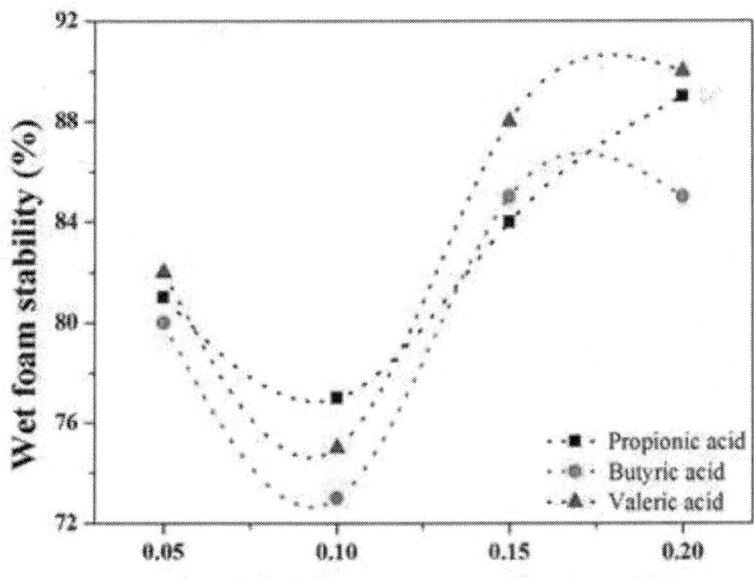

**Figure 13.** Wet foam stability of suspension with respect to different concentration of amphiphiles.[4]

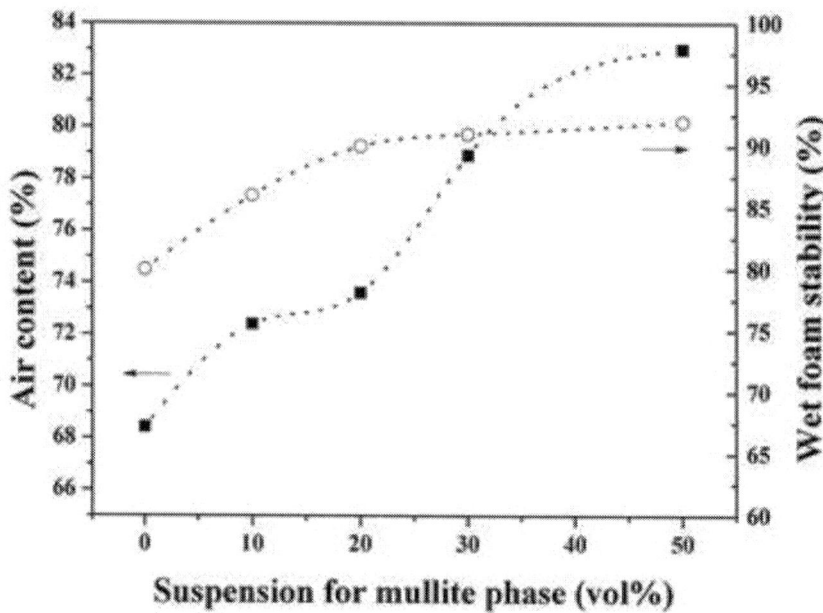

**Figure 14.** Air content and foam stability of $Al_2O_3$-$TiO_2$ equimolar suspension, with respect to different vol% of 3:2 mole ratio of $Al_2O_3$-$SiO_2$ suspension added for the mullite phase.[32]

In Fig. 15, the wet foam stability can be determined by observing the average bubble size with respect to the time after foaming. The foams stabilized with butyric acid and valeric acid show no significant bubble growth unlike the foam stabilized with propionic acid which shows a little coarsening. We can attribute the first two cases of remarkable resistance to the irreversible adsorption of the partially hydrophobized particles at the air-water interface. Therefore, the bubble size remains almost constant with the increase of time up to 6 hours of foaming. The foams stabilized with propionic acid are prone to bubble coarsening due to the pressure difference between two bubbles of different radius which leads to Ostwald ripening. This thermodynamically driven spontaneous process occurs because the internal pressure of a particle is indirectly proportional to the radius of the particle. Large particles, with their lower surface to volume ratio, result in a lower energy state, whereas the smaller particles exhibit higher surface energy. As the system tries to lower its overall energy, molecules on the surface of a small particle tends to detach. It diffuses through colloidal solution and attaches to the surface of larger particle. Therefore, the number of smaller particles continues to shrink, while larger particles continue to grow [17].

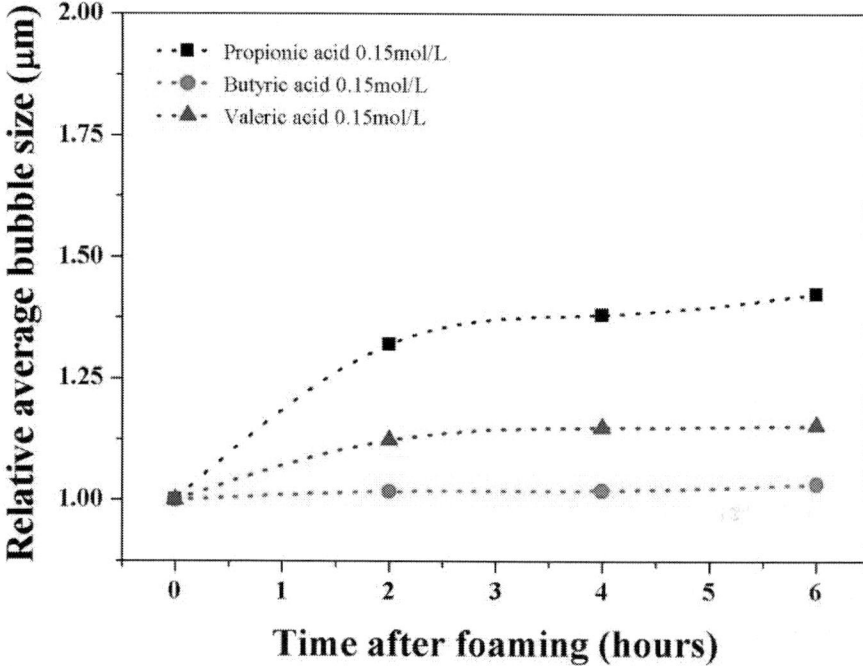

**Figure 15.** Relative average bubble size of suspension with respect to time after foaming.[4]

## 4.3. Adsorption Free Energy

The adsorption free energy plays an important role in stabilizing foams. Particles attached to the gas-liquid interfaces of foams lower the system free energy, by replacing part of the gas-liquid interfacial area. According to Equation (1), G (the Gibbs free energy) is greatest when $\theta$ is 90°; however, the foam stabilization of particles readily occurs when $\theta$ is between 50° and 90°.

Fig. 16 shows the change in the adsorption energy corresponding to the different mole ratio of $SiO_2$ content used to stabilize the suspension. An $Al_2O_3$ loading of 30 vol.% in the suspension was taken as a standard, and experiments were performed with 0.01 mol L$^{-1}$ amphiphiles for stabilization of the particles. The calculations show that the energy level decreases with the nanoparticle size and with increase in $SiO_2$ content. However, after the middle value (0.75) of the $SiO_2$ loading, the van der Waals attraction force between the particles gradually increases, forcing the suspension to destabilize and finally decrease the wet foam stability from 87% to 68%. A higher energy of adsorption of $1.7\times10^8$ kTs could be achieved in the initial suspension without $SiO_2$ content. The adsorption free energy decreases with the increasing concentration. Higher contact angle of 62°-75° with a lower interfacial energy of $1.7\times10^8$ kTs were seen at $SiO_2$ mole ratio of 0.25 giving an interfacial tension of 42-45 mNm$^{-1}$.

Fig. 17 establishes the relationship between adsorption free energy corresponding to the foam stability, with respect to the different vol.% of suspension added for the mullite phase. Low adsorption free energy resulting from the spontaneous bubble growth leads to foam instability. The investigated samples exhibit much higher adsorption free energy of about $2.2 \times 10^{-13}$ J to $2.7 \times 10^{-13}$ J at the interface, resulting in irreversible adsorption of particles at the air-water interface, which leads to outstanding stability.

In Fig. 18, a relationship between adsorption free energy corresponding to the concentrations of different chain length of amphiphile has been established. Stable and unstable zones have been described relating to the data obtained by the wet foam stability graph. Low adsorption free energy (e.g., $2.05 \times 10^{-13}$ J to $3.78 \times 10^{-13}$ J) results from the spontaneous bubble growth leads to foam instability. However, higher adsorption free energy of about $4.52 \times 10^{-13}$ J to $8.22 \times 10^{-13}$ J at the interface results in irreversible adsorption of particles at the air-water interface which leads to outstanding stability.

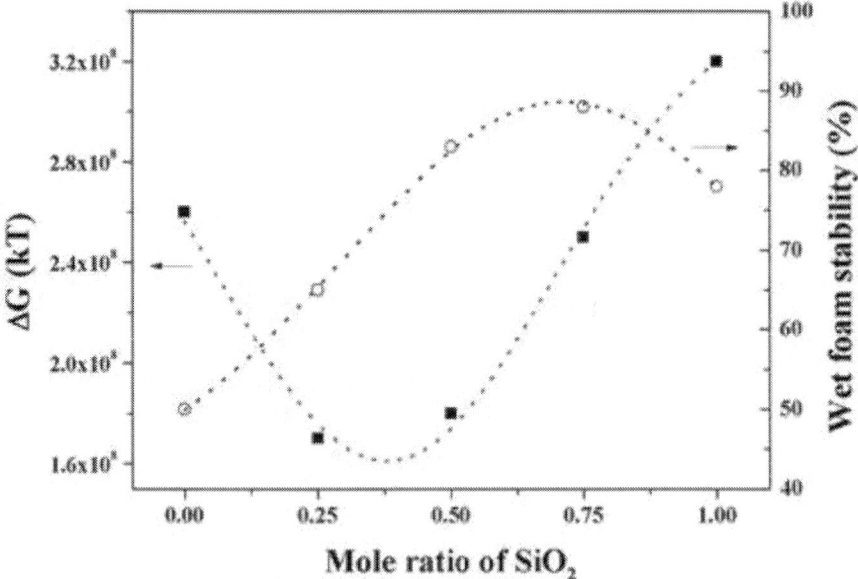

**Figure 16.** Free energy and wet foam stability with respect to the different mole ratio of $SiO_2$.

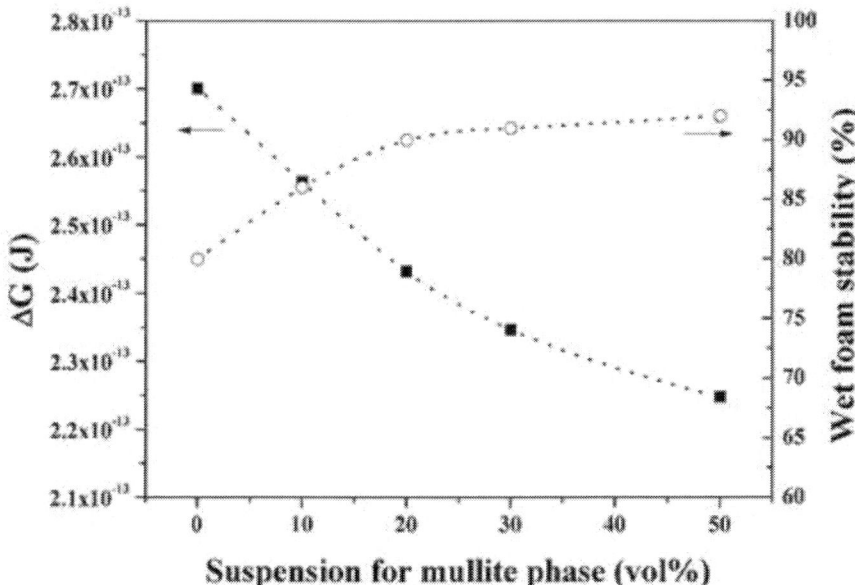

**Figure 17.** Adsorption free energy and foam stability of $Al_2O_3$-$TiO_2$ equimolar suspension, with respect to different vol% of 3:2 mole ratio of $Al_2O_3$-$SiO_2$ suspension added for the mullite phase.[32]

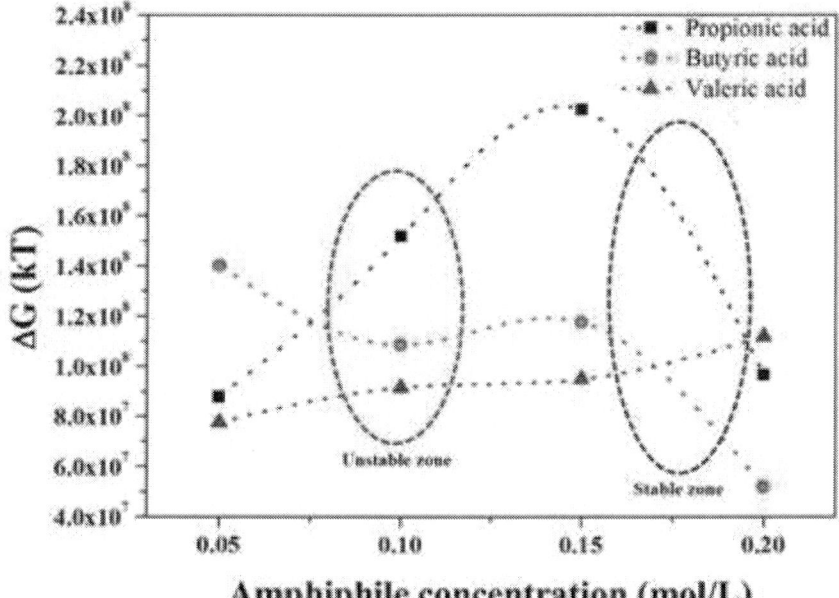

**Figure 18.** Adsorption free energy of suspension with respect to different concentration of amphiphiles.[4]

## 4.4. Laplace Pressure and Bubble Size

Fig. 19 shows the wet foam stability corresponding to the pressure exerted by the bubbles (Laplace pressure) with respect to the different mole ratio of $SiO_2$ content. The Laplace pressure increases with the increase in $SiO_2$ concentration. This behavior can be attributed to the fact that high silica content requires a large volume of water in the suspension, which subsequently lowers the outer pressure of the bubble. The wet foam stability suddenly decreases due to high Laplace pressure when the mole ratio of $SiO_2$ reached at 0.60. The wet foams were stable at the pressure difference between 20 and 25 mPa, which corresponds to the $SiO_2$ mole ratio content of 0.25-0.50. The stability increased to more than 80% at a $SiO_2$ mole ratio of 0.75.

Fig. 20 plots the graph of the Laplace pressure and average bubble size of all evaluated suspensions, with respect to the various vol.% of suspension for the mullite phase. As we can see, instability occurs when the Laplace pressure is too low. Wet foam stability occurs when the Laplace pressure is about 1.5-2.2 mPa. The degree of particle hydrophobization influences the average bubble size of the resultant foams. Fig. 20 shows that the average bubble size decreases with increasing particle concentration and particle hydrophobicity. This is due to the decrease in surface tension and increase in foam viscosity that result from higher particle concentrations. This reduces the resistance of air bubbles against rupture and thus leads to the production of foams with average bubble sizes.

In Fig. 21, the Laplace pressure of all evaluated suspensions has been plotted in a graph with respect to the various concentration of different chain length of amphiphile. As we can see, the instability occurs when the Laplace pressure is too low as in case of 0.10 mol/L of amphiphile concentration. Wet foam stability occurs when Laplace pressure is about 0.8-1.4 mPa. Valeric acid, having the longest chain length, exhibits high Laplace pressure resulting in outstanding stability of wet foam.

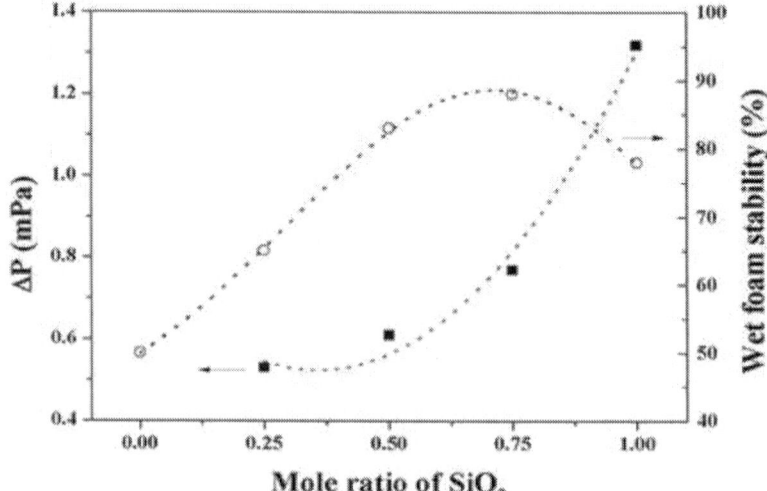

**Figure 19.** Laplace pressure and wet foam stability with respect to the different mole ratio of $SiO_2$ content.

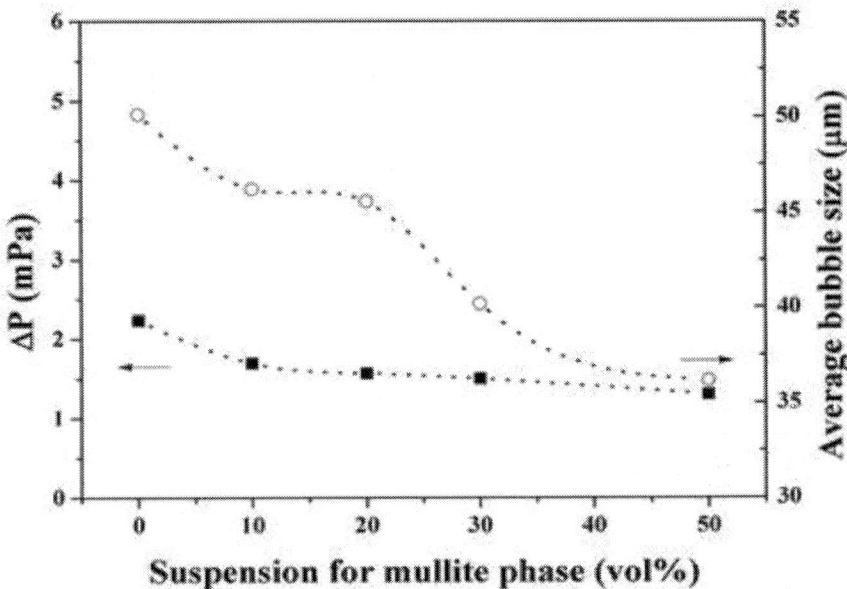

**Figure 20.** Laplace pressure and bubble size of $Al_2O_3$-$TiO_2$ equimolar suspension, with respect to different vol% of 3:2 mole ratio of $Al_2O_3$-$SiO_2$ suspension added for the mullite phase.[32]

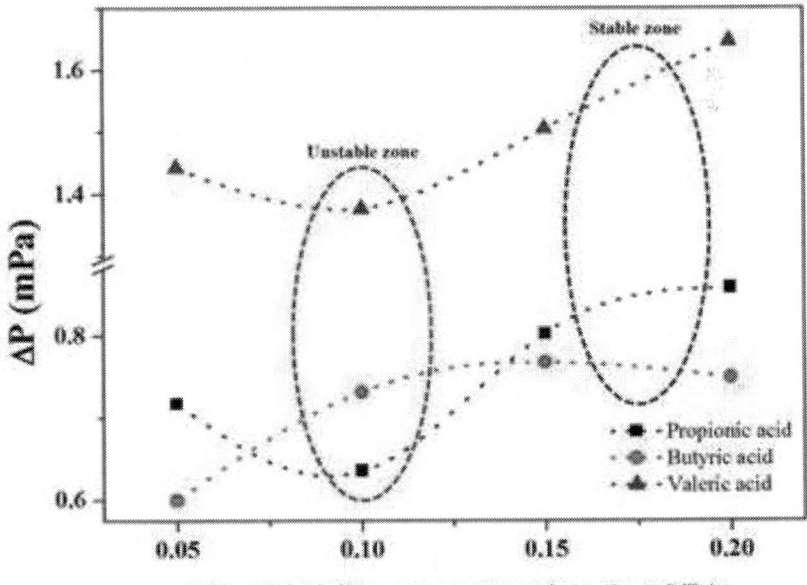

**Figure 21.** Laplace pressure of suspension with respect to different concentration of amphiphiles.[4]

The degree of particle hydrophobization, which is directly related to the concentration of amphiphile, influences the average bubble size of the resultant foams. Fig. 22 shows the bubble size of the suspension and the pore size by thin film or struts formed after the foaming of the particle stabilized suspension and sintering. The average bubble size for these types of stabilized foams was 98-140 μm. The required partial hydrophobization of the particles occurs at this point, which leads to porous ceramics with porosity greater than 80% and pore size of about 108 μm after sintering at 1300°C for 1 hour.

In Fig. 23, it can be seen that the average bubble size decreases with increasing amphiphile concentration and particle hydrophobicity. This is due to the decrease in surface tension and increase in foam viscosity because of higher amphiphile concentrations. This decreases the resistance of air bubbles against rupture and thus leads to produce foams with average bubble sizes. It is interesting to note that valeric acid, having the longest amphiphilic chain, produces very small sized bubbles of about 35-25 μm. This can be attributed to the greater hydrophobicity, which results in enhanced stability of particle stabilized foams against bubble coalescence and Ostwald ripening [see Fig. 26(c)].

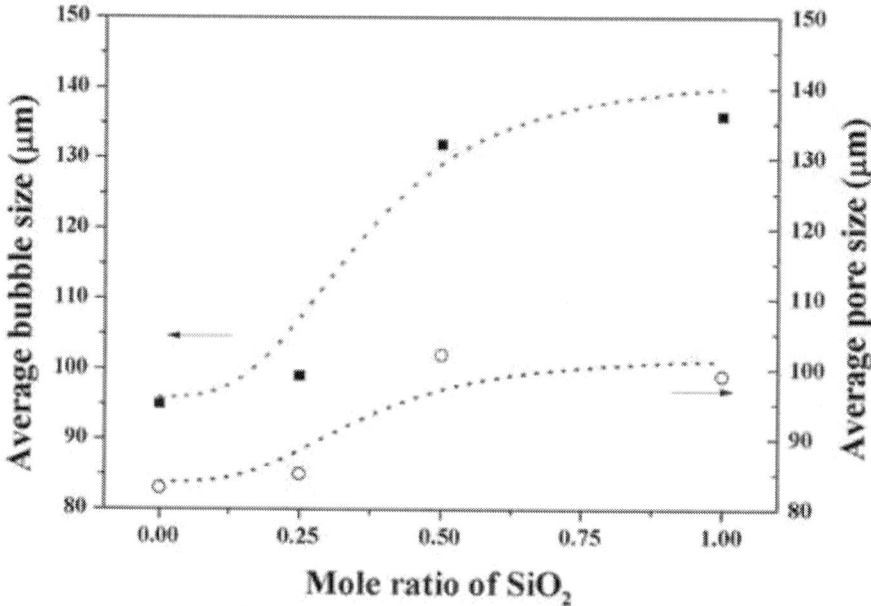

**Figure 22.** Bubble size and pore size with respect to the $SiO_2$ content of the wet foam before and after sintering at 1300°C for 1 hour.

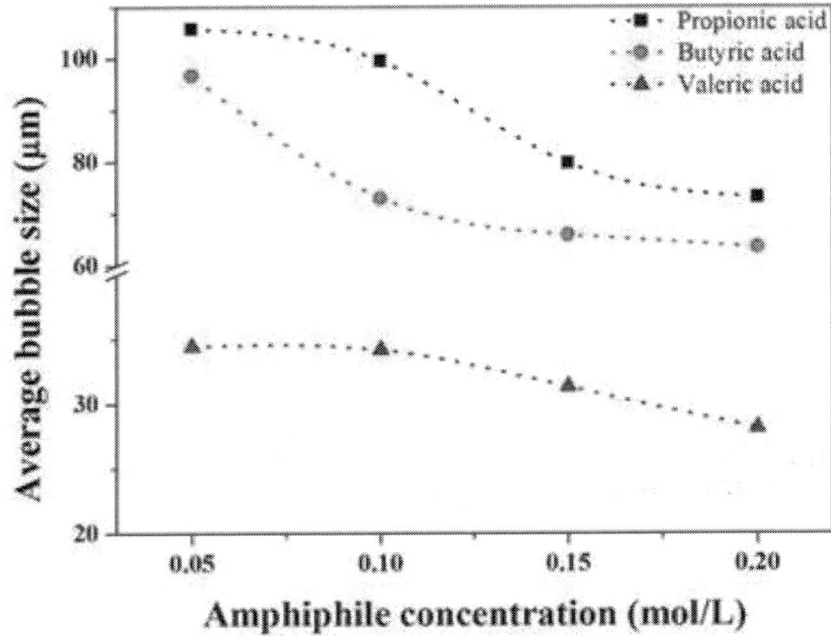

**Figure 23.**Average bubble size of suspension with respect to different concentration of amphiphiles.[4]

## 4.5. Microstructure Analysis

The microstructures are described in Fig. 24, where tailored, open and closed, interconnected pores can be seen. Also, it can be seen that the larger and smaller pores are uniformly distributed. In Fig. 24a-d, different compositions of $Al_2O_3/SiO_2$ with well-developed and narrow pore size distribution can be seen. It shows a hierarchical pore distribution with porosities up to 80% from larger to smaller pores and thick struts (films in wet foams). It leads to produce more stable foams sintered to form porous ceramics with high mechanical strength.

Fig. 25 shows the microstructures of porous (a) AT, (b) ATM1, (c) ATM3, and (d) ATM5, sintered at 1500°C for 1 hour. The microstructures obtained generally consist of open, interconnected pores with a narrow pore size distribution. The composition without addition of mullite (Fig. 25(a)) shows the characteristic microstructure of $Al_2TiO_5$: an open porous and microcracked $Al_2TiO_5$ matrix phase, with the presence of abnormal grain growth. These grains can be attributed to unreacted $Al_2O_3$ and $TiO_2$ due to the formation reaction kinetics, which is a process led by the nucleation and growth of $Al_2TiO_5$ grains, and finally the diffusion of the reactants through the matrix. It is evident from Fig. 25b-d that the addition of mullite has a beneficial effect on grain growth control.

The scanning electron microscope images of 30 Vol% $Al_2O_3$-$SiO_2$ porous ceramics sintered at 1300°C with different chain length amphiphile of concentration 0.15 mol/L are shown in Fig. 26. The microstructures obtained are generally consists of closed pores. It is interesting to note that at the same concentration of amphiphile, the shortest chain carboxylic acid, i.e., propionic acid, produces relatively large pore size than the longest chain carboxylic acid, i.e., valeric acid. This can be attributed to the fact that greater hydrophobicity is achieved with the aid of long carbon chain present in valeric acid which results in small and uniform pore size. The smaller cell sizes result from the high stability of the foams in the wet state, which impedes bubble coarsening. The dense struts as shown in the inset of Fig. 26a-c plays vital role for improving the mechanical strength of the porous ceramics.

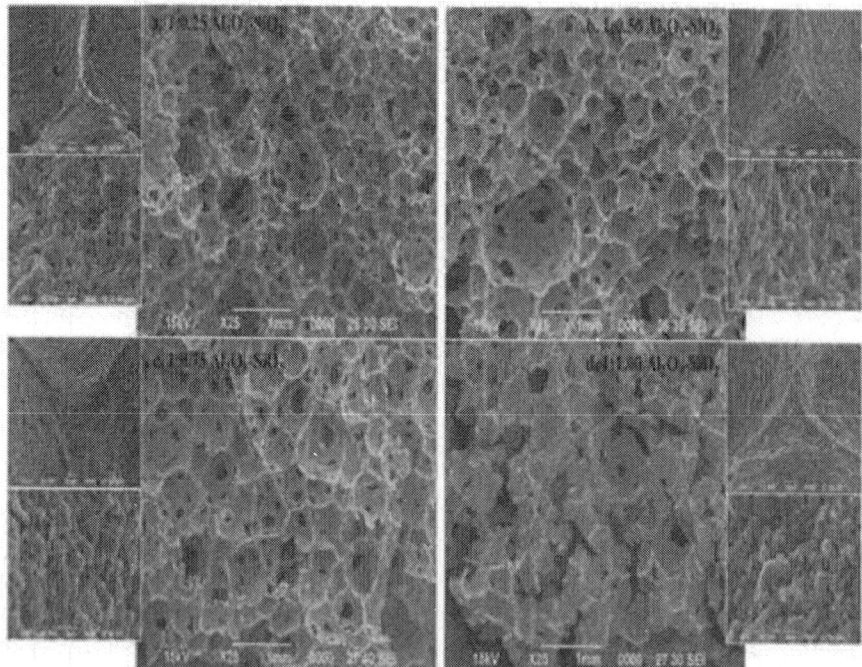

**Figure 24.**Microstructures and thin film (inner cell) of porous ceramics sintered at 30 vol.% $Al_2O_3$ with respect to the different mole ratio of $SiO_2$ content.

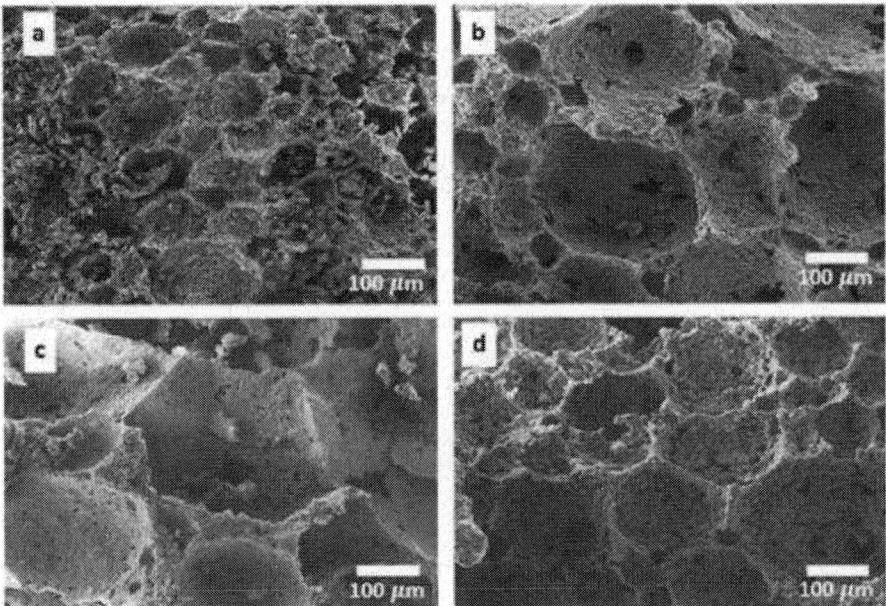

**Figure 25.**Microstructures of (a) AT, (b) ATM1, (c) ATM3, and (d) ATM5 porous ceramics, sintered at 1500°C for 1 hour.[32]

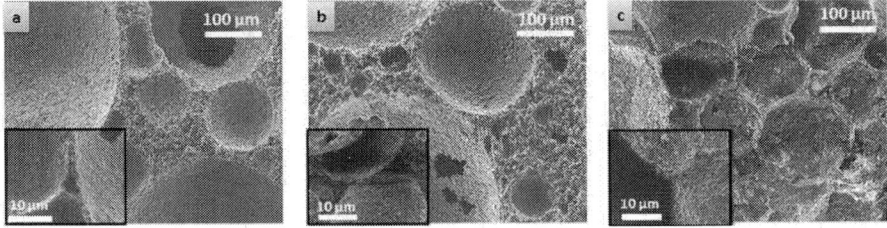

**Figure 26.**Microstructures of porous ceramics using 0.15 mol/L of (a) propionic acid, (b) butyric acid, and (c) valeric acid; the inset in (a), (b), and (c) show single dense struts obtained with direct foaming method.[4]

## 5. CONCLUSIONS

Porous ceramics' microstructures and properties are affected by their method of synthesis. Direct foaming can simply, inexpensively and quickly prepare macroporous ceramics. Open or closed porosities of 45%-85% having been demonstrated. The pores produced by this method result from the direct incorporation of air bubbles into a ceramic suspension, eliminating the need for pyrolysis before sintering. Cellular structures prepared by direct foaming are generally stronger than those prepared by replica synthesis due mainly to the absence of flaws in the cell struts. Given the importance of the chosen synthetic method, this review examines currently available processes for forming porous ceramics. Direct foaming is a simple and versatile process for the low-cost manufacture of porous ceramics for various applications. Its continuous study will result in further improvements of its method and wider applicability of its products. Examination of the literature led to the proposal of an equation describing the inverse proportionality of wet foam stability to the surface of the liquid-air interface.

$$W_{fS} \infty (\frac{1}{\gamma})$$

$W_{fS}$ = Wet foam stability
$\gamma$ = surface tension

# 6. ACKNOWLEDGEMENTS

This research was financially supported by the Ministry of Education, Science Technology (MEST), and The National Research Foundation of Korea (NRF) through the Human Resource Training Project for Regional Innovation and Hanseo University.

# REFERENCES

1. M. Scheffler and P. Colombo, Cellular Ceramics: Structure, Manufacturing, Properties and Applications. p 645. Weinheim, Wiley-VCH, 2005.
2. A. R. Studart, U. T. Gonzenbach, E. Tervoort, and L. J. Gauckler, "Processing routes to macroporous ceramics—a review," J. Am. Ceram. Society (2006).
3. J. Banhart, "Manufacturing Routes for Metallic foams," JOM (2000).
4. N. Sarkar, J. G. Park, S. Mazumder, A. Pokhrel, C. G. Aneziris, and I. J. Kim, Effect of amphiphile chain length on wet foam stability of porous ceramics, Ceram. Int., 41 [3] (2015) 4021-4027.
5. J. Banhart, "Manufacture, classification and application of cellular metals and foams," Prog. Mater. Sci., 46 (2001) 559-632.
6. W. Ramsden, "Separation of solids in the surface-layers of solutions and 'suspensions'," Proc. R. Soc. Lond., 72 (1903) 156.
7. Ya. Guzman, "Certain principles of formation of porous ceramic structures, properties and applications—a review," Glass Ceram., 9 (2003) 28-31.
8. P. Colombo and J.R. Hellmann, "Ceramic foams from preceramic polymers," Mat Res. Innovat., 6 (2002) 260-272.
9. H. M. Princen and A. D. Kiss, "Rheology of foams and highly concentrated Emulsions," J. Colloid. Interface Sci., 128 [1] (1989) 176-187.
10. W. D. Kingery, H. K. Bowen, and D. R. Uhlmann, Introduction to Ceramics, 2nd edition. Wiley Interscience Publication, 1975.
11. P. Colombo, "Conventional and novel processing methods for cellular ceramics," Phil. Trans. R. Soc. A., 364 (2006), 109-124.

12. B. Neirinck, J. Fransaer, O. V. der Biest, JefVleugels, "A novel route to produce porous ceramics," J. Euro. Cerm. Soc., 29 (2009) 833-836.

13. B. P. Binks, "Particles as surfactants—similarities and differences," Curr. Opin. Colloid Interface Sci., 7 (2002) 21-41.

14. Brent S. Murray, "Stabilization of bubbles and foams," Curr. Opin. Colloid Interface Sci., 12 (2007) 231-241.

15. T. S. Horozov, "Foams and foam films stabilized by solid particles," Curr. Opin. Colloid Interface Sci., 13 (2008) 134-140.

16. P. C. Hidber, T. J. Graule, and L. J. Gauckler, "Influence of the dispersant structure on properties of electrostatically stabilized aqueous alumina suspension," J. Eur. Ceram. Soc., 17 [2-3] (2002) 239-249.

17. A. Pokhrel, J. G. Park, J. S. Nam, D. S. Cheong, and I. J. Kim, "Stabilization of wet foams for porous ceramics using amphiphilic particles," J. Kor. Ceram. Soc., 48 [5] (2011) 463-466.

18. N. D. Denkov, I. B. Ivanov, P. A. Kralchevsky, and D. T. Wasan, "A possible mechanism of stabilization of emulsions by solid particles," J. Colloid. Interface Sci., 150, [2] (1992) 589-593.

19. L. J. Gauckler, Th. Graule, and F. Baader, "Ceramic forming using enzyme catalyzed reactions," Mater. Chem. Phys., 61 (1999) 78-102.

20. A. Pokhrel, Zhao Wei, and I. J. Kim, "Wet foam stabilized by amphiphiles to tailor the microstructures of porous ceramics," Key Eng. Mater., 512-515 (2012) 288-292.

21. A. B. Subramaniam, C. Mejean, M. Abkarian, and H. W. Stone, "Microstructure, morphology and lifetime of armored bubbles exposed to surfactants," Langmuir, 22 [14] (2006) 5986-5990.

22. U. T. Gonzenbach, A. R. Studart, D. Steinlin, E. Tervoort, and L. J. Gauckler, "Processing of particle-stabilized wet foams into porous ceramics," J. Am. Ceram. Soc., 90 [11] (2007) 3407-3414.

23. U. T. Gonzenabach, A. R Studart, E. Tervoort, and L. J. Gauckler, "Stabilization of foams with inorganic colloidal particles," Langmuir, (2006) 10983-10988.

24. T. N. Hunter, R. J. Pugh, G. V. Fanks, and G. J. Jameson, "A role of particles in stabilizing foams and emulsions," Adv. Colloid. Inter. Sci., 137 (2008) 57-81.

25. U. T. Gonzenbach, A. R. Studart, E. Tervoort, and L. J. Gauckler, "Macroporous ceramics from particle-stabilized wet foams," J. Am. Ceram. Soc., 90 [1] (2007) 19-22.

26. P. J.Wilde, "Interface: their role in foam and emulsion behavior" Curr. Opin. Colloid Interface Sci., 5 (2000) 176-181.

27. G. Morris, M. R. Pursell, S. J. Neethling, and J. J. Cilliers, "The effect of particle hydrophobicity, separation distance and packing patterns on the stability of a thin film," J. Colloid Interface Sci., 327 (2008) 138-144.

28. U. T. Gonzenbach, A. R. Studart, E. Tervoort, and L. J. Gauckler, "Tailoring the microstructure of particle-stabilized wet foams," Langmuir, 23[3] (2007) 1025-1032.

29. I. Akartuna, A. R. Studart, E. Tervoot, U. T. Gonzenbach, and L. J. Gauckler, "Stabilization of oil-in-water emulsions by colloidal particles modified with short amphiphiles," Langmuir, 24 (2008) 7161-7168.

30. A. R. Studart, U. T. Gonzenbach, I. Akartuna, E. Tervoort, and L. J. Gauckler, "Materials from foams and emulsions stabilized by colloidal particles," J. Mater. Chem., (2007) 3283-3289.

31. A. Pokhrel, J. G. Park, G. H. Jho, J. Y. Kim, and Ik Jin Kim, "Controlling the porosity of particle stabilized Al2O3 based ceramics," J. Kor. Ceram. Society., 48 [6] (2011) 600-603.

32. N. Sarkar, J. G. Park, S. Mazumder, A. Pokhrel, C. G. Aneziris, and I. J. Kim, "Al2TiO5-mullite porous ceramics from particle stabilized wet foam," Ceram. Int., 41 [5] (2015) 6306-6311.

33. I. Aranberri, B. P. Binks, J. H. Clint, and P. D. I. Fletcher, "Synthesis of macroporous silica from solid-stabilised emulsion templates," J. Porous. Mater.,16 (2009) 429-437.

34. U. T. Gonzenabach, A. R Studart, E. Tervoort, and L. J. Gauckler, "Ultrastable particle-stabilized foams," Angew. Chem. Int. Ed., 45 (2006) 3526-3530.

35. E. Dickinson, R. Ettelaie, T. Kostakis, and B. S. Murray, "Factors controlling the formation and stability of air bubbles stabilized by partially hydrophobic silica nanoparticles," Langmuir, 20 (2004) 8517-8525.

36. T. Fukasawa and M. Ando, "Synthesis of porous ceramics with complex pore structure by freeze-dry processing," J. Am. Ceram. Soc., 84 [1] (2001) 230-232.

37. T. Fukasawa, Z. Y. Deng, M. Ando, T. Ohji, and Y. Goto, "Pore structure of porous ceramics synthesized from water-based slurry by freeze-dry process," J. Mater. Sci., 36 (2001) 2523-2527.

38. I. Akartuna, A. R. Studart, E. Tervoot, and L. J. Gauckler, "Macro porous ceramics from particle-stabilized emulsions," Adv. Mater., 20 (2008) 4714-4718.

39. I. Akartuna, E. Tervoot, A. R. Studart, and L. J. Gauckler, "General route for the assembly of functional inorganic capsules," Langmuir 25[21], (2009) 12419-12424.

40. A. Pokhrel, J. G. Park, W. Zhao, and I. J. Kim, "Functional porous ceramics using amphiphilic molecule," J. Ceram. Process. Res., 13 [4] (2012) 420-424.

41. Brent. S. Murray and Ettelaie, "Foam stability: proteins and nanoparticles," Curr. Opin. Colloid Interface Sci., 9 (2004) 314-320.

42. D. M. Alguacil, E. Tervoort, C. Cattin, and L. J.Gauckler, "Contact angle and adsorption behavior of carboxylic acids on α-Al2O3 surfaces," J. Colloid Interface Sci., 353 (2011) 512-518.

43. G. Kaptay, "On the equation of the maximum capillary pressure induced by solid particles to stabilize emulsions and foams and on the emulsion stability diagrams," Colloids Surf. A: Physicochem. Eng. Aspects., 282-283 (2006) 387-401.

44. S. Barg, C. Soltmann, M. Andrade, D. Koch, and G. Grathwohl, "Cellular ceramics by direct foaming of emulsified ceramic powder suspensions," J. Am. Ceram. Soc., 91 [9] (2008) 2823-2829.

45. C. Tuck and J. R. G. Evans, "Porous ceramics prepared from aqueous foams," J. Mater. Sci. Lett., 18 (1999) 1003-1005.

46. F. Schuth and W. Schmidt, "Microporous and mesoporous materials" Adv. Eng. Mater., 4 [5] 92005) 269-279

47. A. R. Studart, R. Libanori, A, Moreno, U. T. Gonzenbach, E. Tervoort, and L.J. Gauckler, "Unifying model for the electrokinetic and phase behavior of aqueous suspensions containing short and long amphiphiles," Langmuir, 27 (2011) 11835-11844.

48. J. C. H. Wong, E. Tervoort, S. Busato, Urs. T. Gonzanbech, A. R. Studart, P. Ermanni, and L. J. Gauckler, "Designing macro porous polymers from particle-stabilized foams," J. Mater. Chem., 20 (2010) 5628-5640.

49. A. R. Studart, Julia Studer, Lei Xu, K. Yoon, H. C. Shum, and D. A. Weitz, "Hierarchical porous materials made by drying complex suspensions," Langmuir, 27 [3] (2011) 955-964.

50. P. C. Hiemenz and R. Ayagopalan, "Principles of colloid and surface chemistry," p. 650. 3rd edition. Marcel Dekker Inc, New York, 1997.

CHAPTER 4

# Micro ARC Oxidation of Wire ARC Sprayed Al-Mg6, Al-Si12, Al Coatings on Low Alloyed Steel

*Levent Cenk Kumruoglu, Fatih Ustel, Ahmet Ozel, Abdullah Mimaroglu*

Faculty of Engineering, University of Sakarya, Esentepe Campus, Sakarya, Turkey

## ABSTRACT

Micro arc oxidation of wire arc sprayed Al-Mg6, Al-Si12 and Pure Al coatings on low carb-on steel has been performed. The coatings have been analyzed using optic microscope, scanning electron microscopy, X-ray diffraction and surface roughness tester. At the same time, voltage and current regimes are investigated during the process. Then after MAO process, uniform $Al_2O_3$ ceramic coatings have been deposited on surface of Al-Mg6, Al-Si12 and Al coated steel. The ceramic coatings are mainly composed of $Al_2O_3$ phase. The compound coatings show high hardness and significant improvement of corrosion resist-ance property.

**Keywords:** Micro Arc Oxidation, Plasma Electrolytic Oxidation, Arc Spray, AISI 1010 Steel, MAO, PEO, SEM, XRD

## 1. INTRODUCTION

Studies focusing on how to improve the protection of steel against degradation by corrosion or wear are the main objective of researchers of processes and materials, always aiming to reduce costs by preserving the structure and increasing its service life.

Cathodic and anodic protection, coatings and painting are some of the methods employed to combat degradation [1]. Micro Arc Oxidation (MAO), also known as Plasma Electrolytic Oxidation, is globally gaining increased popularity as a novel means of deposition dense, thick, ultra-hard ceramic coatings on metals like Al, Mg, Ti, Zr and Nb and their alloys with an objective to improve the wear, thermal, chemical, corrosion, oxidation resistance. MAO process employs Electro-chemical and Electro-thermal oxidation in alkaline electrolytic medium with an alternating and direct current-high voltage surface multiple discharges [2,3]. During oxidation, many visible spark or micro arc spots move rapidly on the metal surface in aqueous solution. The local instantaneous temperature and pressure inside these micro arc discharge channels can reach 103 - 104 K and 102 - 103 MPa, respectively [4].

In the Arc Spray Process a pair of electrically conductive wires such as Al, Al-Si12, Al-Mg6 are melted by means of an electric arc. The molten material is atomized by compressed air and propelled towards the substrate surface. The impacting molten particles on the substrate rapidly solidify to form a coating. Compared to other thermal spray methods, wire arc spray is considered as a simple, low cost, efficient coating process [5-12].

The purpose of this study was to explore the potential of the wire arc spray process in applying Al-Mg6, Al-Si12 and Pure Al coatings on low carbon steel and examination of micro arc oxidation of arc spray coatings. This paper discusses the coating properties and microstructure through changes in spray processes and micro arc oxidation of coated samples

## 2. EXPERIMENTAL PROCEDURE

Sticks made of low carbon steel (AISI 1010) of 17.5 mm and length of 50 mm were used as substrates. Aluminum alloys, Al-Mg6, Al-Si12 and Pure Al, were used as the arc spraying material. The current employed in arc spraying were 100, 200 and 300 A respectively at the spraying voltage of 28, 30 and 32 V, and the distance of spraying between the work piece and spraying gun were 150 and 200 mm at the compressed air pressure of 3 and 4 Bar. All wire arc sprayed coatings were carried out using robot arm to obtain homogeny coating thickness. Wires were sprayed using a electric wire arc spray gun and robotic system (Sulzer-Metco). Spray parameters used in tests are listed in **Table 1**.

The thicknesses of all sample details of arc sprayed coatings were measured with Positector 6000. Also knoop and micro hardness was measured using hardness tester. Vickers microhardness measurements on all the coating cross-sections and top surface of coating have been carried out, all measurements have been performed at random locations on each sample.

Then micro arc oxidation studies were carried out for all arc sprayed samples. The micro arc oxida-tion processes were conducted on a high-power DC electrical source, a plexiglass container with a sample holder as the electrolyte cell, and a stirring and cooling system (**Figure 1**).

Aqueous solutions of 5 g/L $Na_2B_4O_7$ (sodium tetraborate), 5 g/L KOH (Potassium hydroxid), 2 g/L $Na_3PO_4$ (sodium phosphate), 20 g/L glycerin and distilled water used as components of the electrolyte. The current density was approximately 12 A/dm$^2$. The solution temperature was controlled at 35°C. After MAO processing, samples were rinsed thoroughly with water and then were dried in cool air. MAO parameters used in tests are listed in **Table 2**.

The surface morphology and cross-sections of the arc spray and micro arc coatings were examined by a JEOL600 scanning electron microscopy (SEM) and optic microcopy. Film phase composition was estimated using a Rigaku X-ray diffractometer (XRD) in the $2\theta = 10° - 90°$ range

**Table 1.** Deposition by the electric arc spray process.

| Sample Code | Wire Type | Current | Voltage | Distance | Pressure |
|:---:|:---:|:---:|:---:|:---:|:---:|
| AS1 | Al-Si12 | 100 A | 28 V | 15 cm | 3 Bar |
| AS2 | Al-Si12 | 200 A | 28 V | 15 cm | 3 Bar |
| AS3 | Al-Si12 | 300 A | 28 V | 15 cm | 3 Bar |
| AS4 | Al-Si12 | 100 A | 28 V | 15 cm | 4 Bar |
| AS5 | Al-Si12 | 200 A | 28 V | 15 cm | 4 Bar |
| AS6 | Al-Si12 | 300 A | 28 V | 15 cm | 4 Bar |
| AS7 | Al-Si12 | 200 A | 28 V | 20 cm | 3 Bar |
| AS8 | Al-Si12 | 200 A | 30 V | 20 cm | 3 Bar |
| AS9 | Al-Si12 | 200 A | 32 V | 20 cm | 3 Bar |
| AM1 | Al-Mg6 | 200 A | 28 V | 15 cm | 3 Bar |
| AM1 | Al-Mg6 | 200 A | 28 V | 20 cm | 3 Bar |
| AM1 | Al-Mg6 | 200 A | 28 V | 15 cm | 4 Bar |
| Al | Al%100 | 200 A | 28 V | 15 cm | 3 Bar |

Note: AS: Al-Si12, AM: Al-Mg6, A: Al %100.

**Table 2.** Experimental parameters of MAO treatment.

| Sample Code | Treatment time | Electrolyte Temperature | pH | C. of E. |
|:---:|:---:|:---:|:---:|:---:|
| AS2 | 150 min | 20°C - 35°C | 13 | 15 mS |
| AS2 | 100 min | 20°C - 35°C | 13 | 15 mS |
| AS2 | 50 min | 20°C - 35°C | 13 | 15 mS |
| AM1 | 150 min | 20°C - 35°C | 13 | 15 mS |
| AM1 | 100 min | 20°C - 35°C | 13 | 15 mS |
| AM1 | 50 min | 20°C - 35°C | 13 | 15 mS |
| Al | 150 min | 20°C - 35°C | 13 | 15 mS |

C. of E.: Conductivity of Electrolyte.

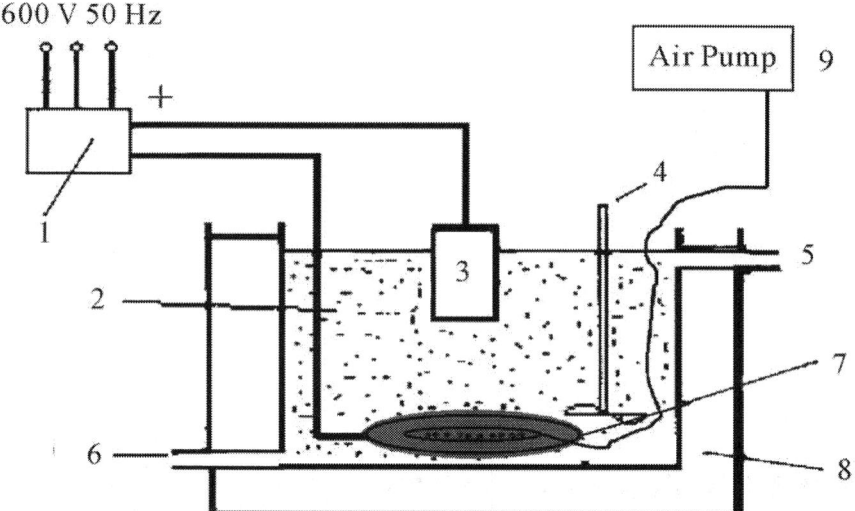

**Figure 1.** Schematic view of micro arc oxidation deposition system: 1, power supply and control system; 2, electrolyte; 3, workpiece; 4, mixer; 5, cooling water inlet; 6, cooling water outlet; 7, stainless-steel anode plate; 8, plexiglass; 9, air pump.

## 3. RESULTS AND DISCUSSIONS

### 3.1. The Effect of Process and Arc Spray Parameters on Microstructure and Microhardness Properties

When a thermal spray coating is applied onto a surface, the process parameter during spraying is known to have a strong effect on coating properties.

Polished cross section of the arc spray deposits were digitized by using an optical microscope with a digital camera. The effect of increasing arc current, arc voltage, operating gas pressure and spray distance were presented in Figures 2-5 respectively.

Figures 6 and 7 show the surface roughness, thickness and microhardness of wire sprayed upper layer of substrate using different process parameters.

**Figure 8** gives the evolution of the microhardnes and coating thickness in function of arc current. As it can be seen in **Figure 8** that high arc voltages cause high thickness and low hardness.

As can be seen in **Figure 9**, increasing the operating gas pressure increases the microhardness of surface. At the same time, the effect of arc current-arc voltage, arc current-spray distance on microhardness and coating thickness, were given in Figures 10 and 11.

Effect of Arc Current (Al12Si, 3 bar, 32 volt, 20 cm)

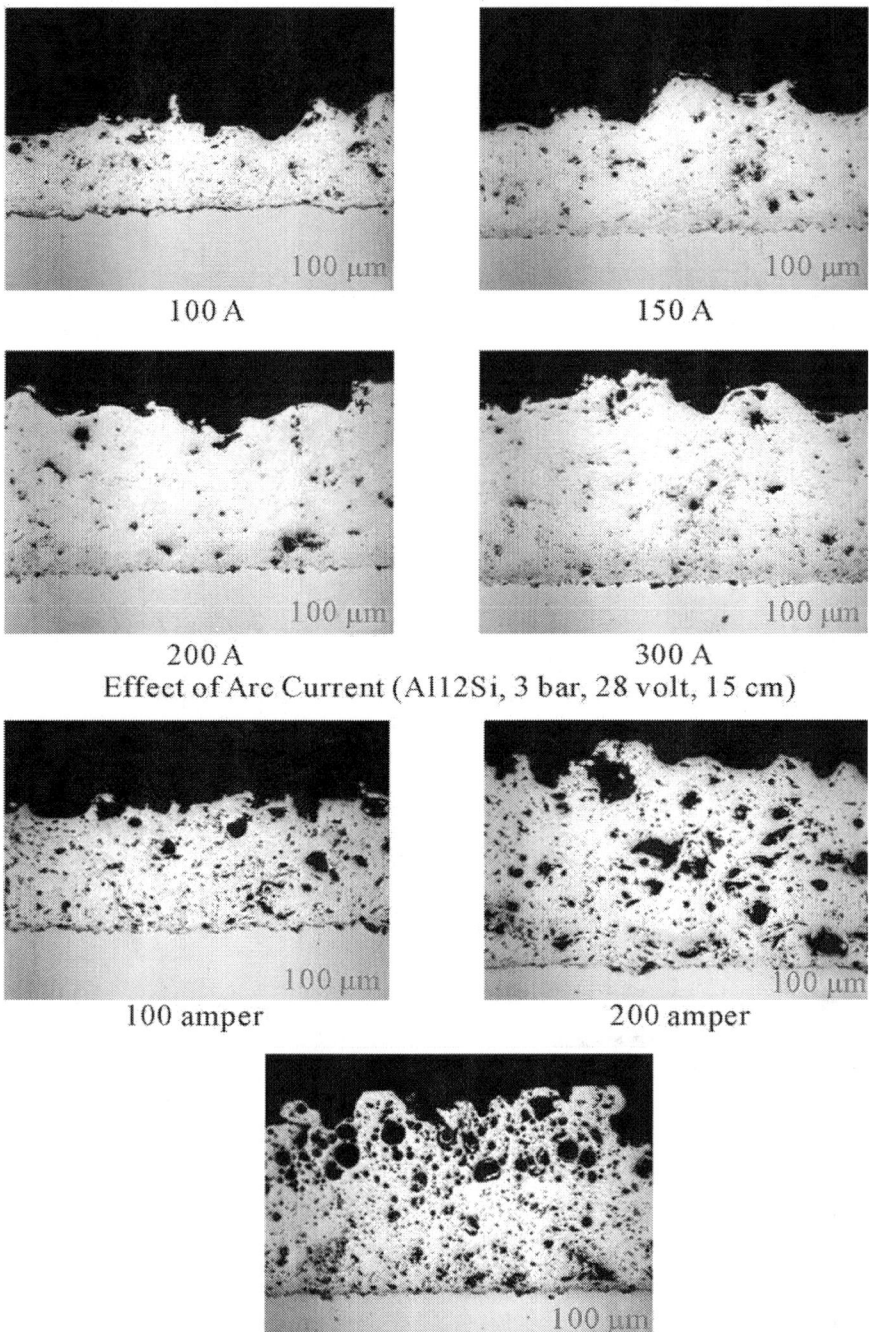

**Figure 2.** Effect of arc current on cross-section microstructure properties of AlSi12 wire coated onto carbon steel.

## 3.2. SEM-EDS and XRD Studies of Arc Spray Coatings and MAO Process

Investigation of coatings microstructure revealed dependence of structure morphology on process parameters.

The experimental studies showed that; the coating thickness of the arc sprayed layer increased and the microhardness of layer decreased with the increase of arc current from 100 A to 300 A. It is claim that, increasing the arc current increases the wire feed speed. The deposition rate is directly tied to the wire feed speed of the arc spraying process.

Effect of Arc Voltage (V) (Al12Si, 100 A, 32 volt, 20 cm)

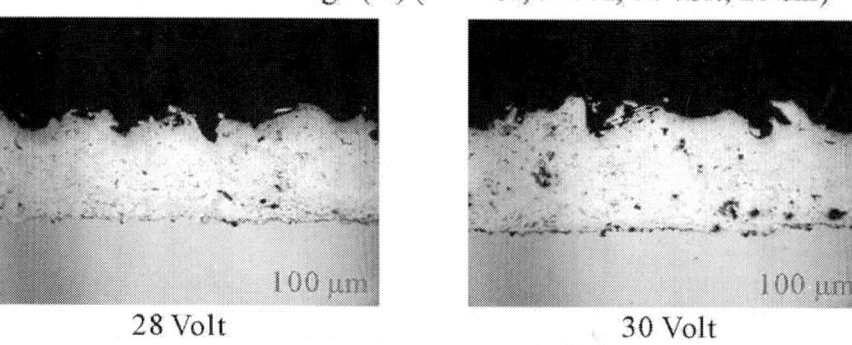

28 Volt                    30 Volt

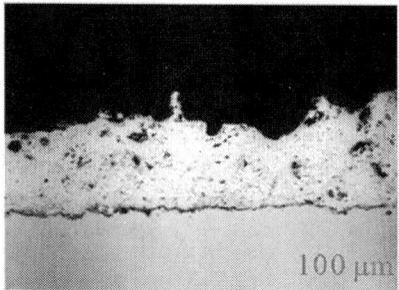

32 Volt

**Figure 3.** Effect of arc voltage on cross-section microstructure properties of Al-Si12 wire coated onto the carbon steel.

Effect of Operating gas pressure (Al12Si, 300 amper, 28 volt, 15 cm)

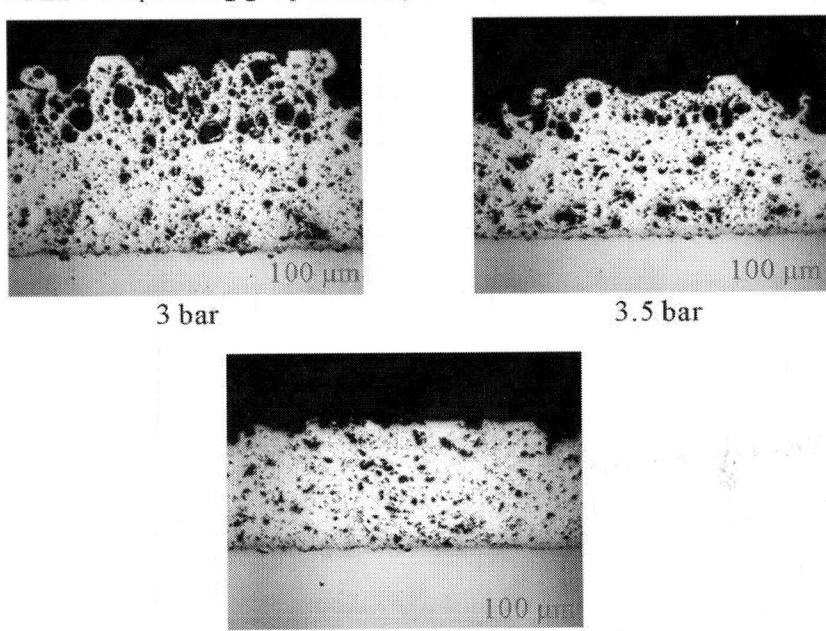

3 bar                                    3.5 bar

4 bar

Effect of Operating gas pressure (Al12Si, 100 amper, 28 volt, 15 cm)

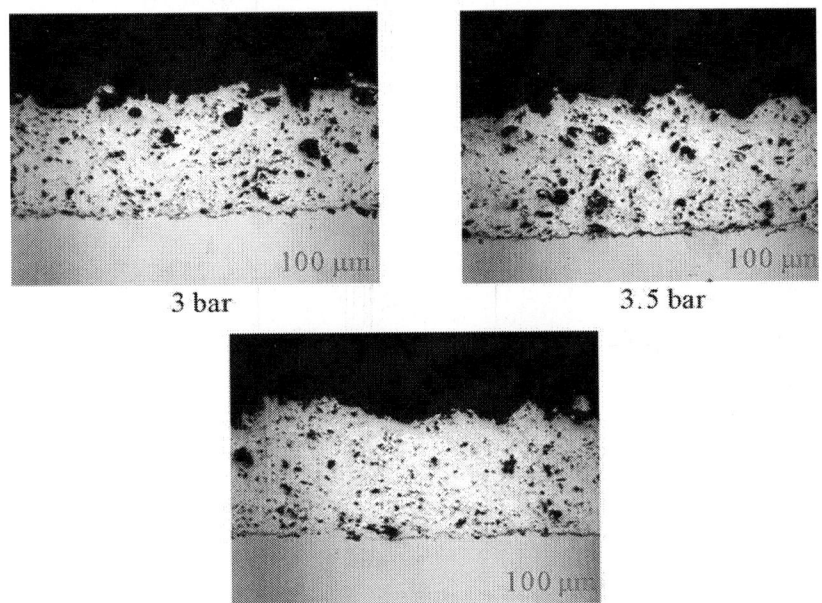

3 bar                                    3.5 bar

4 bar

**Figure 4.** Effect of operating gas pressure and arc current on microstructure. (a) (Al-12Si, 300 A, 28 V, 15 cm); (b) (Al-12Si, 100 A, 28 V, 15 cm).

Increasing the operating gas pressure produces moredense and lower thickness of coated layer. Increase in operating gas pressure influences the micro hardness of deposited layer. As the operating gas pressure increased, the hardness of arc spray coatings increases Figures 6-12. The average Knoop hardness of Al-Mg6, Al-Si12 and Al were given in **Figure 12**.

The hardness test results revealed that the hardness of the coatings increased and thickness of coatings decreasedwith increasing spray distance, in the all coating fabricated with the wire arc spray. This occurs because with the increase in the spray distance, the temperature of the sprayed particles decreases, the amount of air taken in thespray increases and, as a result, the cooling rate of the coating increases. It also found that the increases of coating thickness were lowered the hardness and enhanced the porosity and the coating roughness.

Effect of spray distance (cm) (Al12Si, 28 volt, 3 bar)

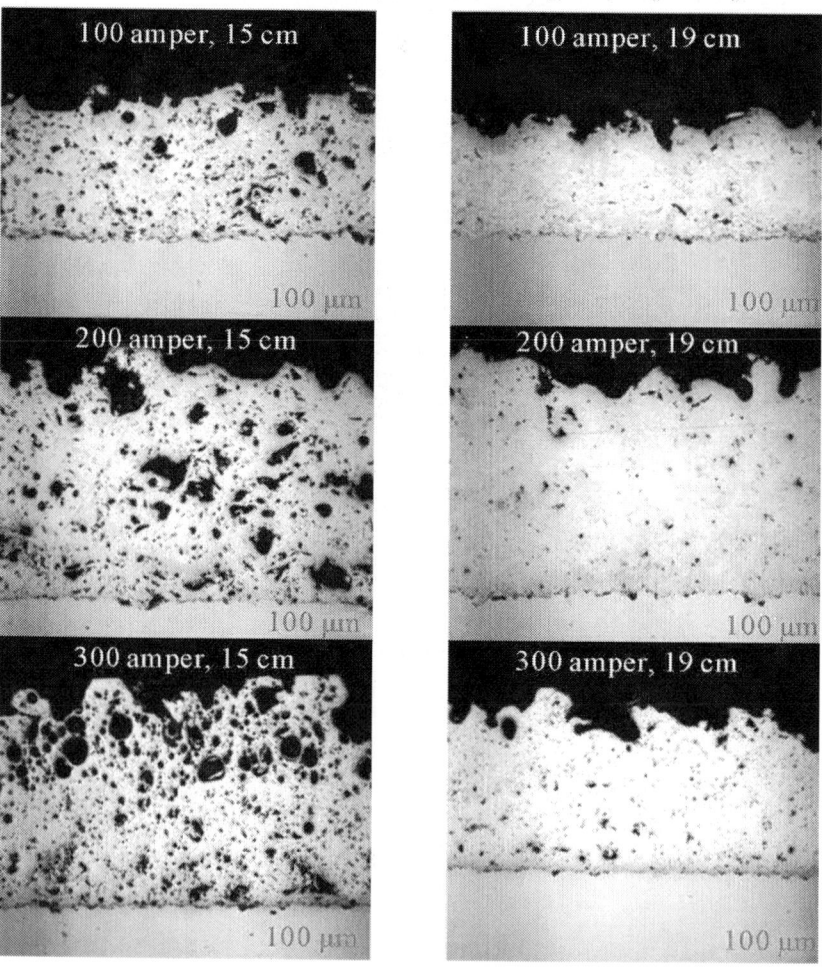

**Figure 5.** Effect of spray distance on microstructure of arc sprayed coating (Al-12Si, 28 V, 3 bar).

Figures 13 and 14 show the cross section view of arc sprayed coating using Al-Mg6 wire. EDS analyses reveals that the atomized particles were oxidized while arc spraying. In the air wire arc spray process, the oxide content is relatively high due to oxidation of the molten particles. The quantitative analysis reveals that the amount of oxides in the coating increases when increasing air-flow rate. During spraying, the effect of air-flow rate causes significant in-flight oxidation of the molten particles because increasing air-flow rate leads to higher gas stream velocities which in turn break up the molten particles into smaller ones [13]. This affects the break voltage level while micro ac oxidation process. Typical micrographs of coatings produced by the MAO technique on an Al-Mg6 arc-sprayed coating, obtained using scanning electron microscopy, are shown in **Figure 15**.

**Figure 6.** Coating thickness and micro hardness properties of arc sprayed coatings (Spray Distance 20 cm).

**Figure 7.** Coating thickness and micro hardness properties of arc sprayed coatings (Spray distance 15 cm)

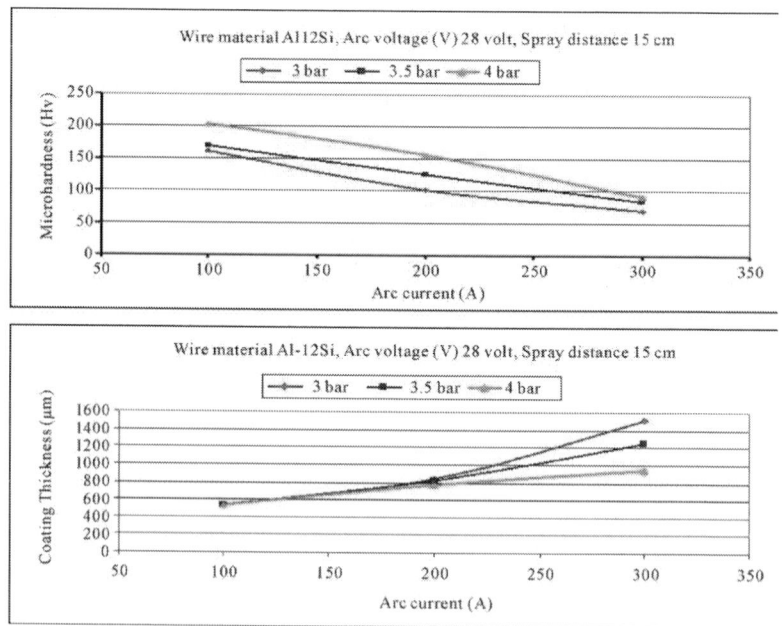

**Figure 8.** Effect of arc current on microhardness and coating thickness of wire arc spray process.

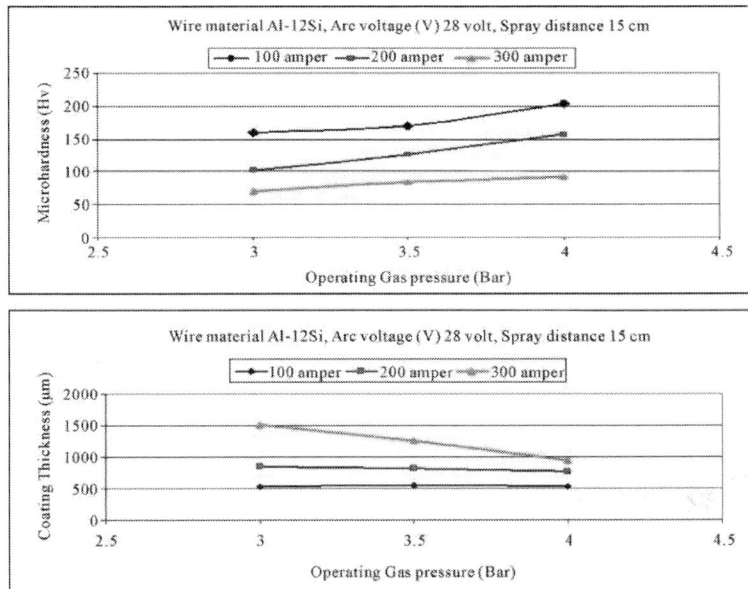

**Figure 9.** Effect of operating gas pressure on microhardness and coating thickness of wire arc spray process.

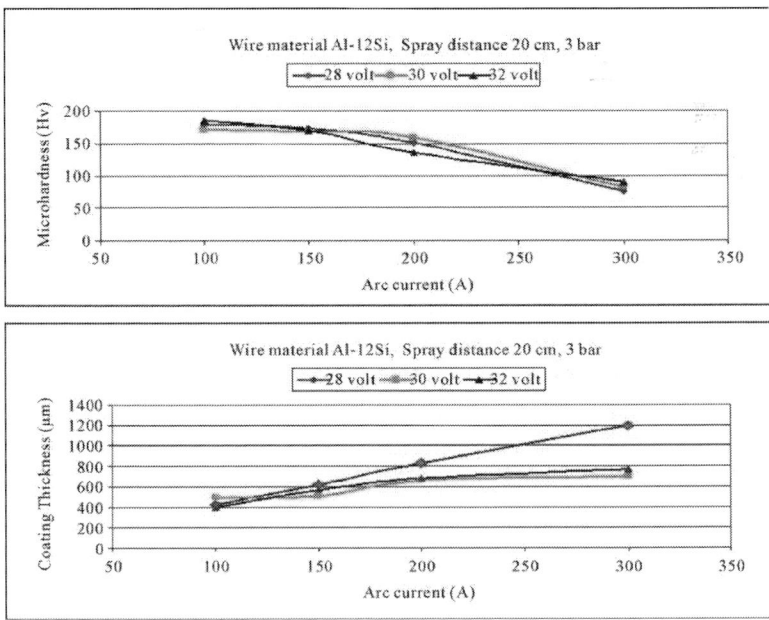

**Figure 10.** Effect of arc current and arc voltage on microhardness and coating thickness of wire arc spray process.

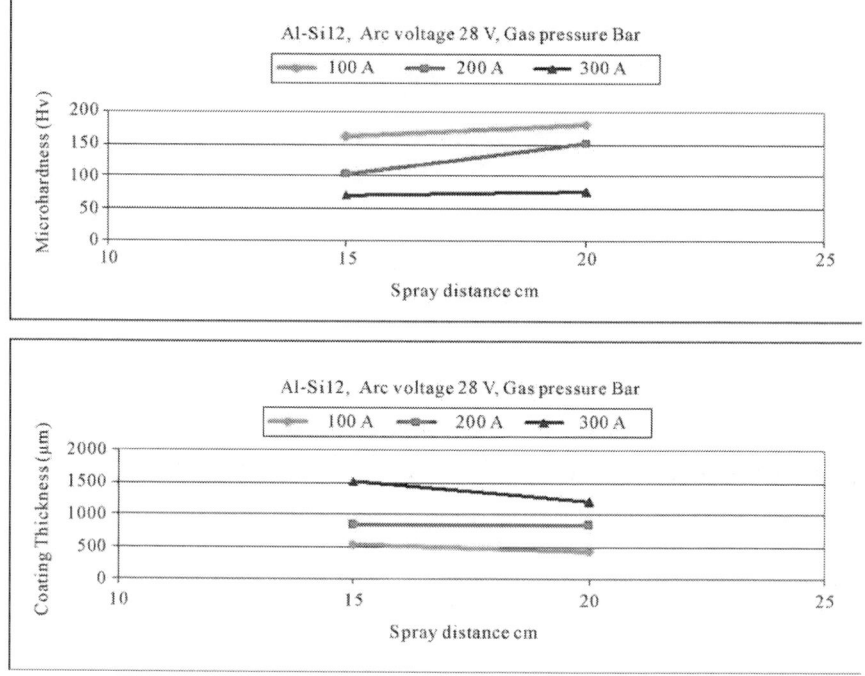

**Figure 11.** Effect of arc current and spray distance on microhardness and coating thickness of wire arc spray process.

In order to understand the composition characteristics, element distributions were measured over the specimen surface by EDS technique. It was found that the specimenwas mainly composed of three elements, magnesium, aluminum, oxygen (Al-Mg6), aluminum, oxygen silicon (Al-Mg12) and aluminum, oxygen (Al). The composition of the anodic coating was mainly crystalline oxide forms of Al and amorphous materials. The kinetics of MAO coatings is interface-controlled and largely dependent on the applied current density; the influence of electrolyte temperature, electrolyte additives and chemical composition of substrate. The final oxide forms of substrate are also largely dependent on chemical composition of substrate, the final phases of micro arc oxides samples were shown in **Figure 16**. The surface roughness of MAO coatings is a linear function of the final coating thickness, and increases with increasing coating thickness.

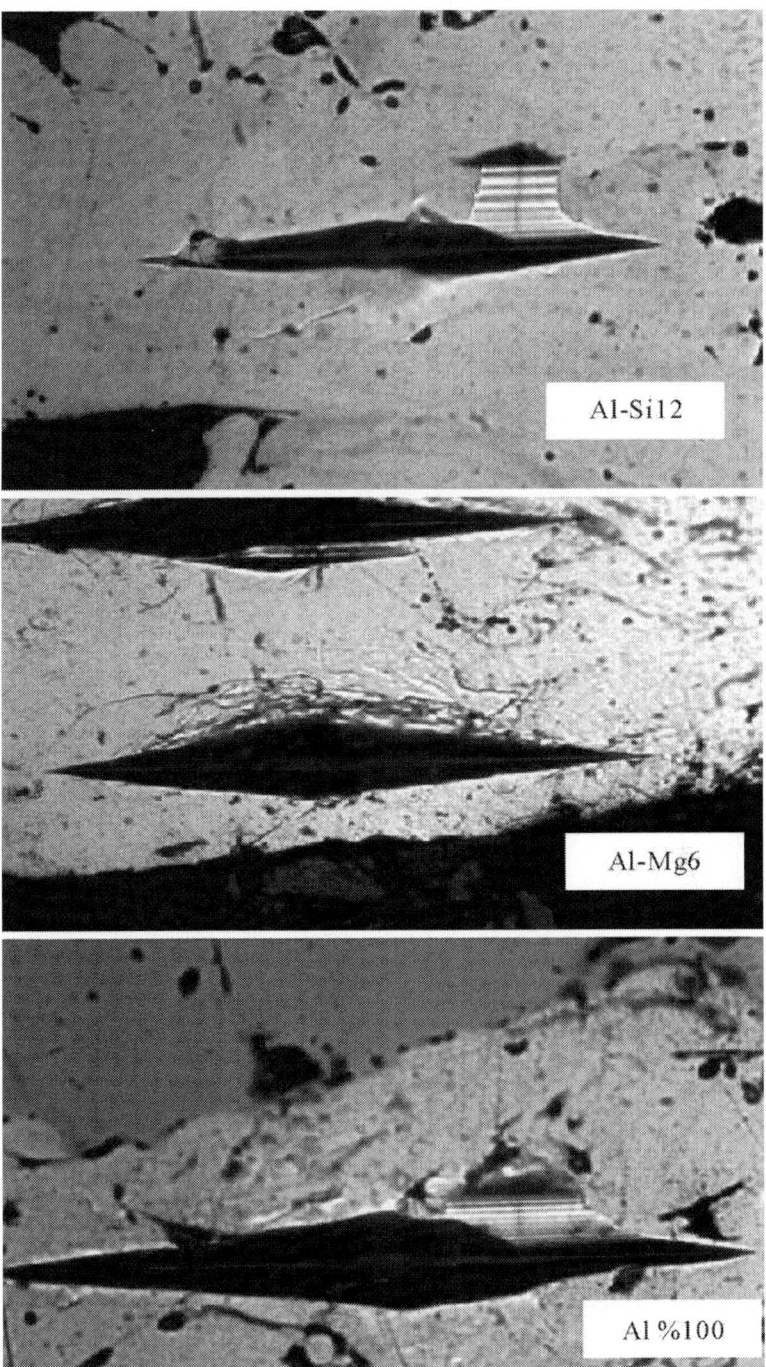

**Figure 12.** Average Knoop hardness values of different arc spray coating on steel substrate (Al-Si12 140 Hk), (Al-Mg6 80 Hk), (Al 100% 45 Hk).

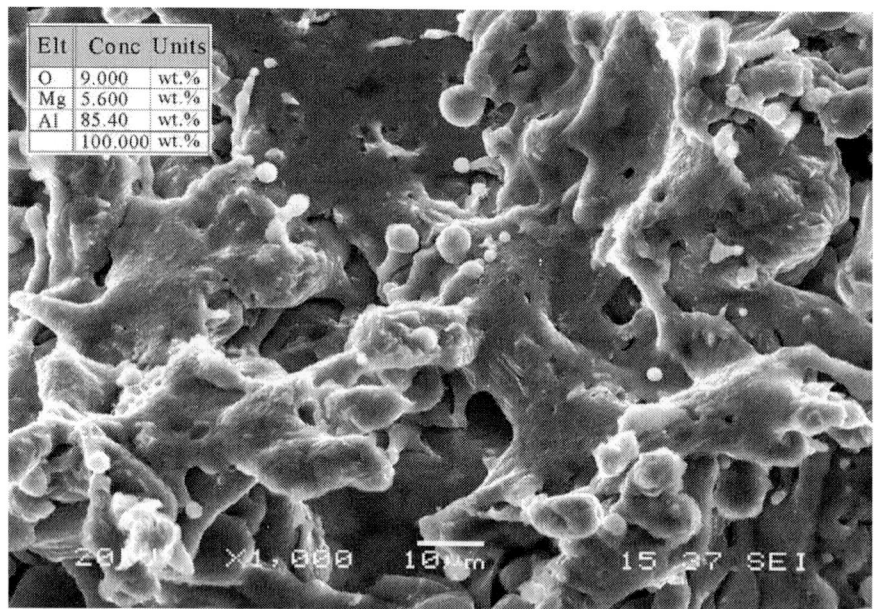

| Elt | Conc | Units |
|-----|------|-------|
| O | 9.000 | wt.% |
| Mg | 5.600 | wt.% |
| Al | 85.40 | wt.% |
| | 100.000 | wt.% |

**Figure 13.** SEM image and Spectrum EDS analysis of arc spray coating (wire material Al-Mg6).

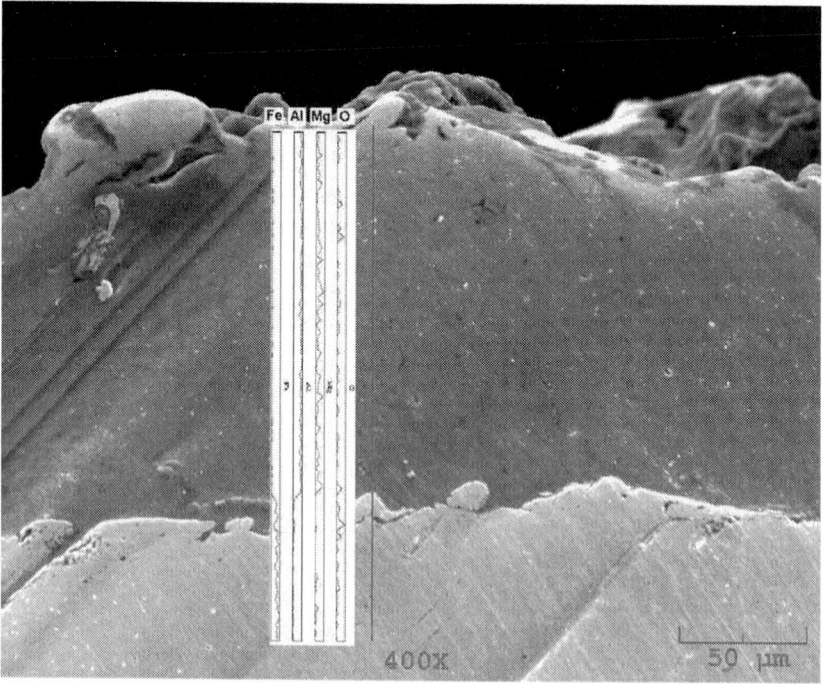

**Figure 14.** Cross section SEM image and line EDS analysis of arc spray coating (wire material Al-Mg6).

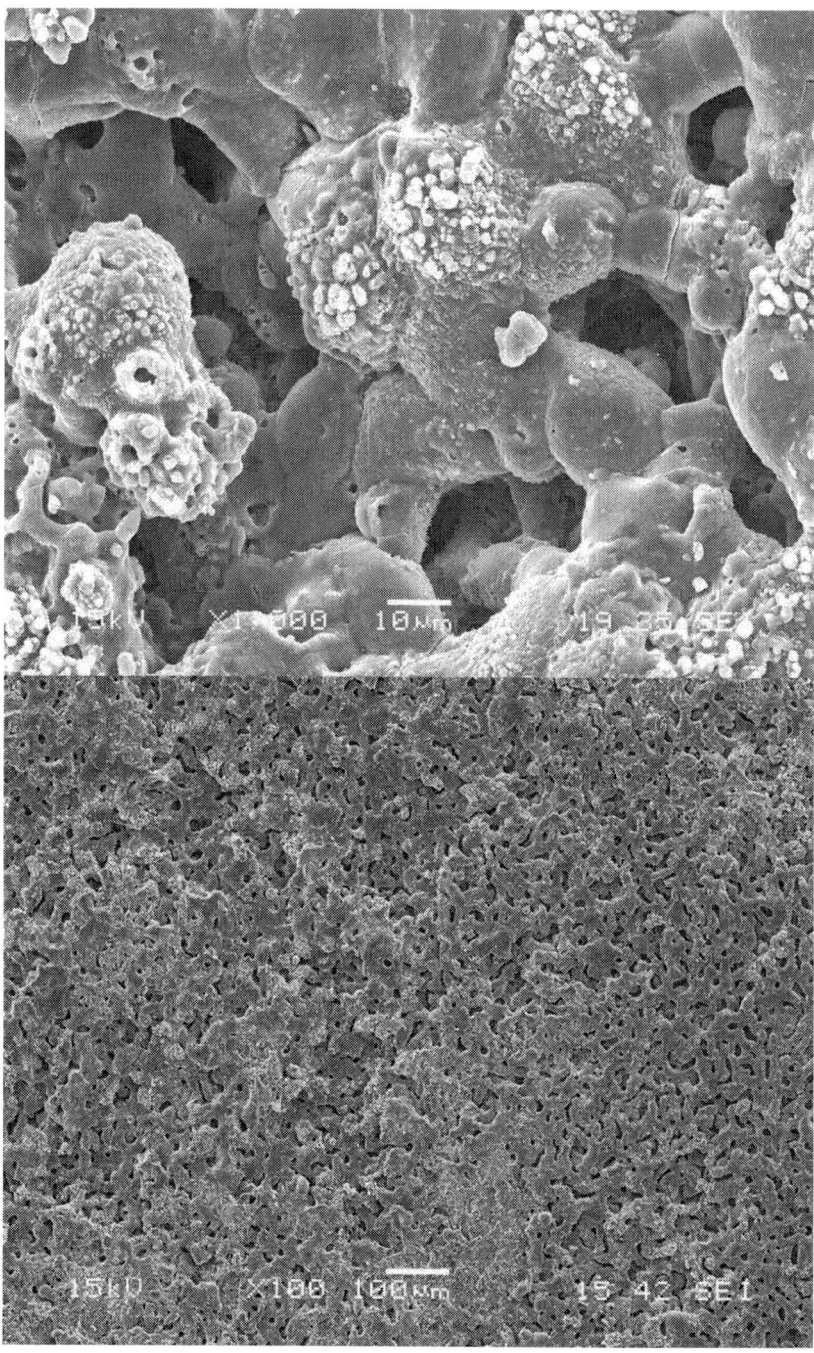

**Figure 15.** SEM micrographs of the specimen after micro arc oxidation treatment on arc sprayed layer using Al-Mg6 wire.

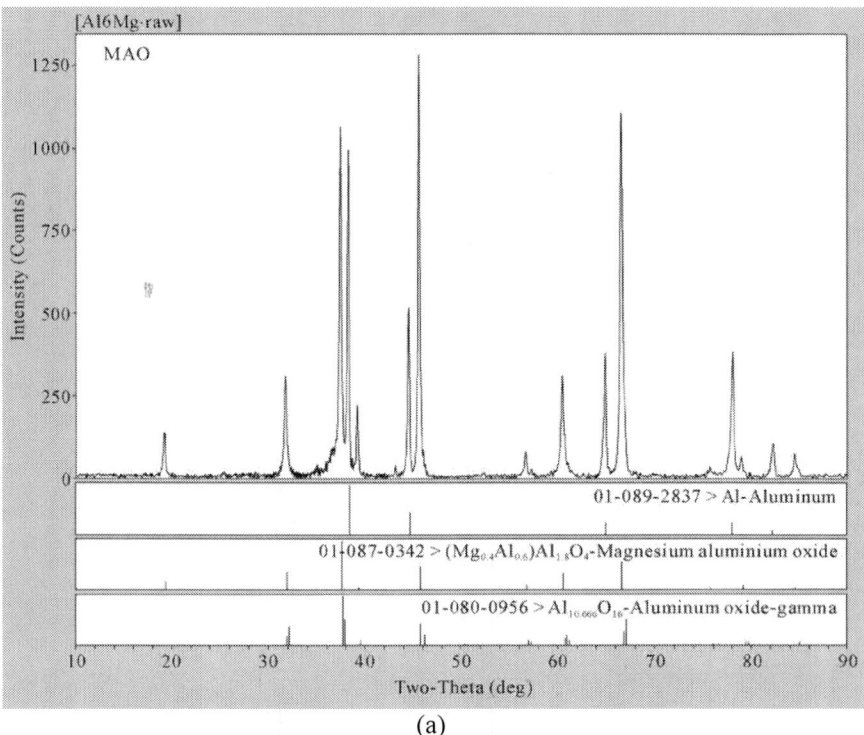

(a)

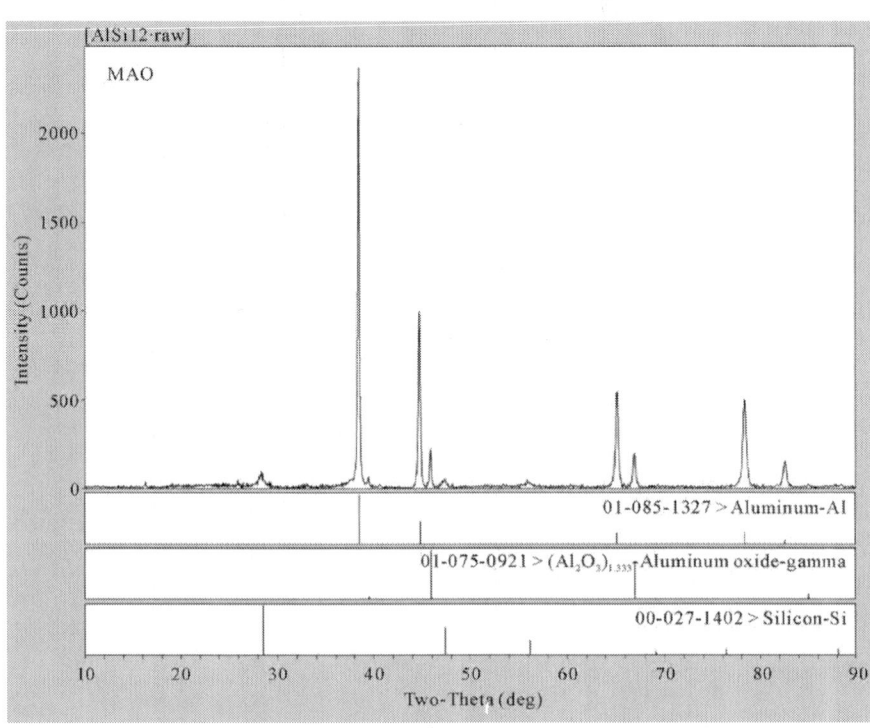

(b)

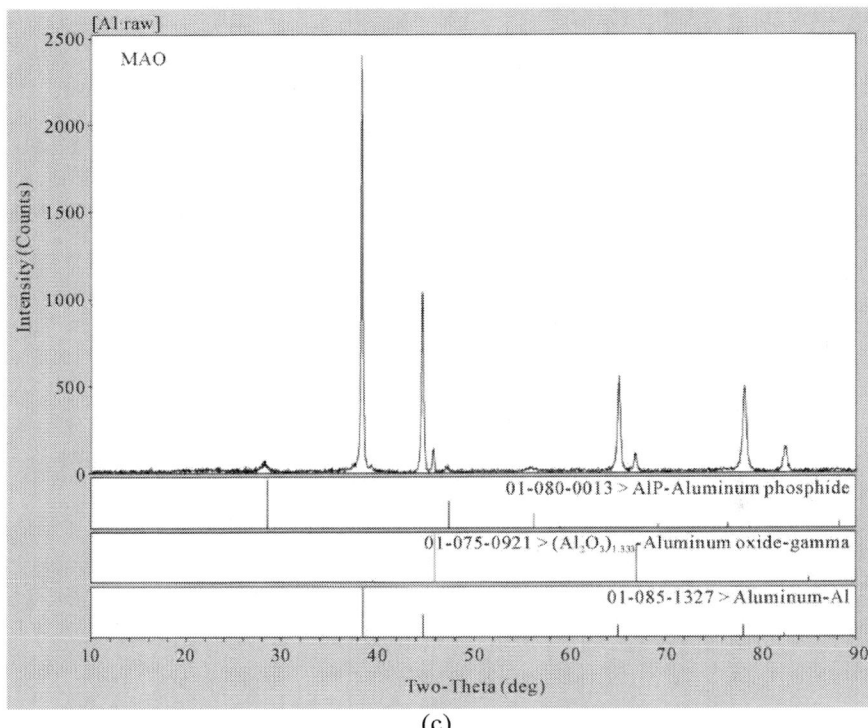

(c)

**Figure 16.** XRD analysis of oxide coatings produced by the MAO technique on arc sprayed aluminum coatings. (a) Al-Mg6 wire material; (b) Al-Si12 wire material; (c) Pure Al wire material.

## 4. CONCLUSIONS

Multi-layer coatings have been performed on low carbon steel using a combined technique of wire arc spraying and micro arc oxidation.

The experimental results of arc spray coatings show that spray distance, arc voltage—current, operating gas pressure are very effective parameters in hardness and coating thickness.

MAO processes were carried out for anodizing materials such as Al-Mg6, Al-Si12, pure Al that coated on steel substrate using wire arc spray technique that are hard to anodize with conventional anodizing processes.

The chemical compositions of the substrate and electrolyte constituents exert a decisive influence on phase composition of the MAO films. XRD analysis results showed that the oxide films composed of Magnesium aluminum oxide, Aluminum oxide and Al for the AlMg6 arc sprayed substrate, Aluminum oxide, Al, Si for the Al-Si12 arc sprayed substrate and Aluminum phosphide, Aluminum oxide, Al for Al arc sprayed substrate.

# 5. REFERENCES

1.  M. H. Regina, et al., "Comparison of Aluminum Coatings Deposited by Flame Spray and by Electric Arc Spray," Surface and Coatings Technology, Vol. 202, No. 1, 2007, pp. 172-179. doi:10.1016/j.surfcoat.2007.05.067

2.  A. L. Yerokhin, X. Nie, A. Leyland, A. Matthews, S. J. Dowey, "Plasma Electrolysis for Surface Engineering," Surface and Coatings Technology, Vol. 122, No. 2-3, 1999, pp. 73-93. doi:10.1016/S0257-8972(99)00441-7

3.  V. Pokhmurskii, et al., "Plasma Electrolytic Oxidation of Arc-Sprayed Aluminum Coatings," Journal of Thermal Spray Technology, Vol, 16, No. 5-6, 2007, pp. 998-1004.doi:10.1007/s11666-007-9104-x

4.  M. D. Klapkiv, "State of Electrolytic Plasma in the Process of Synthesis of Oxides Based on Aluminum," Materials Science, Vol. 31, No. 4, 1995, pp. 494-499.doi:10.1007/BF00559144

5.  C. Moreau, P. Gougeon and M. Lamontagne, "Influence of Substrate Preparation on the Flattening and Cooling of Plasma-Sprayed Particles," Journal of Thermal Spray Technology, Vol. 4, No. 1, 1995, pp. 25-33. doi:10.1007/BF02648525

6.  S. Shakeri and S. Chandra, "Splashing of Molten Tin Droplets on a Rough Steel Surface," International Journal of Heat and Mass Transfer, Vol. 45, No. 23, 2002, 4561-4575.doi:10.1016/S0017-9310(02)00170-9

7.  M. Fukumoto, E. Nishioka and T. Matsubara, "Flattening and Solidification Behavior of a Metal Droplet on a Flat Substrate Surface Held at Various Temperatures," Surface and Coatings Technology, Vol. 120-121, 1999, pp. 131-137. doi:10.1016/S0257-8972(99)00349-7

8.  N. Mehdizadeh, S. Chandra and J. Mostaghimi, "Formation of Fingers around the Edges of a Drop Hitting a Metal Plate with High Velocity," Journal of Fluid Mechanics, Vol. 510, 2004, pp. 353-373. doi:10.1017/S0022112004009310

9.  M. Pasandideh-Fard, V. Pershin, S. Chandra and J. Mostaghimi, "Splat Shapes in a Thermal Spray Coating Process: Simulations and Experiments," Journal of Thermal Spray Technology, Vol. 11, No. 2, 2002, pp. 206-217. doi:10.1361/105996302770348862

10. A. McDonald, M. Lamontagne, C. Moreau and S. Chandra, "Impact of Plasma-Sprayed Metal Particles on Hot and Cold Glass Surfaces," Thin Solid Films, Vol. 514, No. 1-2, 2006, pp. 212-222. doi:10.1016/j.tsf.2006.03.010

11. M. Fukumoto, E. Nishioka and T. Matsubara, "Flattening and Solidification Behavior of a Metal Droplet on a Flat Substrate Surface Held at Various Temperatures," Surface and Coatings Technology, Vol. 120-121, 1999, pp. 131-137. doi:10.1016/S0257-8972(99)00349-7

12. A. Nobuyuki, et al., "Surface Modification of Al Sprayed Coatings by Direct Diode Laser Remelting Process," Transactions of JWRI, Vol. 35, No. 2, 2006),

13. M. P. Planche, et al., "Relationships between in-Flight Particle Characteristics and Coating Microstructure with a Twin Wire Arc Spray Process and Different Working Conditions," Surface and Coatings Technology, Vol. 182, No. 2-3, 2004, pp. 215-226.doi:10.1016/S0257-8972(03)00873-9

# Potentiality of Clay Raw Materials from Northern Morocco in Ceramic Industry: Tetouan and Meknes Areas

*M. El Ouahabi[1]\*, L. Daoudi[2], F. De Vleeschouwer[3,4], R. Bindler[5], N. Fagel[1]*

[1]UR Argile, Géochimie et Environnement Sédimentaires (AGEs), Département de Géologie, Université de Liège, Liége, Belgium
[2]Département de Géologie, Faculté des Sciences et Techniques, Marrakech, Maroc
[3]Université de Toulouse; INP, UPS; EcoLab (Laboratoire Ecologie Fonctionnelle et Environnement); ENSAT, Castanet Tolosan, France
[4]CNRS; EcoLab; Castanet Tolosan, France
[5]Department of Ecology and Environmental Sciences, Umeå University, Umeå, Sweden

## ABSTRACT

This study aims at evaluating the potential suitability of Tetouan and Meknes (central Morocco) clay material as raw materials in various ceramic applications by investigating their textural, chemical, thermal and firing characteristics. Textural properties were identified by specific surface area, cation exchange capacity (CEC) and bulk density ($\rho s$). Chemical and thermal properties were assessed using XRF and TG/DTA techniques, respectively. Firing characteristics at temperatures from 800°C to 1100°C were determined by linear firing shrinkage, loss on weight and water absorption capacity. The Meknes clays are characterised by medium cation exchange capacity (CEC) and specific surface area (SSA) values due to their moderate smectite content. The Tetouan clays have medium to low CEC and medium SSA values. The main oxides in the clayey samples are $SiO_2$ (35 - 54.3 wt%), $Al_2O_3$ (20.6 - 43.9 wt%), and $Fe_2O_3$ (9.7 - 22.4 wt%). The amount of CaO in Meknes clays ranges from 8 to 12 wt%, whereas CaO is only present in some Tetouan clay (TE4, TE7, TN4 and TN5). A significant densification of ceramic behaviour could be noticed for most of Tetouan clays at firing temperatures above 1000°C. Meknes clays show earlier densification from 800°C. The chemical, textural and ceramic properties of Tetouan and Meknes clays indicate their suitability as raw materials for the production of structural ceramics. The high amount of $Fe_2O_3$ in all clays makes them inappropriate in fine ceramics.

**Keywords:**Clay Materials, Ceramic Properties, Ceramic Suitability, Morocco

# 1. INTRODUCTION

Throughout the world, clays are the main raw materials exploited in the fabrication of various ceramic products for building construction. Due to inherently complex physical, chemical and mineralogical characteristics, clays have unique properties related to their own natural genesis [1] - [4] . In most cases, for economic reasons, the ce- ramics industry relies on clays from nearby deposits. As a consequence, the clay characterization is important for the technical performance of local products [5] [6] .

Pure clays do not occur in nature, they contain mixtures of different clay and associated minerals [7] - [9] . At present, many ceramic tiles are manufactured from mixtures of mineral raw materials, composed essentially of clays and materials such as quartz, feldspar and carbonates. In the fabrication process, the raw materials are mixed in variable proportions taking into account the influence of each component on the properties of the final products [10] [11] . The components that play fundamental roles for optimum processing, and hence perfor- mance of the final products, are clay fraction (<2 μm) for plasticity, feldspar for fluxing and silica as filler ma- terial [12] - [14] . The selection of the appropriate raw materials is based on specific criteria, which are related ei- ther to the behaviour during the different stages of manufacturing or to the overall chemical composition. The microstructure and properties of any ceramic depends on the characteristics of the raw materials and processing parameters [15] to assure the quality of ceramic products.

Morocco is among the world top 20 producers and consumers of clayey building          materials          (http://www.lematin.ma/reader-2007/files/lematin/2011/01/30). The industry of ceramic tiles and bricks is most frequent in northern Morocco. Natural clayey materials are abundant in this region. They play an important economic role with a national production of about 45% for building materials (bricks, ceramic tiles and refracto- ries). However, national ceramic production is still insufficient, and to fill this deficit the Moroccan government has to import ceramic products mainly from Spain, Italy and Egypt.

The Tetouan and Meknes area have large construction and development projects to meet the population growth in particular for social and luxury housing construction, which is experiencing a significant economic and tourism development. Meknes and Tetouan clays are currently used for the traditional production of small- scale ceramic factories. Meknes clays are used in small traditional pottery factories using local clays. Artisanal and semi-industrial exploitation of these clays, without any prior study, causes various problems during the manufacturing process, such as deformations and breakage of products.

The aim of this study is to evaluate the viability of the Tetouan and Meknes clay deposits as an industrial mineral resource by comparing their textural, chemical and thermal properties as well as drying and firing be- haviours.

## 2. MATERIAL

The studied clayey raw materials were collected in the Tetouan and Meknes areas. The Tetouan clay deposit is located in north-western Morocco a few kilometres from the town of Tetouan, in the external Rif domain [16] [17] (Figure 1). A total of nineteen clay samples were collected in this area. There are two categories of samples: 10 clay samples collected from the exploited quarries labelled (TE) and 9 clays sampled from clay deposits, which are not yet exploited, labelled (TN). The Meknes clay deposit is located in north-central Morocco, about 15 km east of the town of Meknes, in the Saïss Basin zone situated in the Pre-Rif domain (Figure 1). Six Miocene yellow sandy clay samples [18] were collected from the quarries of pottery and artisanal ceramic.

A preliminary characterization of those clay materials was done in a previous study [19] . Tetouan clays are characterized by diversified mineralogical assemblages (in particular a variable proportion of clay, quartz and calcite) in contrast with to Meknes clays (high clay content, quartz and calcite). The clay fraction of Tetouan is dominated by illite and kaolinite with variable contributions of chlorite. Meknes clays are illite and kaolinite, associated with smectite. The studied clay materials consist generally of fine particles with medium to high plas- ticity and low organic matter content [19] .

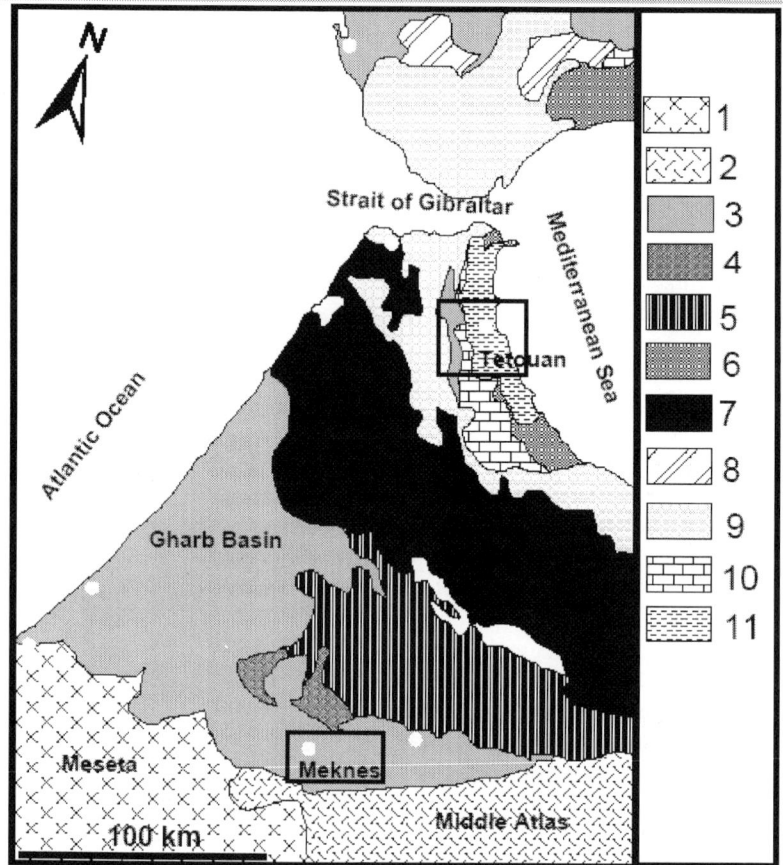

**Figure 1.** Structural sketch map of the northern occidental part of the Rif Chain and the Northern central of Morocco (modified after [45] ). 1—Foreland basement, 2—Meseta and Atlas cover series; 3—Foredeep basins; 4—Detached Atlasic cover at Prerif front; 5—Prerif; 6—Alpujarrides-Sebtide nappes; 7—Intrarif, Mesorif, Rif nappes; 8—Sud-Betic Zone; 9—Mag- hrebian Flyschs; 10—Dorsale calcaire units, 11—Malaguide- Ghomaride nappes

# 3. ANALYTICAL METHODS

## 3.1. Textural Properties

The specific surface area of the samples was characterized by the analysis of nitrogen adsorption-desorption isotherms, performed at 77°K. The measurements were performed using a sorptomatic Carlo Erba 1900, con-trolled by a computer (Industrial Chemistry, Department of Chemistry, ULg). The analysis of the isotherms was performed according to the methodology of Vallée, Keller [20] , which provided specific surface area ($S_{BET}$), micropore volume calculated by the Dubinin-Radushkevich equation (VDUB), and total pore volume calculated from the adsorbed volume at saturation (Vp). The bulk density ($\rho$s) was obtained by helium pycnometry on the powdered sample, using Micromeritics Accupyc 1330. Textural analyses were done at the laboratory of Indus- trial Chemistry (Department of Chemistry, ULg).

Cation Exchange Capacity (CEC) was measured by the Schollenberger method [21] . The samples were first saturated by ammonium acetate (1 N), and then the ammonium ions in the supernatant were deprotonated into ammonia with sodium hydroxide solution (0.1 N). The ammonia content was then determined by distillation into a known amount of acid and back-titrated by the Kjeldahl method [22] .

## 3.2. Chemical and Thermal Properties

The chemical composition of selected elements (Si, Al, Fe, Ca, Mn, Mg, Na, K, Ti, P and S) was measured as oxides on 2 g of dried and homogenized powder of clayey samples using a Bruker S8 Tiger wavelength-disper- sive X-Ray Fluorescence (WD-XRF) spectrometer equipped with a Rh anticathode (Department of Ecology and Environmental Sciences, Umeå University). Calibration was made using 35 commercially available certified reference materials of similar matrix (sedimentary rocks, river, lake and marine sediments, sands and soils). The accuracy ranges from 3% to 7% except for S (25%) and P (20%). Reproducibilites are above 99% except for S (89%) and P (97%). More details about the method and the calibration can be found in [23] . The same powdered samples were heated to 1000°C for 2 h to determine the Loss On Ignition (L.O.I).

Differential scanning calorimetry (DSC) and thermogravimetry (TG) were conducted simultaneously using a NETZSCH STA 409 PC instrument (Industrial Chemistry, Department of Chemistry, ULg). Samples were heated from room temperature to 800°C at 10°C min$^{-1}$ under atmospheric air [24] .

## 3.3. Ceramic Properties

Industrial tests were carried out as part of the ceramic evaluation process. Clay samples were dried at 105°C for 24 hr and ground to a fine powder and then sieved using a mesh-size of 100 μm. Each clay sample was wetted in order to achieve the proper plasticity for modeling. The samples obtained with these shaping techniques were 4 cm long, 2 cm wide and 2 cm thick. The drying was done in a shaded and ventilated room. For different drying times, the mass and the value of the length were measured to calculate the linear shrinkage. The dried samples (24 h at shaded room plus 12 h at 105°C in oven) were kiln-fired at different temperatures (800°C, 850°C, 900°C, 950°C, 1000°C, 1050°C and 1100°C) over 1 h. The fired samples were tested for loss on ignition, shrinkage and water absorption capacity after firing. The linear shrinkage (LS) was determined following the conventional techniques. The water absorption capacity (WAC) was determined according to standard procedure UNE 67-027 [25] , in fired clay pieces after each cooking cycle [26] . After preliminary measurements at the end of firing and cooling, the specimens of each batch were kept dry in an oven until their submission to water absorption. Each dry and cool specimen was weighed (P1) and the three were then immersed into clean water at 25°C for 24 h. The specimens were removed from the water, their surfaces were wiped off and the weight (P2) of each was measured immediately. The water absorption capacity (WAC) was calculated as WAC = (P2 − P1)/P1 × 100%.

# 4. RESULTS AND DISCUSSION

## 4.1. Textural Properties

Specific surface area ($S_{BET}$), pore volume (Vp), micropore volume (VDUB) and cation exchange capacity (CEC) of the samples are listed in Table 1. The surface area of the Meknes samples ranges from 33.3 to 37.9 m²·g⁻¹. Sample M1 has the highest specific surface area and M3 the lowest. According to these specific surface area values, M1 has less illite and smectite content, knowing that the specific surface area of kaolinite is usually smaller (10 - 20 m²·g⁻¹) with respect to that of illite (80 - 100 m²·g⁻¹) [27] , but similar values for smectite (19 m²·g⁻¹) [28] . The Tetouan clays have specific surface area ($S_{BET}$) values ranging between 17 and 36 m²·g⁻¹. Among them TE1, TE3 and TN9 display the highest *SSA* values due to their illite, kaolinite and interstratified clay contents [29] .

The CEC values of the Meknes samples (9.0 - 20.6 meq·100⁻¹g) are higher than the Teouan clays (7.1 - 18.5 meq·100⁻¹g). The high CEC values of the Meknes samples may be explained by their smetite content [19] , a mineral characterized by higher CEC value ranging between 80 and 150 meq·100⁻¹g [7]
.

The highest $S_{BET}$ and CEC of Meknes and some Tetouan clays (TN and TE) reflect their fine grain size, which is represented by the high amount of clay fraction. The highest $S_{BET}$ and CEC of those clays could cause a diffi- culty to dry and lead to the development of cracks in the drying process. The bulk density (ρs) of all clay sam- ples are almost similar, ranging from 2.7 to 3.2 g·cm⁻³.

## 4.2. Chemical and Thermal Properties

The most abundant oxides in the studied samples are $SiO_2$, $Al_2O_3$, CaO and $Fe_2O_3$ (Table 2). According to the major element abundances, the samples are divided into two groups (Figure 2). The first group is characterized by very low CaO (<1.9%), slightly high $Fe_2O_3$ (from 12% to 22%) and high $Al_2O_3$ (from 29% to 44%) in com- parison to the second group. On the basis of their low CaO content, the first group is qualified as a non-cal- careous clay group, which is mainly represented by the Tetouan clays (TE and TN). The second group displays higher CaO (12% - 22%), slightly lower $Al_2O_3$ (20% - 25%) and lower $Fe_2O_3$ (8% - 12%) concentrations and is labelled as a moderate calcareous group. Group 2 consists of all the Meknes samples and TE4, TE7 from the exploited Tetouan samples and TN4 and TN5 from unexploited Tetouan clays. The higher contents of $Fe_2O_3$ (>5%) in the samples give the reddish color of the clay-based products after firing. The L.O.I at 1000°C (Table 2) ranges from 6% to 24%, which is associated with the presence of clay minerals, hydroxides and organic matter7] [30] . These material losses are confirmed by the thermal analysis and organic matter content as indicated in Figure 3 and Table 2.

**Table 1.** Textural properties of the clay samples. SBET: specific surface area; Vp: pore volume; VDUB: micropore volume; CEC: specific surface area; ρs: bulk density.

| Samples | $S_{BET}$ $(m^2 \cdot g^{-1})$ | Vp $(cm^3 \cdot g^{-1})$ | VDUB $(cm^3 \cdot g^{-1})$ | CEC $(meq\ 100^{-1} \cdot g)$ | ρs $(g \cdot cm^{-3})$ |
|---|---|---|---|---|---|
| | ±5 | ±0.05 | ±0.01 | | |
| Meknes | | | | | |
| M1 | 37.9 | 0.055 | 0.015 | 12.9 | 2.7 |
| M2 | 36.3 | 0.055 | 0.015 | 9.1 | 2.7 |
| M3 | 33.3 | 0.053 | 0.015 | 9.7 | 2.7 |
| M4 | 36.3 | 0.053 | 0.015 | 10.1 | 2.7 |
| M5 | 34.7 | 0.049 | 0.015 | 20.6 | 2.7 |
| M6 | 35.2 | 0.094 | 0.015 | 12.5 | 2.7 |
| Unexploitable Tetouan clays | | | | | |
| TN1 | 32.4 | 0.069 | 0.015 | 18.5 | 2.7 |
| TN2 | 32.9 | 0.072 | 0.015 | 17.1 | 2.8 |
| TN3 | 31.0 | 0.072 | 0.015 | 17.2 | 2.8 |
| TN4 | 34.6 | 0.08 | 0.02 | 8.5 | 2.8 |
| TN5 | 26.2 | 0.061 | 0.014 | 10.0 | 2.8 |
| TN6 | 27.9 | 0.064 | 0.015 | 18.3 | 2.7 |
| TN7 | 26.2 | 0.086 | 0.012 | 13.1 | 2.8 |
| TN8 | 30.2 | 0.091 | 0.012 | 9.3 | 2.8 |
| TN9 | 36.4 | 0.101 | 0.015 | 17.9 | 2.8 |
| Exploitable Tetouan clays | | | | | |
| TE1 | 35.5 | 0.077 | 0.015 | 13.8 | 2.8 |
| TE2 | 34.5 | 0.075 | 0.015 | 14.6 | 2.8 |
| TE3 | 36.4 | 0.078 | 0.015 | 10.3 | 2.8 |
| TE4 | 31.0 | 0.068 | 0.014 | 10.7 | 2.9 |
| TE5 | 26.0 | 0.069 | 0.011 | 7.6 | 2.9 |
| TE6 | 17.7 | 0.047 | 0.008 | 12.0 | 2.8 |
| TE7 | 18.1 | 0.048 | 0.01 | 7.1 | 2.8 |
| TE8 | 21.3 | 0.056 | 0.01 | 13.7 | 3.0 |
| TE9 | 18.2 | 0.048 | 0.008 | 18.1 | 3.2 |
| TE10 | 27.0 | 0.062 | 0.015 | 17.6 | 2.8 |

The DSC/TG (Figure 3(a) and Figure 3(b)) show at least 4 peaks. The first endothermic peak is observed at 90°C - 150°C for Tetouan samples and between 96°C and 113°C for Meknes samples. Such peaks can be attrib- uted to the removal of adsorbed and interlayered water. The associated mass lost ranges from 0.4% to 0.8% for Meknes clays and between 0.4% and 1.3% for Tetouan clays. A large exothermic peak is observed between 200°C - 450°C. Due to organic matter decomposition [31] , it was observed in all studied clays, but was espe- cially well marked at 450°C for M6. A broad endothermic band, sintered at 520°C to 550°C for Meknes samplesand at 523°C to 568°C for Tetouan area samples, is due to clay mineral dehydroxylation and α quartz ® β quartz transformation [25] . A substantial loss of weight (5% - 7%) is associated with this endothermic peak for Tetouan samples. As an exception, some Tetouan samples (TN4, TN5, TE7 and TE4) lost less than 3% weight. Meknes samples lost about 4%, demonstrated by elevated clay and quartz content in this area. Small additional endo- thermicpeaks occurred in all samples at 700°C, which are due to carbonate decomposition [32]

**Table 2**. Chemical composition of the clay samples.

| | $SiO_2$ | $Al_2O_3$ | $Fe_2O_3$ | CaO | MnO | MgO | $Na_2O$ | $K_2O$ | $TiO_2$ | $P_2O_5$ | $SO_2$ | Loss On Ignition |
|---|---|---|---|---|---|---|---|---|---|---|---|---|
| | | | | | | Meknes | | | | | | |
| M1 | 44.9 | 24.83 | 14.20 | 11.38 | 0.06 | 2.69 | 0.81 | 4.12 | 0.75 | 0.28 | 0.22 | 17.85 |
| M2 | 42.7 | 24.83 | 12.16 | 12.67 | 0.05 | 2.34 | 0.78 | 4.07 | 0.75 | 0.25 | 0.08 | 17.37 |
| M3 | 41.3 | 21.05 | 15.63 | 8.15 | 0.04 | 2.67 | 0.84 | 3.37 | 0.65 | 0.24 | 0.42 | 18.77 |
| M4 | 43.1 | 22.67 | 13.18 | 9.95 | 0.05 | 2.42 | 0.84 | 3.71 | 0.70 | 0.24 | 0.08 | 17.61 |
| M5 | 41.0 | 23.84 | 14.83 | 11.70 | 0.07 | 1.82 | 0.67 | 3.61 | 0.73 | 0.22 | 0.02 | 18.29 |
| M6 | 43.1 | 24.15 | 12.40 | 10.47 | 0.05 | 2.55 | 0.92 | 3.83 | 0.72 | 0.26 | 0.52 | 16.77 |
| | | | | | Unexploitable Tetouan clays | | | | | | | |
| TN1 | 45.5 | 30.53 | 15.84 | 0.69 | 0.08 | 2.06 | 0.94 | 3.57 | 0.90 | 0.13 | 0.10 | 9.90 |
| TN2 | 47.7 | 34.58 | 18.96 | 0.39 | 0.03 | 2.44 | 0.92 | 3.49 | 1.05 | 0.13 | 0.00 | 9.24 |
| TN3 | 47.1 | 35.48 | 22.39 | 0.25 | 0.05 | 2.59 | 1.16 | 3.30 | 1.03 | 0.14 | 0.02 | 8.93 |
| TN4 | 35.9 | 20.63 | 9.75 | 20.94 | 0.14 | 2.14 | 1.05 | 3.54 | 0.53 | 0.17 | 0.84 | 23.90 |
| TN5 | 35.9 | 20.90 | 9.86 | 22.16 | 0.15 | 2.24 | 1.05 | 3.52 | 0.52 | 0.15 | 0.68 | 24.41 |
| TN6 | 47.6 | 32.69 | 14.53 | 0.48 | 0.06 | 1.94 | 0.86 | 3.08 | 0.97 | 0.12 | 0.00 | 8.87 |
| TN7 | 54.3 | 30.95 | 13.38 | 0.73 | 0.04 | 1.94 | 1.16 | 2.84 | 0.90 | 0.11 | 0.02 | 7.48 |
| TN8 | 51.1 | 30.46 | 12.24 | 1.90 | 0.03 | 1.72 | 1.29 | 2.89 | 0.93 | 0.13 | 0.10 | 7.94 |
| TN9 | 44.2 | 29.66 | 15.07 | 4.55 | 0.10 | 1.63 | 0.89 | 2.89 | 1.05 | 0.27 | 0.02 | 11.65 |
| | | | | | Exploitable Tetouan clays | | | | | | | |
| TE1 | 48.6 | 32.39 | 17.01 | 0.45 | 0.03 | 1.68 | 1.78 | 4.36 | 0.97 | 0.18 | 0.00 | 7.20 |
| TE2 | 52.7 | 31.44 | 18.01 | 0.45 | 0.05 | 2.36 | 1.35 | 2.91 | 0.83 | 0.22 | 0.08 | 6.82 |
| TE3 | 54.3 | 31.48 | 11.49 | 0.42 | 0.03 | 1.76 | 1.73 | 3.95 | 0.85 | 0.16 | 0.02 | 7.05 |
| TE4 | 36.7 | 25.39 | 9.92 | 11.84 | 0.10 | 6.00 | 0.94 | 5.13 | 0.67 | 0.10 | 0.02 | 19.52 |
| TE5 | 51.9 | 33.97 | 18.10 | 0.45 | 0.03 | 1.89 | 1.91 | 4.43 | 1.00 | 0.19 | 0.00 | 8.72 |
| TE6 | 51.5 | 33.78 | 17.47 | 0.49 | 0.03 | 1.76 | 1.91 | 4.41 | 1.00 | 0.18 | 0.06 | 6.57 |
| TE7 | 35.0 | 20.63 | 8.61 | 17.67 | 0.18 | 5.36 | 0.81 | 3.88 | 0.55 | 0.11 | 0.02 | 21.35 |
| TE8 | 44.3 | 43.95 | 19.87 | 0.38 | 0.03 | 1.19 | 0.84 | 6.91 | 0.78 | 0.08 | 0.02 | 11.13 |
| TE9 | 48.3 | 42.40 | 16.64 | 0.31 | 0.02 | 0.85 | 1.13 | 6.07 | 0.72 | 0.05 | 0.02 | 8.95 |
| TE10 | 47.6 | 34.46 | 20.87 | 0.41 | 0.05 | 2.40 | 1.11 | 3.32 | 1.02 | 0.14 | 0.02 | 8.96 |

## 4.3. Ceramic Properties

### 4.3.1. Drying Behaviour

Bigot's curves (Figure 4) were used as preliminary indicators in the choice of raw materials [33] for the ce- ramic industry. The drying is accompanied by shrinkage of the clay materials, due to porewater loss. The Bigot's curves exhibit the two characteristic phases of the drying process: an initial weight loss with shrinkage and successive weight loss with no further shrinkage. The behaviour of all Meknes clay samples was somewhatsimilar (Figure 4), although there were some small differences in shrinkage, ranging from 12% to 17%, accom- panied by a loss of weight from 19% to 22%. Tetouan clay samples showed a low drying shrinkage (2% - 6%), in contrast to the Meknes clay samples. Tetouan clays showed a greater increase of the loss on weight (20% - 23%). The highest linear shrinkage of Meknes clays can be explained by its smectite content and its high PI value [34] . Consequently Meknes clays have more problematic drying behaviour. By contrast, Tetouan (TN and TE clays) showed a low drying shrinkage, and they have more suitable behaviour.

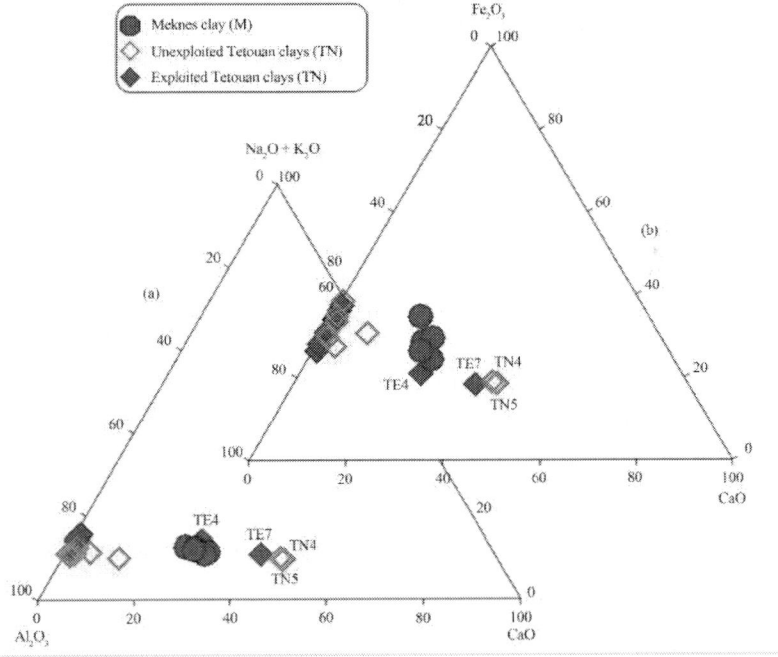

**Figure 2.** Ternary plot of: (a) Al2O3-CaO-Na2O + K2O; and (b) Al2O3-CaO-Fe2O3 (all in wt%) for studied clay samples.

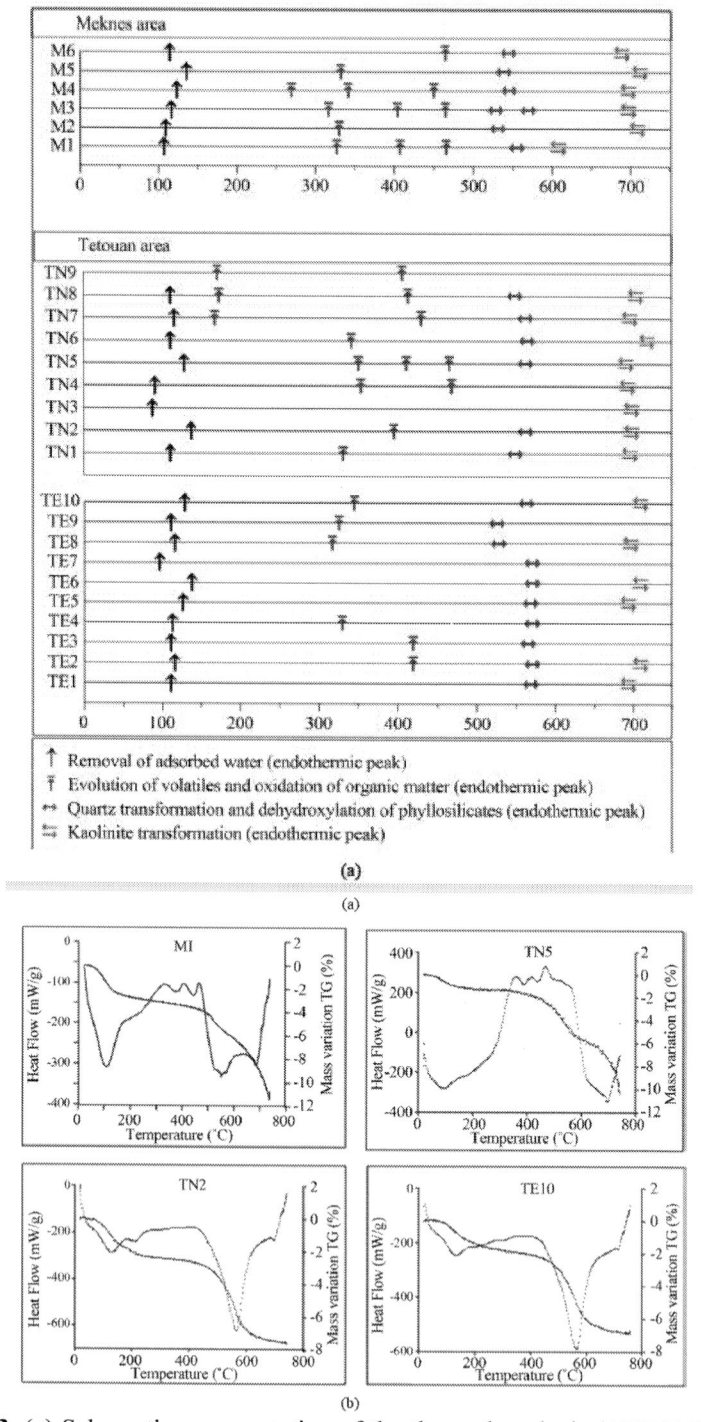

**Figure 3**. (a) Schematic representation of the thermal analysis (ATD/TG) of the clay samples; (b) Some examples of TG/ DSC curves.

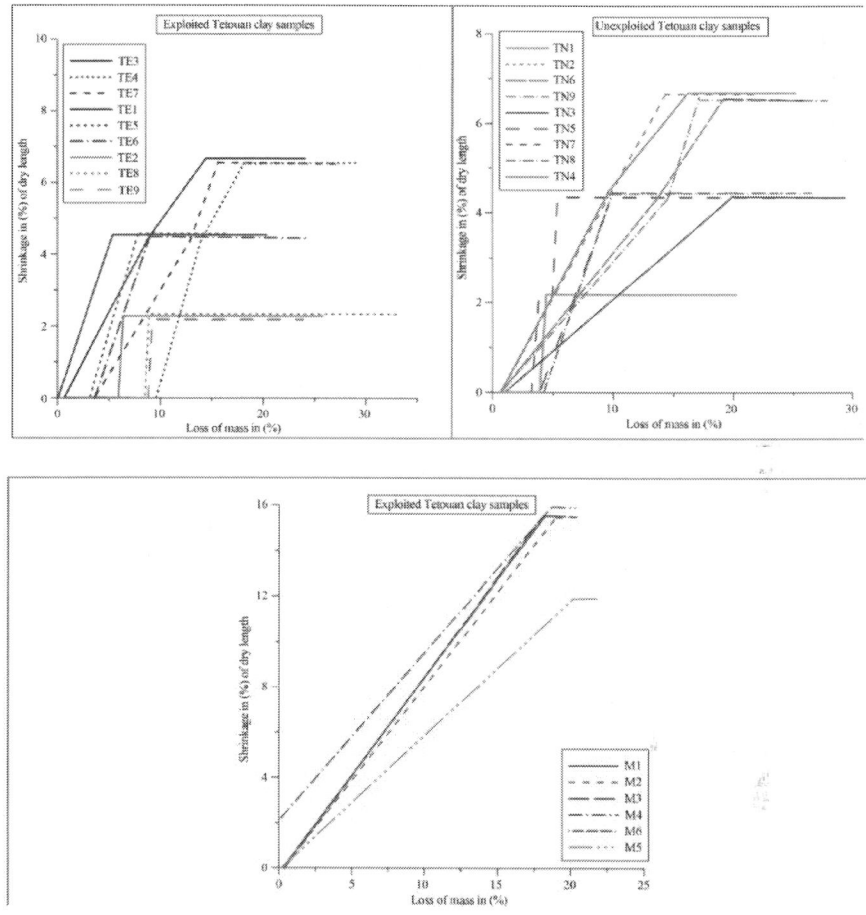

**Figure 4.** Bigot's diagram of the examined clay samples.

### 4.3.2. Firing behaviour

The results for linear firing shrinkage, loss on mass and water absorption of the fired clays are displayed in Table 3 and Figure 5, respectively. Linear firing shrinkage and water absorption have been frequently used as quality and process control parameters in the development and manufacturing stages of the production of struc- tural ceramics such as floor and wall tiles [35] [36] . The linear firing shrinkage could be used as a direct meas- ure of the extent of densification [37] . The linear firing shrinkage increased gradually from 1000°C to 1110°C for most of the Tetouan clays (TN and TE). This evolution reflects densification of the fired clays. As an excep- tion, some fired clays (TE5, TE7 and TE8) do not show any shrinkage. At 800°C, Meknes clays shrank from 17% to 29% and their shrinkage increased again gradually from 850°C to 1110°C. Meknes clays are characterized by an early densification from 800°C. This is due to the formation of a glassy phase, because they contain large amounts of fluxing agents ($Na_2O$, $K_2O$ and MgO) (Table 2). Loss on weight stability is indicated from 1000°C for most of the sintered clay bodies (Table 3). It can be seen that an increase in the shrinkage with a decrease in water absorption are associated with an increase in firing temperature. This trend is found especially above 1000°C. Shrinkage is related to the formation of high temperature crystalline phases, and the rates of shrinkageare directly related to the development of glassy material. Formation of gases during vitrification can also have a significant effect on the net dimensions of fired clay minerals. The greatest shrinkage above 1000°C is due to a more significant liquid phase formation. During liquid phase formation, the liquid surface tension and capillarity help to bring particles close together and reduce porosity [38] .

A significant decrease of the water absorption is observed for Tetouan clays (TE and TN) from 1000°C (Fig- ure 5). As an exception, TE4 and TN4 present the highest water absorption capacity (about 18%). The Meknes clays have high water absorption capacity, ranging from 9% to 21%. At 1100°C, the water absorption of Te- touan clays decreased to 0.4% and 2.6% for TE6 and TN8, respectively, indicating the total vitrification of those clays. For Meknes clays, the water absorption decreased from 6% to 13%, with the exception of M1, which had 20% (Figure 5).

## 4.4. Suitability for Ceramics Applications

The suitability of raw clay for their use in the manufacture of ceramic products is determined by mineralogy,chemistry and physical properties of the material. These factors will determine the behaviour of the clay during forming, drying and firing with direct influence on the final product [39] . In order to evaluate the suitability of the studied clays for different ceramic products, a ternary diagram from [40] based on XRD results [19] is shown in Figure 6(a). Two groups can be identified: those with high quartz contents and those that are rich in oxides, carbonates and feldspars. The mineralogical data for the samples (Figure 6(b)) also suggest that most of the Meknes clays can be used for vitrified red floor tile making. However Tetouan clays can be used for differ- ent ceramic applications due to their high quartz amount and their low amount of iron oxide. TN7 and TE9 may be used for making clinker products and TE7 for vitrified red floor tiles (Figure 6(c)). Some Tetouan samples (TN5, TN9 and TE7) are located close to the porous light-coloured field; they can therefore be used for porous light-coloured wall tile product.

The ternary diagram ($Fe_2O_3$ + CaO + MgO)/$Al_2O_3$/($Na_2O$ + $K_2O$) (Figure 6(c)) is used to classify raw claymaterials and industrial ceramic bodies [40] . All Meknes clays plot in the field of red ceramics, close to the field for porous light-coloured ceramics. Most of the Tetouan clays were located outside of the red ceramics field, with the exception of TE4, TE7, TN4 and TN5, which are close to the field for porous light-coloured ceramics. The ideal composition for an optimum white body product ($SiO_2$ = 72 wt%, $Al_2O_3$ and total oxides = 8 wt%)was estimated by Fiori et al., [40] , Wilson [41] and Baccour et al., [42] , who stated that clays containing 8% or more of $Fe_2O_3$ are red-firing clays. In this context, none of the Meknes and Tetouan clays can be used for the production of fine ceramics, due to their high $Fe_2O_3$ amount (>8%). Such application would require processing to reduce this iron oxide content. However, Tetouan clays could be considered as raw materials for structural ceramic products.

**Table 3**. Linear shrinkage (%) of the studied clays during firing process

| | 800°C | 850°C | 900°C | 950°C | 1000°C | 1050°C | 1100°C |
|---|---|---|---|---|---|---|---|
| Meknes clays | | | | | | | |
| M1 | 29.8 | 23.9 | 21.7 | 21.5 | 21.6 | 21.4 | 20.3 |
| M2 | 19.3 | 15.3 | 10.2 | 9.7 | 10.2 | 9.6 | 8.8 |
| M3 | 21.8 | 16.0 | 14.4 | 13.9 | 14.2 | 14.1 | 13.8 |
| M4 | 18.6 | 13.7 | 9.3 | 9.0 | 9.1 | 9.0 | 7.4 |
| M5 | 18.2 | 14.2 | 9.2 | 8.5 | 8.8 | 8.0 | 7.2 |
| M6 | 17.4 | 12.9 | 9.1 | 8.8 | 9.2 | 8.6 | 6.7 |
| Unexploitable Tetouan clays | | | | | | | |
| TN1 | 0 | 0 | 0 | 0 | 3.2 | 3.2 | 3.2 |
| TN2 | 0 | 0 | 3.1 | 3.1 | 3.1 | 3.1 | 6.3 |
| TN3 | 0 | 0 | 3.1 | 3.1 | 3.1 | 3.1 | 3.1 |
| TN4 | 0 | 0 | 2.9 | 2.9 | 2.9 | 2.9 | 2.9 |
| TN5 | 0 | 0 | 0 | 0 | 3.1 | 3.1 | 3.1 |
| TN6 | 0 | 0 | 0 | 0 | 0 | 0 | 3.2 |
| TN7 | 0 | 0 | 0 | 0 | 0 | 3.1 | 3.1 |
| TN8 | 0 | 0 | 0 | 0 | 3.2 | 3.2 | 3.2 |
| TN9 | 0 | 0 | 3 | 3 | 3 | 6.1 | 6.1 |
| Exploitable Tetouan clays | | | | | | | |
| TE1 | 0 | 0 | 3.1 | 3.1 | 3.1 | 3.1 | 6.3 |
| TE2 | 0 | 0 | 3.3 | 3.3 | 3.3 | 3.3 | 3.3 |
| TE3 | 0 | 0 | 3.1 | 3.1 | 3.1 | 3.1 | 6.3 |
| TE4 | 0 | 0 | 3.3 | 3.3 | 3.3 | 3.3 | 3.3 |
| TE5 | 0 | 0 | 0 | 0 | 0 | 0 | 0 |
| TE6 | 0 | 0 | 3.1 | 3.1 | 3.1 | 6.3 | 6.3 |
| TE7 | 0 | 0 | 0 | 0 | 0 | 0 | 0 |
| TE8 | 0 | 0 | 0 | 0 | 0 | 0 | 0 |
| TE9 | 0 | 0 | 0 | 0 | 3.1 | 3.1 | 6.3 |

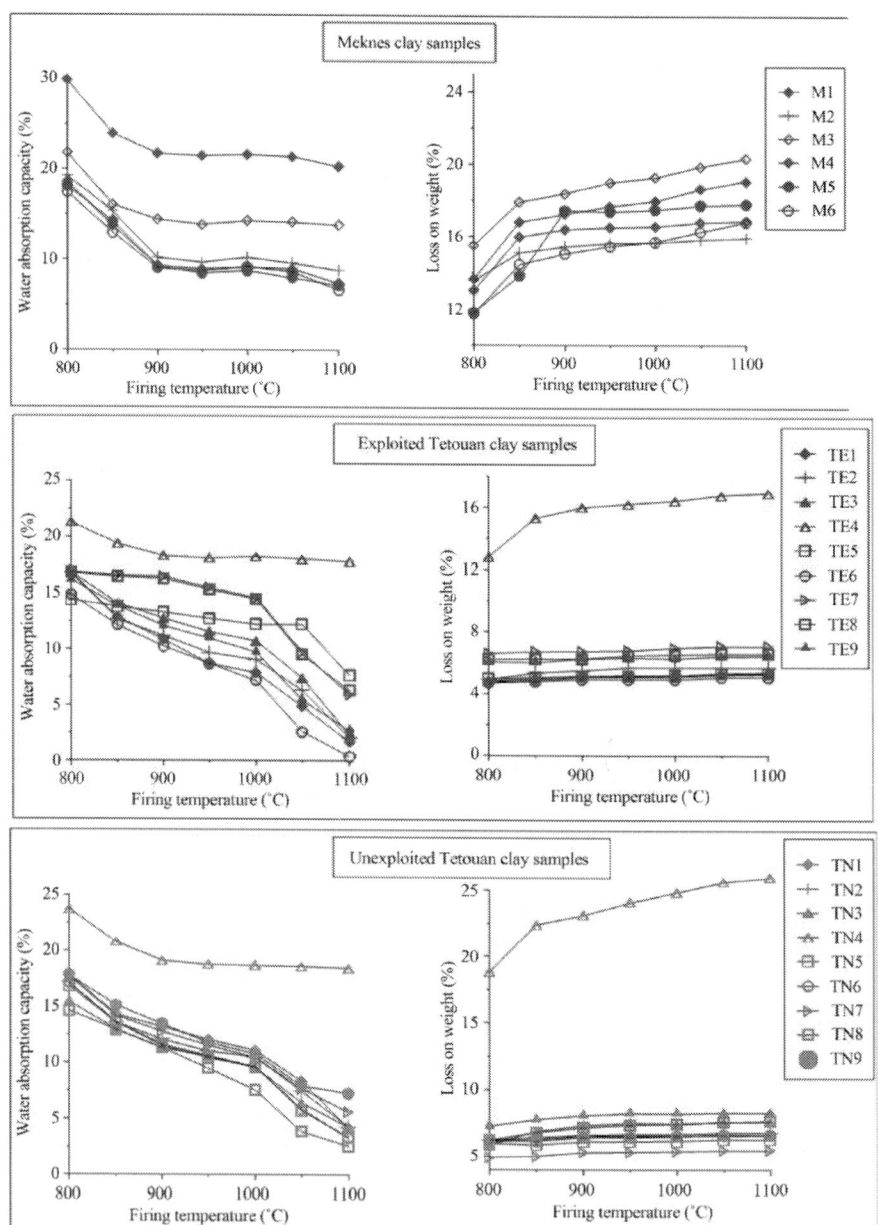

**Figure 5**. Influence of firing temperature on the variation of linear shrinkage and loss on weight.

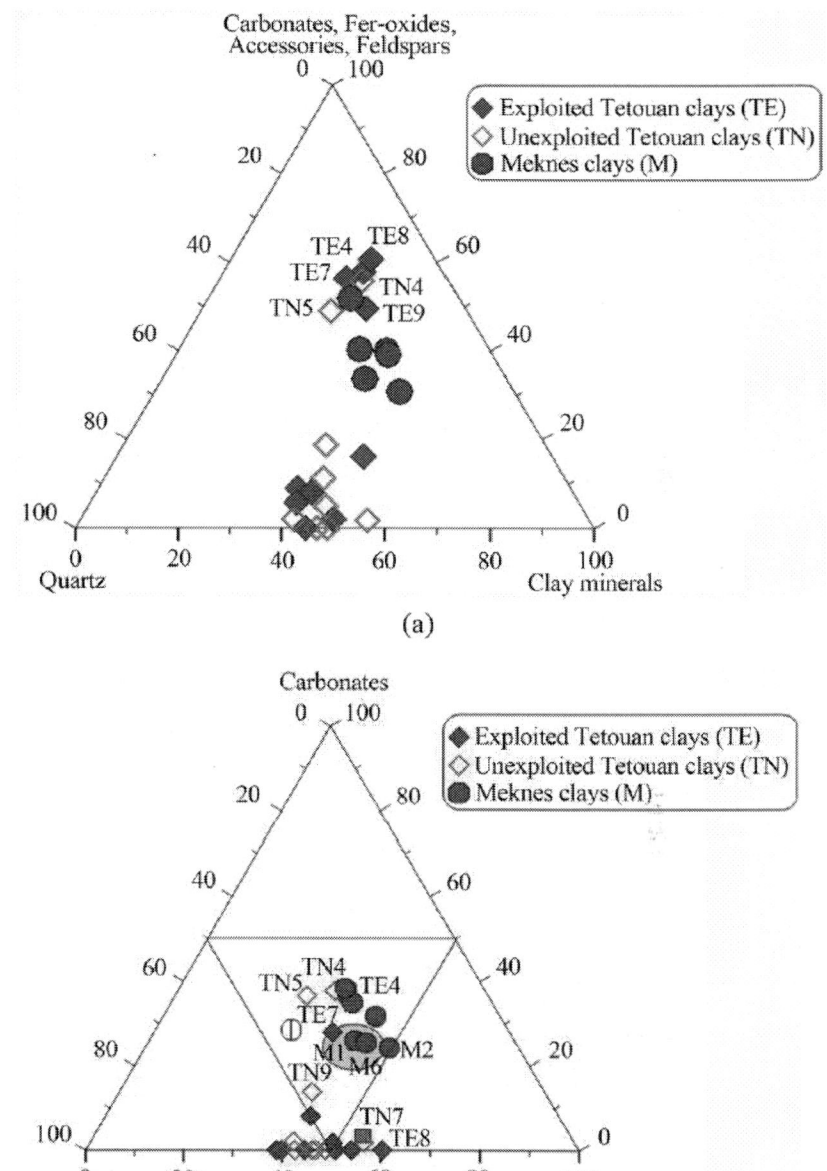

(a)

(b)

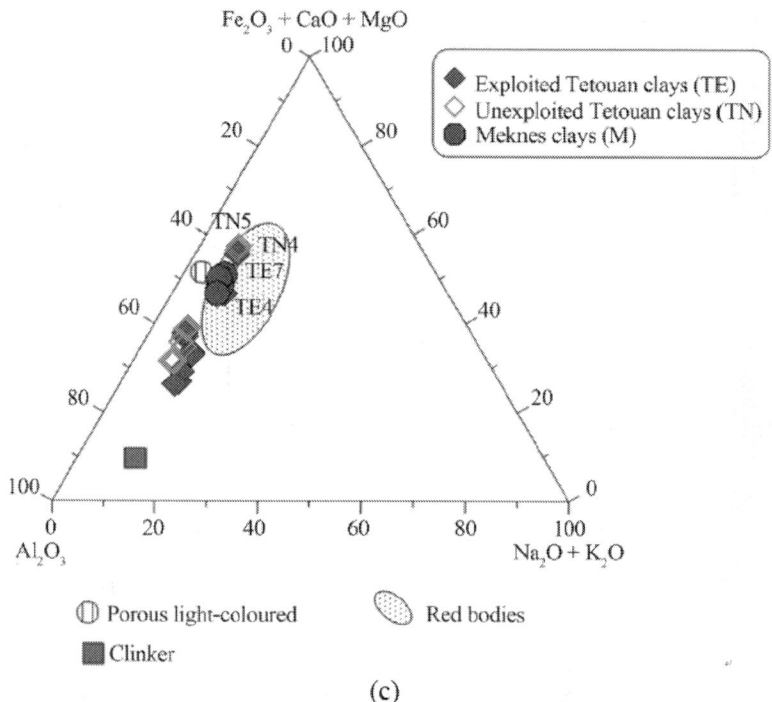

(c)

**Figure 6**. (a) Ternary diagram: quartz/carbonates + Fe-oxides + accessories + feldspars/clays minerals for the studied clays, after [42] ; (b) Triangular chart: Quartz + feldspars/calcite + dolomite/phyllosilicates for the studied clay materials, after [42] . (c) Ternary diagram of Fiori et al., [42] for clays studied: $Fe_2O_3 + CaO + MgO/Al_2O_3/(Na_2O + K_2O)$

The high firing shrinkage for M1 and M3 from Meknes area (>10%) make them inappropriate to make ce- ramics, due to the risk of dimensional defects. Water absorption can restrict the use of ceramics as building ma- terial [37] [44] . Ceramics used as flooring should have a water absorption <1%, roofing tiles <20%. Some Te- touan (TE1, TE2, TE3, TE6, TE9 and TN8) clays can also be used as flooring. Meknes clays and most of unex- ploited Tetouan clays show adequate water absorption for a use as roofing tiles.

## 5. CONCLUSIONS

Tetouan and Meknes clay materials are composed mainly of $SiO_2$, $Al_2O_3$, and $Fe_2O_3$. Meknes clays are charac- terized by a higher CaO content.

The ceramic behaviours of Tetouan and Meknes clays were interpreted by drying shrinkage, linear firing shrinkage, loss on weight and water absorption. Meknes clays are characterized by a high drying shrinkage, and therefore there will be more difficulty to dry the samples. To solve this problem, sand could be added to reduce their plasticity. Tetouan clays show optimal drying shrinkage. All the fired ceramic properties show moderate changes from 800°C to 1000°C.

However, significant changes are observed above 1000°C. An increase in the firing shrinkage and associated with a decrease in the water absorption are especially observed from 1100°C.

Meknes clays and most of Tetouan clays (TE1, TE2, TE3, TE6, TE9 and TN8) can be used on flooring pro- duction. In addition some exploited Tetouan clays and currently unexploited Tetouan clays are also suitable for making roofing tiles. Meknes clays and most of the Tetouan clays can be used for different ceramic applications due to their high quartz amount and their small iron oxide amount. Only TN7 and TE9 are being used for mak- ing clinker products.

This study demonstrates that exploited Tetouan and Meknes raw clay are suitable for the manufacture of structural ceramic products, with or without additives depending on the ceramic type. New clay outcrops such as unexploited Tetouan clays are also suitable. They constitute a potential ceramic raw material for growing Mo- roccan ceramic tile industry, and are thus not limited to artisanal production.

## ACKNOWLEDGEMENTS

The WD-XRF data acquisition was made possible at Umeå University through a research grant from the Kempe Foundation to Richard Bindler. The authors would like to thank René Pirard (Laboratoire de Génie Chimique, Université de Liège, Belgium) for the Calorimetry (ATD-TG) and specific surface area (SSA) analyses. We would like to thank Anne Iserentant (Université catholique de Louvain) for the CEC analysis. Finally, our thanks go to Dekayir (Université de Marrakech, Morocco) for his help on the field.

## REFERENCES

1.  Bauluz, B., Mayayo, M.J., Fernández-Nieto, C., Cultrone, G. and González López, J.M. (2003) Assessment of Tech- nological Properties of Calcareous and Non-Calcareous Clays Used for the Brick-Making Industry of Zaragoza (Spain). Applied Clay Science, 24, 121-126. http://dx.doi.org/10.1016/S0169-1317(03)00152-2

2.  Gomes, C. (1989) Argilas. O que sãoe para que servem. Fundação Calouste Gulbenkian, Lisboa.

3.  Hajjaji, M., Kacim, S. and Boulmane, M. (2002) Mineralogy and Firing Characteristics of a Clay from the Valley of Ourika (Morocco). Applied Clay Science, 21, 203-212. http://dx.doi.org/10.1016/S0169-1317(01)00101-6

4.  Verduch, A.G. (1995) Características de las Arcillas Empleadas en la Fabricación de Ladrillos. Técnica Cerámica, 232, 214-228.

5.  Fabbri, B. (1994) Quality Assurance of Ceramic Clays—A Deeper Understanding. Schmid, Freiburg, Allemagne.

6. Sánchez, E., Ginés, F., Agramunt, J. and Monzó, M. (1998) Control de calidad de las arcillas empleadas en la fabricación de los soportes de baldosas cerámicas. Proceedings Qualicer, 98, 97-112.

7. Grim, R.E. (1968) Clay Mineralogy. McGraw-Hill, New York.

8. Moore, D.M. and Reynolds, R.C. (1997) X-Ray Diffraction and the Identification and Analysis of Clay Minerals. Oxford University Press, Oxford, New York.

9. Bergaya, F. and Lagaly, G. (2006) Chapter 1. General Introduction: Clays, Clay Minerals, and Clay Science. In: Faïza Bergaya, B.K.G.T. and Gerhard, L., Eds., Developments in Clay Science, 1-18.

10. Sousa, S.J.G. and Holanda, J.N.F. (2005) Development of Red Wall Tiles by the Dry Process Using Brazilian Raw Materials. Ceramics International, 31, 215-222.http://dx.doi.org/10.1016/j.ceramint.2004.05.003

11. Sousa, S. and Holanda, J. (2007) Thermal Transformations of Red Wall Tile Pastes. Journal of Thermal Analysis and Calorimetry, 87, 423-428.http://dx.doi.org/10.1007/s10973-006-7030-7

12. Kamseu, E., Leonelli, C., Boccaccini, D.N., Veronesi, P., Miselli, P., Pellacani, G. and Melo, U.C. (2007) Characteri- sation of Porcelain Compositions Using Two China Clays from Cameroon. Ceramics International, 33, 851-857.http://dx.doi.org/10.1016/j.ceramint.2006.01.025

13. Alcântara, A.C.S., Beltrão, M.S.S., Oliveira, H.A., Gimenez, I.F. and Barreto, L.S. (2008) Characterization of Ceramic Tiles Prepared from Two Clays from Sergipe-Brazil. Applied Clay Science, 39, 160-165. http://dx.doi.org/10.1016/j.clay.2007.05.004

14. Celik, H. (2010) Technological Characterization and Industrial Application of Two Turkish Clays for the Ceramic In- dustry. Applied Clay Science, 50, 245-254.http://dx.doi.org/10.1016/j.clay.2010.08.005

15. Kingery, W., Bowen, H. and Uhlmann, D. (1976) Introduction to Ceramics. John Wiley & Sons, New York.

16. Durand Delga, M., Hottinger, L., Marcáis, J., Mattauer, M., Lilliard, Y. and Suter, G. (1960-1962) Livre à la mémoire du Professeur Paul Fallot consacré à l'évolution paléogéographique et structurale des domaines méditerranéens et alpins d'Europe. Vol. 1, Société géologique de France, Paris, 399-422.

17. Wildi, W. (1983) La chaîne tello-rifaine (Algérie, Maroc, Tunisie): Structure, stratigraphie et évolution du Trias au Miocène. Revue de Géologie Dynamique et Géographie Physique, 24, 201-297.

18. Aït Brahim, L. (1991) Tectonique cassante et états de contraintes récentes au nord du Maroc. Thèse de doctorat de l'Université Mohamed, 360.

19. El Ouahabi, M., Daoudi, L. and Fagel, N. (2014) Preliminary Mineralogical and Geotechnical Characterization of Clays from Morocco: Application to Ceramic Industry. Clay Minerals, 49, 1-17.

20. Lecloux, A.J. (1981) Texture of Catalysts. In: Anderson, J.R. and Boudart, M., Eds., Catalysis, Science and Techno- logy, Vol. 2, Springer, Berlin, 171-230.

21. Schollenberger, C.J. and Simon, R.H. (1945) Determination of Exchange Capacity and Exchangeable Bases in Soil- Ammonium Acetate Method. Soil Science, 59, 13-24.http://dx.doi.org/10.1097/00010694-194501000-00004

22. Mackenzie, R.C. (1952) A Micromethod for Determination of Cation Exchange Capacity of Clay. Clay Minerals Bulletin, 1, 203-204.http://dx.doi.org/10.1180/claymin.1952.001.7.03

23. Vleeschouwer, F.D., Renson, V., Claeys, P., Nys, K. And Bindler, R. (2011) Quantitative WD-XRF Calibration for Small Ceramic Samples and Their Source Material. Geoarchaeology, 26, 440-450. http://dx. doi.org/10.1002/gea.20353

24. Abajo, M. (2000) Manual sobre Fabricación de Baldosas. Tejas y Ladrillos. In: Beralmar, S.A., Ed., Barcelona.

25. U. 67-027-84 (1984) Determinacion de la absorcion de agua de ladrillos de arcilla. Instituto Espanol de Normalizacion, AENOR.

26. I. 10545-3 (1995) International Standard for Ceramic Tiles—Part 3. Determination of Water Absorption, Apparent Porosity, Apparent Relative Density and Bulk Density. International Organization for Standardization.

27. Ferrari, S. and Gualtieri, A.F. (2006) The Use of Illitic Clays in the Production of Stoneware Tile Ceramics. Applied Clay Science, 32, 73-81.http://dx.doi.org/10.1016/j.clay.2005.10.001

28. Ravichandran, J. and Sivasankar, B. (1997) Properties and Catalytic Activity of Acid-Modified Montmorillonite and Vermiculite. Clays and Clay Minerals, 45, 854-858.http://dx.doi.org/10.1346/CCMN.1997.0450609

29. Omotoso, O., Mikula, R.J. and Stephens, P.W. (2002) Surface Area of Interstratified Phyllosilicates in Athabasca oil Sands from Synchrotron XRD. Advances in X-Ray Analysis, 45, 363-391.

30. Milheiro, F.A.C., Freire, M.N., Silva, A.G.P. and Holanda, J.N.F. (2005) Densification Behaviour of a Red Firing Brazilian Kaolinitic Clay. Ceramics International, 31, 757-763.http://dx.doi.org/ 10.1016/j. ceramint.2004.08.010

31. Baccour, H., Medhioub, M., Jamoussi, F. and Mhiri, T. (2009) Influence of Firing Temperature on the Ceramic Properties of Triassic Clays from Tunisia. Journal of Materials Processing Technology, 209, 2812-2817.http://dx.doi.org/10.1016/j.jmatprotec.2008.06.055

32. Yariv, S. (2004) The Role of Charcoal on DTA Curves of Organo-Clay Complexes: An Overview. Applied Clay Science, 24, 225-236.http://dx.doi.org/10.1016/j.clay.2003.04.002

33. Štubna, I., Chmelík, F., Trník, A. and Šín, P. (2012) Acoustic Emission Study of Quartz Porcelain during Heating up to 1500°C. Ceramics International, 38, 6919-6922.http://dx.doi.org/10. 1016/j. ceramint. 2012.05.021

34. Maitra, S., Choudhury, A., Das, H.S. and Pramanik, M. (2005) Effect of Compaction on the Kinetics of Thermal Decomposition of Dolomite under Non-Isothermal Condition. Journal of Materials Science, 40, 4749-4751. http://dx.doi.org/10.1007/s10853-005-0843-0

35. Dondi, M., Fabbri, B. and Guarini, G. (1998) Grain-Size Distribution of Italian Raw Materials for Building Clay Products: A Reappraisal of the Winkler Diagram. Clay Minerals, 33, 435-442. http://dx.doi. org/10.1180/000985598545732

36. Manning, D.A.C. (1995) Introduction to Industrial Minerals. Chapman & Hall, London.http://dx.doi.org/10.1007/978-94-011-1242-0

37. Correia, S.L., Curto, K.A.S., Hotza, D. and Segadães, A.M. (2004) Using Statistical Techniques to Model the Flexural Strength of Dried Triaxial Ceramic Bodies. Journal of the European Ceramic Society, 24, 2813-2818.http://dx.doi.org/10.1016/j.jeurceramsoc.2003.09.009

38. Correia, S.L., Hotza, D. and Segadães, A.M. (2004) Simultaneous Optimization of Linear Firing Shrinkage and Water Absorption of Triaxial Ceramic Bodies Using Experiments Design. Ceramics International, 30, 917-922.http://dx.doi.org/10.1016/j.ceramint.2003.10.013

39. Shepard, F.P. (1954) Nomenclature Based on Sand-Silt-Clay Ratios. Journal of Sedimentary Petrology, 24, 151-158.

40. Monteiro, S.N. and Vieira, C.M.F. (2004) Influence of Firing Temperature on the Ceramic Properties of Clays from Campos dos Goytacazes, Brazil. Applied Clay Science, 27, 229-234. http://dx.doi.org/10. 1016/j.clay.2004.03.002

41. Mardare, C.C., Mardare, A.I., Fernandes, J.R.F., Joanni, E., Pina, S.C.A., Fernandes, M.H.V. and Correia, R.N. (2003) Deposition of Bioactive Glass-Ceramic Thin-Films by RF Magnetron Sputtering. Journal of the European Ceramic Society, 23, 1027-1030.http://dx.doi.org/10.1016/S0955-2219(02)00278-9

42. Fiori, C., Fabbri, B., Donati, G. and Venturi, I. (1989) Mineralogical Composition of the Clay Bodies Used in the Italian Tile Industry. Applied Clay Science, 4, 461-473.http://dx.doi.org/10.1016/0169-1317(89)90023-9

43. Murray, H.H. (2007) Applied Clay Mineralogy, Vol. 2, Developments in Clay Science. Elsevier B.V., Amsterdam.

44. Machado, A.T., Valenzuela-Diaz, F.R., de Souza, C.A.C. and de Andrade Lima, L.R.P. (2011) Structural Ceramics Made with Clay and Steel Dust Pollutants. Applied Clay Science, 51, 503-506. http://dx.doi.org/10.1016/j.clay.2011.01.004

45. Saadi, M., Hilali, E.A. and Boudda, A. (1980) Carte Géologique de la chaîne Rifaine—Echelle 1:500.000. Ministere de l'Energie et des Mines, Direction de Géologie, Service de Géologie du Maroc. Notes et Mémoires, No. 245.

CHAPTER 6

# Corrosion Performance of Atmospheric Plasma Sprayed Alumina Coatings on AZ31B Magnesium Alloy Under Immersion Environment

*D. Thirumalaikumarasamy, K. Shanmugam[1], V. Balasubramanian[2]*

[1,2] Department of Manufacturing Engineering, Annamalai University, Annamalainagar, 608 002 Chidambaram, Tamil Nadu, India

## ABSTRACT

Plasma sprayed ceramic coatings are successfully used in many industrial applications, where high wear and corrosion resistance with thermal insulation are required. The alumina powders were plasma sprayed on AZ31B magnesium alloy with three different plasma spraying parameters. In the present work, the influence of plasma spray parameters on the corrosion behavior of the coatings was investigated. The corrosion behavior of the coated samples was evaluated by immersion corrosion test in 3.5 wt% NaCl solution. Empirical relationship was established to predict the corrosion rate of plasma sprayed alumina coatings by incorporating process parameters. The experiments were conducted based on a three factor, five-level, central composite rotatable design matrix. The developed relationship can be effectively used to predict the corrosion rate of alumina coatings at 95% confidence level. The results indicate that the input power has the greatest influence on corrosion rate, followed by stand-off distance and powder feed rate.

## Abbreviations

- APS, atmospheric plasma spraying;
- $P$, power;
- $S$, stand-off distance;
- $F$, powder feed rate;

- RSM, response surface methodology;
- CR, corrosion rate

**Keywords:**  Corrosion; Plasma spraying; Alumina coating; Mg alloy

# 1. INTRODUCTION

Magnesium alloys are the lightest of all metals used as the basis for constructional alloys. Lightness, high strength-to-weight ratio and its demonstrated versatility make this material a great choice for application in aerospace, transportation and in civil and military applications [1] and [2]. However, several drawbacks restrict the application of unprotected magnesium alloys, especially their low wear and corrosion behavior. The use of coatings is one of the most effective strategies to protect these light alloys against corrosion and wear.

Ceramic coating, for example, has shown its ability to be used where corrosion and wear co-exist. As far as anticorrosion and antiwear applications are concerned, the most frequently used coating materials are oxide ceramic coatings [3]. Among the oxide materials, aluminum oxide, $Al_2O_3$, which is more often referred as alumina, is an exceptionally important ceramic material which has many technological applications. Alumina ceramic can retain up to 90% of their strength even at 1100 °C. Because of the excellent properties of alumina ceramics, they are widely used in many refractory materials, grinding media, cutting tools, high temperature bearings, a wide variety of mechanical parts, and critical components in chemical process environments, where materials are subjected to aggressive chemical attack, increasingly higher temperatures and pressures [4]. It is reported that the corrosion resistance of alumina coatings is higher than that of cermet and metallic coatings [5]. Thermally sprayed $Al_2O_3$ is widely employed to protect components against sliding, abrasive and erosive wear in a number of applications including rolls, pump bodies and plungers, industrial machinery parts (such as packaging and food processing equipment), etc. [6]. Alumina ($Al_2O_3$) is the most widely established coating material, and these coatings have been used for many applications in textile, electronic, aerospace, and aircraft industries because of their dielectric and wear resistance properties [7]. Alumina coatings could be prepared by techniques such as chemical vapor deposition (CVD), physical vapor deposition (PVD), sol–gel, plasma electrolytic oxidation (PEO) and plasma spray [8]. Among these techniques, plasma spraying is one of the most popular methods because it does not cause deterioration of the substrate, and by using this, comparatively thick coatings can be formed at a low cost and high deposition rate. Since the 1960s atmospheric plasma spraying (APS) has been widely used in industry and found applications in many fields such as automotive, aeronautical, medical, and paper milling [9]. As a result of the spraying process, ceramic coatings usually have open porosity and incomplete bonding between lamellae, which are deleterious when the coatings have to perform in an aggressive environment. The porosity allows a path for electrolytes from the outer surface to the substrate [10].

Therefore, reduction of porosity of the sprayed coatings plays a key role in improving the corrosion resistance of the coatings.

Celik et al. [11] studied the effect of grit blasting of substrate on the corrosion behavior of plasma-sprayed $Al_2O_3$ coatings. The results of their investigation showed that the corrosion resistance of the alumina coatings increased with decreasing porosity and coating thickness. Aruna et al. [12] reported that out of the three coatings prepared at different critical plasma spray parameters (CPSP), the coating deposited with a moderate CPSP exhibited the maximum wear and corrosion resistance, moderate microhardness and <10% spallation with 700 thermal cycles. Zhongshan et al. [13] used a heat treatment to aluminum arc sprayed coatings to improve the behavior of AZ31 alloy immersed in 3.5% NaCl. Chiu et al. [14] used hot pressing and anodizing as post-treatment of the thermally sprayed coating to improve its protection effectiveness against corrosion. Arrabal et al. [15] and Carboneras et al. [16] used oxy-acetylene flame spraying to deposit Al/SiCp coating on different magnesium alloys and a cold pressing post-treatment to reduce the porosity values of the coating.

All the above literature focused on the corrosion behavior of thermal sprayed coatings. To the best of our knowledge, there is no information available on the role of spraying parameters such as power, stand-off distance, and powder feed rate on the corrosion behavior of plasma sprayed ceramic coatings on magnesium alloy in open literature. Hence, the present investigation was carried out to develop an empirical relationship to predict the corrosion rate of plasma sprayed alumina coatings. The effect of input power, stand-off distance and powder feed rate on corrosion behavior of alumina coating is reported in this paper.

# 2. METHODOLOGY

## 2.1. Finding the working limits of the parameters

The chemical compositions of the AZ31B alloy used in this study are as follows (in wt%): Al 3.0, Zn 0.1, Mn 0.2 and Mg balance. The dimensions of the coated specimen are shown in Fig. 1a. Plasma spraying of the alumina powder was carried out using an in-house system. Details about this system are reported elsewhere [10] and [17]. Coating thickness for all the deposits was maintained at $200 \pm 15$ μm. This was achieved using the following method. The values of the factors were set in the machine as prescribed by a run in the design matrix, after which one single layer was made. The gun traverse rate was constant for all experiments; it was 300 mm s$^{-1}$, and the coating track overlap was 30%. The thickness of the single layer made after the run was measured using a digital micrometer (Model DIGIMATIC MDC-25SB). The number of passes required to achieve 200 μm thickness was determined, and the spray run was carried out to achieve it. The microstructure of the coatings for fixing the working range of parameters is presented inTable 1.

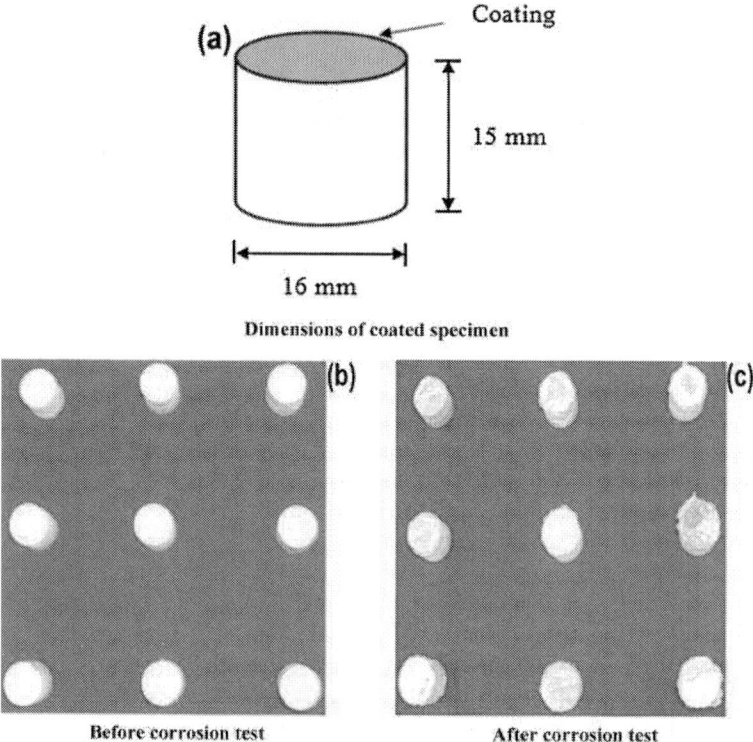

**Figure. 1**. Experimental details.

**Table 1.**Microstructure observation for fixing the working range of parameters.

| SI no. | Parameters | Parameter range | Microstructure | Name of the defect |
|---|---|---|---|---|
| 1 | Power | <18 kW | 50 μm | Coating delamination |
| 2 | Power | >18 kW | 50 μm | Poor deposition efficiency |
| 3 | Stand-off distance | <10 cm | 50 μm | Peeling-off of the coating |
| 4 | Stand-off distance | >13 cm | 50 μm | Poor coating adhesion |

## 2.2. Developing the Design Matrix

With a view to study the effects of the considered process parameters on the corrosion rate, statistically designed experiments, based on a factorial technique, were used to reduce the cost and time and to obtain the required information pertaining to the main and the interaction effects of the parameters on the response. Table 2 presents the ranges of factors considered, and Table 3 shows the 20 sets of coded conditions used to form the design matrix.

**Table 2.** Important APS process parameters and their levels.

| Factors | Notations | Units | Levels | | | | |
|---|---|---|---|---|---|---|---|
| | | | -1.682 | -1 | 0 | +1 | 1.682 |
| Power | P | kW | 18 | 19.4 | 21.5 | 23.6 | 25 |
| Stand-off distance | S | cm | 10 | 10.6 | 11.5 | 12.4 | 13 |
| Powder feed rate | F | gpm | 15 | 20 | 25 | 30 | 35 |

**Table 3.** Design matrix and experimental results.

| Spray condition | Coded values | | | Original value | | | Porosity (vol%) | Corrosion rate (mm/year) |
|---|---|---|---|---|---|---|---|---|
| | P | S | F | P (kW) | S (cm) | F (gpm) | | |
| 1 | -1 | -1 | -1 | 19.4 | 10.6 | 20 | 12 | 15.9172 |
| 2 | 1 | -1 | -1 | 23.6 | 10.6 | 20 | 7 | 7.92 |
| 3 | -1 | 1 | -1 | 19.4 | 12.4 | 20 | 14 | 14.3527 |
| 4 | 1 | 1 | -1 | 23.6 | 12.4 | 20 | 6 | 9.9182 |
| 5 | -1 | -1 | 1 | 19.4 | 10.6 | 30 | 10 | 11.2399 |
| 6 | 1 | -1 | 1 | 23.6 | 10.6 | 30 | 9 | 8.902 |
| 7 | -1 | 1 | 1 | 19.4 | 12.4 | 30 | 18 | 15.8024 |
| 8 | 1 | 1 | 1 | 23.6 | 12.4 | 30 | 13 | 15.568 |
| 9 | -1.682 | 0 | 0 | 18 | 11.5 | 25 | 14 | 19.02 |
| 10 | 1.682 | 0 | 0 | 25 | 11.5 | 25 | 5 | 12.9808 |
| 11 | 0 | -1.682 | 0 | 21.5 | 10 | 25 | 9 | 9.8012 |
| 12 | 0 | 1.682 | 0 | 21.5 | 13 | 25 | 15 | 15.344 |
| 13 | 0 | 0 | -1.682 | 21.5 | 11.5 | 15 | 8 | 6.122 |
| 14 | 0 | 0 | 1.682 | 21.5 | 11.5 | 35 | 12 | 6.8973 |
| 15 | 0 | 0 | 0 | 21.5 | 11.5 | 25 | 5 | 5.0241 |
| 16 | 0 | 0 | 0 | 21.5 | 11.5 | 25 | 6 | 4.9898 |
| 17 | 0 | 0 | 0 | 21.5 | 11.5 | 25 | 5 | 5.5344 |
| 18 | 0 | 0 | 0 | 21.5 | 11.5 | 25 | 6 | 4.892 |
| 19 | 0 | 0 | 0 | 21.5 | 11.5 | 25 | 5 | 5.9254 |
| 20 | 0 | 0 | 0 | 21.5 | 11.5 | 25 | 5 | 4.9994 |

## 2.3. Conducting The Experiments

The porosity analysis was carried out using an optical microscope equipped with an image analysing system (Make: MEIJI, Japan; Model: MIL-7100) [18]. Customary metallographic procedures were adopted to polish the cross-section of the coatings. A 200 μm square area was selected on the polished cross section of the coating, and the image was analyzed. The same procedure was repeated at five random locations to find out the average percentage volume of porosity. It was explained in Fig. 2.

The porosity analysis results of coatings produced are summarized in Table 4. In the case of APS coatings, porosity can be inter-lamellar or intra lamellar. Inter-lamellae pores mostly form from the random build-up of splats and the volume change during solidification results in intra lamellae pores [19]. Further, several possible sources of porosity in a coating have also been identified including: curling up of splats due to thermal stresses; incomplete filling of interstices during deposition; presence of unmelted particles in the spray; satellite droplets formed by splat break-up at the time of impact; overshooting of liquid over solidified splats during droplet spreading; entrapment of gas between splats and the presence of an oxide layer on the spray particles. Precursor powder can also be trapped by the molten splats. It should be noted that not all powder particles are fully molten when they land on the substrate. Some of them may be molten droplets without enough kinetic energy to form splats. These droplets re-solidify and get attached to the substrate by landing over the solidifying splats [20]. All of the trapped and attached spherical particles are integrated into the coating after being covered by subsequent deposits. In this investigation, different types of pores (Table 4) were observed in the coatings. Pore size and pores are classified from 0 to 10 μm as type A and designated as A02, A04, A06 and A08. Pores classification in the range from 10 to 25 μm as type B and designated as B02, B04, B06 and B08. Type A pores may be formed as a result of interaction between material particles and the gaseous medium. The type A pores are formed owing to placement in a layer of particles which are either liquid or rehardened in the plasma jet, containing within them gas bubbles which remain in place when the granule is deposited. As a result, voidscan be localized in the material of a size generally commensurate with large micropores. Otherwise, these types of pores may be caused by gas dissolution from the liquid material being deposited. The type B pores are caused by the splashing of particles on impact with deposited material; or it may be due to voids resulting from the poor deformation of partially melted particles. These pores can have different sizes and exceedingly intricate shapes [21]. Moreover, results show that coating consists of type A and type B pores and distributed as A08 and B08 forms which are evidence of the characteristics of plasma-sprayed coating. The microstructure of plasma-sprayed alumina coatings may be altered with change in plasma spray conditions. It can be explained by changes in total porosity, shape/size of the inter-splat pores and unmelted particles.

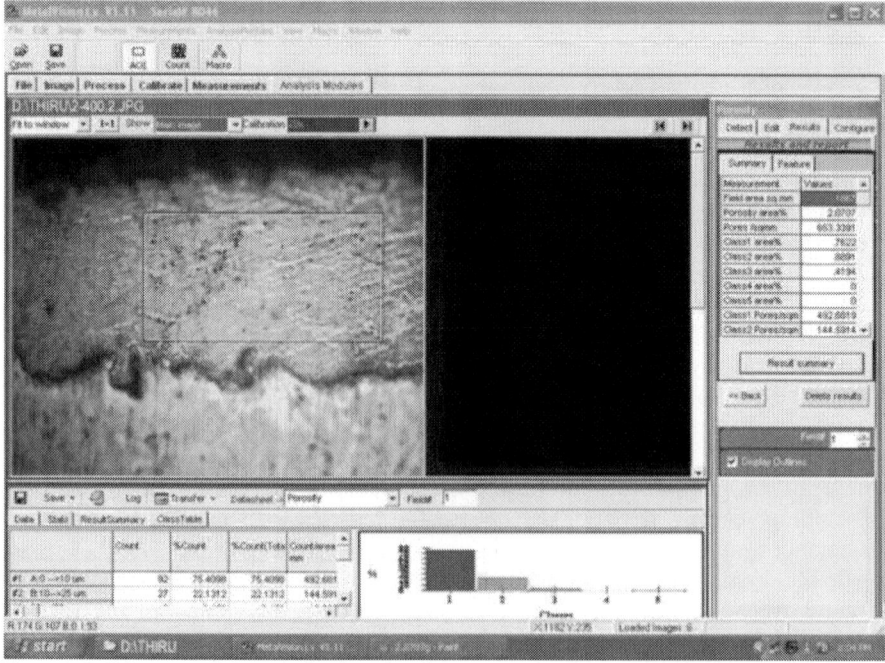

**Figure 2.** Measure the porosity level using computer image analyzing software.

**Table 4.** Results summary of porosity analysis.

| Total area measured | | | 0.0432 mm$^2$ |
|---|---|---|---|
| Porosity | 8.6% | Pores/mm$^2$ | 2796.8 |
| Classes | Area % | Pores/mm$^2$ | Type |
| 0 > 10 μm | 1.8 | 2098.69 | A08 |
| 10 > 25 μm | 3.9 | 642.52 | B08 |
| 25 > 75 μm | 3.5 | 49.87 | – |
| 75 > 125 μm | 0 | 0 | – |
| >125 μm | 0 | 0 | – |

The presence of pores, cracks, and localized compositional variations could degrade corrosion resistance [22]. The pores and microcracks would be preferential corrosion initiation sites when exposed to the electrolyte. Chloride ions penetrated into the pores and microcracks easily and promoted localized corrosion, such as pitting. With the progress of corrosion, the pores and microcracks joined together and formed macrocracks and finally led to the removal of loosened platelets of the coating (Fig. 3).

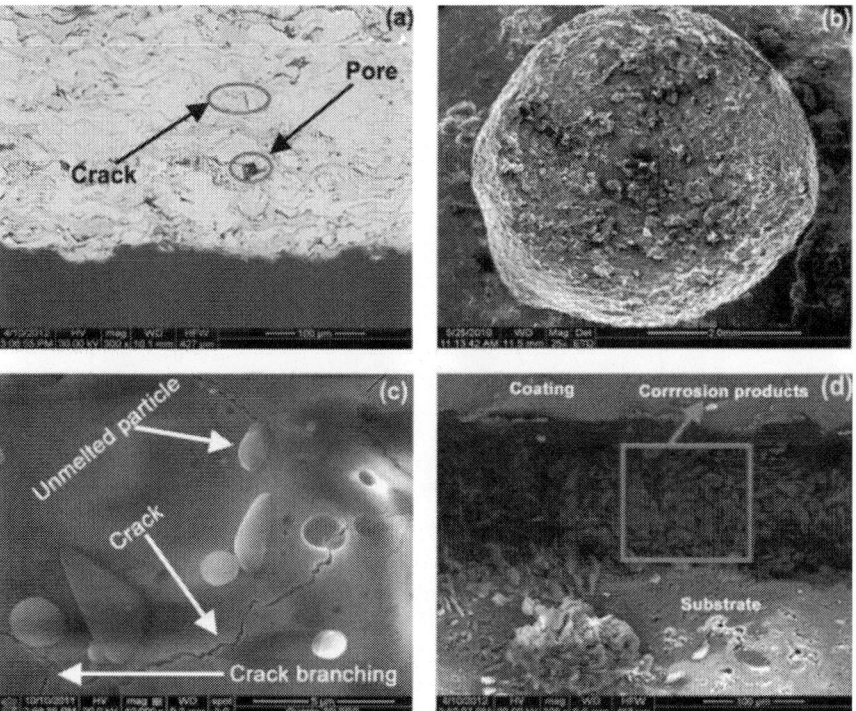

**Figure 3.** (a) SEM image of alumina coating, (b and c) high magnification images of pore and crack and (d) cross-section of the alumina coated specimen after immersion in NaCl solution.

The corrosion behavior of the coatings was evaluated by conducting immersion corrosion test in a 3.5% NaCl solution with a constant pH value of 7 and exposure time of 6 h. For each experimental condition two coated specimens were prepared and tested.Fig. 1b and c presents the corrosion test samples. The specimens were ground with 500#, 800#, 1200#, and 1500# grit SiC papers washed with distilled water and dried by warm flowing air. The corrosion rates of the as coated specimens were estimated through the weight loss measurement. The original weight ($W_O$) of the specimen was recorded and then immersed in the solution of 3.5% NaCl solution for 6 h. Finally, the corrosion products were removed by immersing the specimens for 1 min in the solution prepared by using 50 g chromium trioxide ($CrO_3$), 2.5 g silver nitrate ($AgNO_3$) and 5 g barium nitrate ($Ba(NO_3)_2$) for 250 ml distilled water. The final weight (wt) of the specimen was measured and the net weight loss was calculated using the following equation [23]:

$$CR = 87.6 * W/(A * D * T)$$

where $W$ = weight loss in mg, $A$ = surface area of the specimen in $cm^2$, $D$ = density of the Mg alloy substrate in $g/cm^3$, $T$ = corrosion time in h.

The Scanning Electron Microscope (Make: Jeol, Japan; Model: 6410-LV) was used to analyze the size and morphology of the parent materials. The powder is fused and then crushed, which gives its characteristic angular shape as shown in Fig. 4. SEM micrographs of the plasma-sprayed alumina coating are shown in Fig. 5a and b. A number of microcracks were observed on the surface of the coating. Fig. 6a and b shows the optical microstructure and SEM images of the alumina coating. From these micro-graphs, it can be seen that the coating microstructure consists of completely melted splat structures, unmelted particulate regions, pores between the splats and cracks within the splats. EDS analysis of the interface area (Fig. 7) detected the presence of Al and O in the coating.

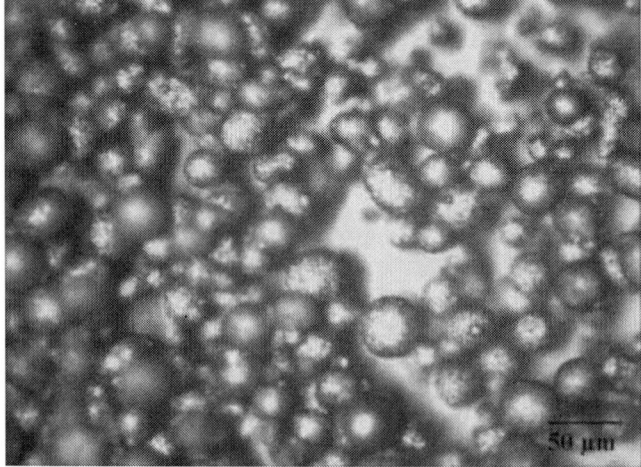

**Figure 4.** Optical micrograph of the $Al_2O_3$ powder.

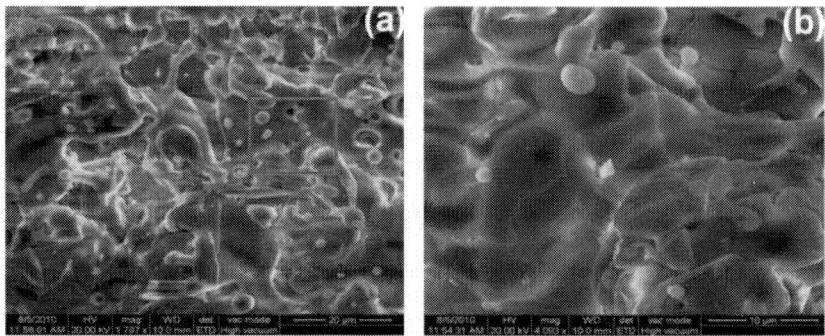

**Figure 5.** SEM surface morphology of alumina coating.

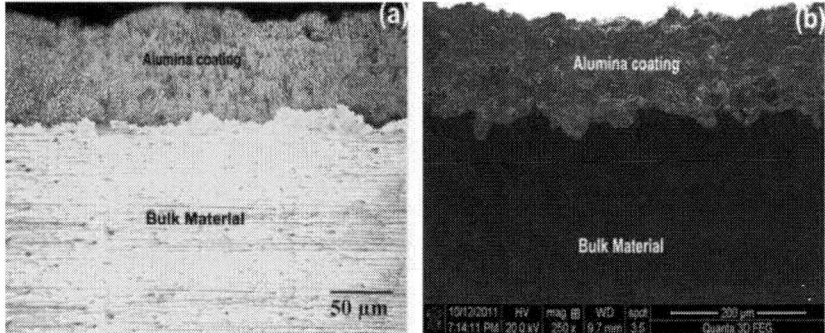

**Figure 6.** Optical microstructure and SEM image of the coating.

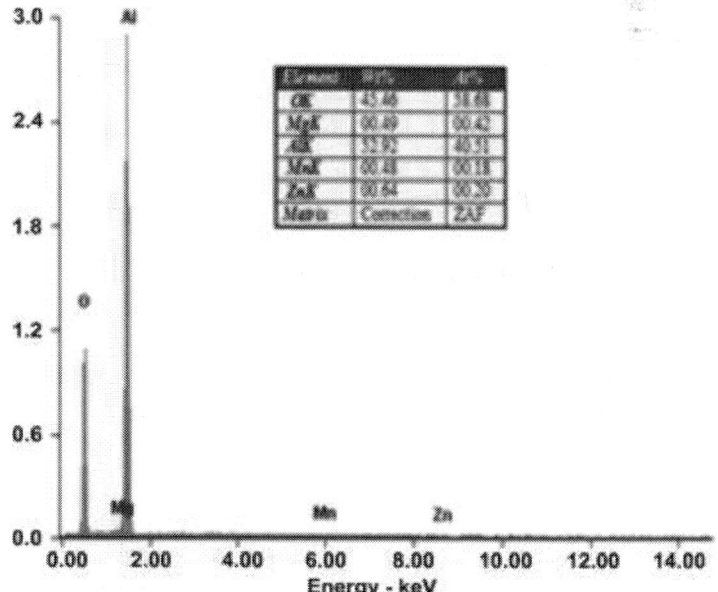

**Figure 7.** EDAX analysis.

## 3. PREDICTIVE STATISTICAL MODEL FOR CORROSION RATE

In this study, a response surface model-building technique was utilized to predict the corrosion rate in terms of the input power, the stand-off distance and the powder feed rate. Details of the model building technique are discussed below.

### 3.1. Response Surface Methodology (RSM)

RSM is an experimental strategy that explores the space of the process independent variables, and an empirical statistical modeling, to develop an appropriate relationship between the responses (output) and the process variables or factors (input). In the present investigation, to correlate the process parameters and the corrosion rates, an empirical relationship was developed to predict the responses based on experimentally measured values. The response is a function of power ($P$), stand-off distance ($S$), powder feed rate ($F$) and hence it can be expressed as

$$CR = f(P, S, F)$$

The empirical relationship chosen includes the effects of the main and interaction effect of all factors. The construction of empirical relationship and the procedure to calculate the values of the regression coefficients can be referred elsewhere [24]. In this work, the regression coefficients were calculated with the help of Design Expert V 8.1 statistical software. After determining the coefficients (at a 95% confidence level), the final empirical relationship was developed using these coefficients. The final empirical relationship to estimate the response is given below:

$$CR = 5.22 - 1.84P + 1.54S + 0.34F + 0.71PS + 1.23PF + 1.35SF$$
$$+ 3.89P^2 + 2.67S^2 + 0.53F^2 \, (mm/year)$$

The analysis of variance (ANOVA) technique was used to find the significant main and interaction factors. The results of second order response surface model fitting in the form of analysis of variance (ANOVA) are given in Table 5. The determination coefficient ($r^2$) indicated the goodness of fit for the model. The Model $F$-value of 249.23 implies the model is significant. There is only a 0.01% chance that a "Model $F$-value" this large could occur due to noise. Values of "Prob > $F$" <0.0500 indicate that model terms are significant. In this case $P$, $S$, $F$, $PS$, $PF$, $SF$, $P^2$, $S^2$ and $F^2$ are significant model terms. Values greater than 0.1000 indicate the model terms are not significant. If there are many insignificant model terms (not counting those required to support hierarchy), model reduction may improve the model. The "Lack of Fit $F$-value" of 1.14 implies the Lack of Fit is not significant relative to the pure error. There is a 44.49% chance that a "Lack of Fit $F$-value" this large could occur due to noise. Non-significant lack of fit is good. The "Pred $R$-squared" of 0.9780 is in reasonable agreement with the "Adj $R$-squared" of 0.9915. "Adeq precision" measures the signal to noise ratio. $P$ ratio greater than 4 is desirable. Our ratio of 46.90 indicates an adequate signal. The normal probability of the corrosion rate shown in Fig. 8 reveals the residuals were falling on the straight line, which meant that the errors were distributed normally. All of these indicated an excellent suitability of the regression model. Each of the observed values compared with the experimental values shown inFig. 9.

**Table 5.** ANOVA test results.

| Source | Sum of squares | df | Mean square | F value | p-ValueProb > F | |
|---|---|---|---|---|---|---|
| Model | 404.847 | 9 | 44.983 | 249.2319 | <0.0001 | Significant |
| P | 46.35474 | 1 | 46.35474 | 256.8322 | <0.0001 | |
| S | 32.24242 | 1 | 32.24242 | 178.6417 | <0.0001 | |
| F | 1.623079 | 1 | 1.623079 | 8.992797 | 0.0134 | |
| PS | 4.013228 | 1 | 4.013228 | 22.23561 | 0.0008 | |
| PF | 12.15097 | 1 | 12.15097 | 67.32342 | <0.0001 | |
| SF | 14.56596 | 1 | 14.56596 | 80.70388 | <0.0001 | |
| $P^2$ | 217.629 | 1 | 217.629 | 1205.791 | <0.0001 | |
| $S^2$ | 103.0545 | 1 | 103.0545 | 570.9817 | <0.0001 | |
| $F^2$ | 4.056546 | 1 | 4.056546 | 22.47561 | 0.0008 | |
| Residual | 1.804865 | 10 | 0.180487 | | | |
| Lack of Fit | 0.961151 | 5 | 0.19223 | 1.139189 | 0.4449 | Not significant |
| Pure error | 0.843715 | 5 | 0.168743 | | | |
| Cor. total | 406.6519 | 19 | | | | |
| Std. dev. | 0.424837 | R-squared | | 0.995562 | | |
| Mean | 10.05754 | Adj R-squared | | 0.991567 | | |
| CV% | 4.224065 | Pred R-squared | | 0.978082 | | |
| PRESS | 8.912924 | Adeq precision | | 46.90273 | | |

df: degrees of freedom; CV: coefficient of variation; F: Fisher ratio; p: probability.

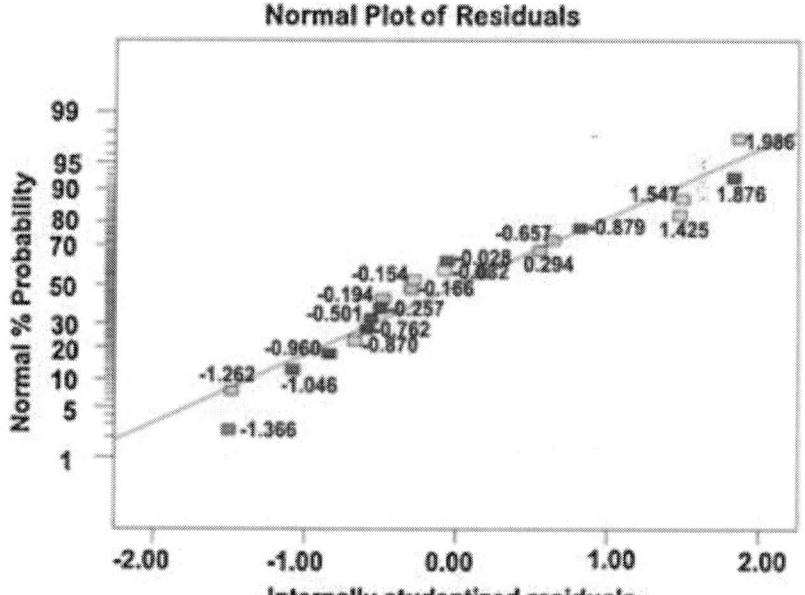

**Figure 8.** Normal probability plot for the response.

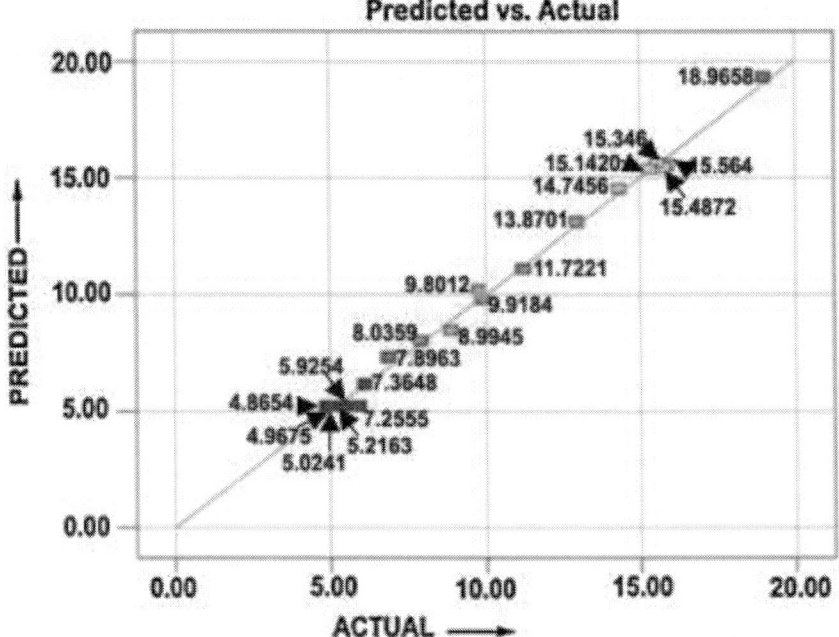

**Figure 9.** Correlation plot for the response.

## 4. RESULTS AND DISCUSSION

The developed mathematical model can be used to predict the range of parameters used in the investigation by substituting their respective values in coded form. Based on these models, the main and interaction effects of process parameter on corrosion rate were computed and plotted in graphical form as shown in Fig. 10 and Fig. 11.

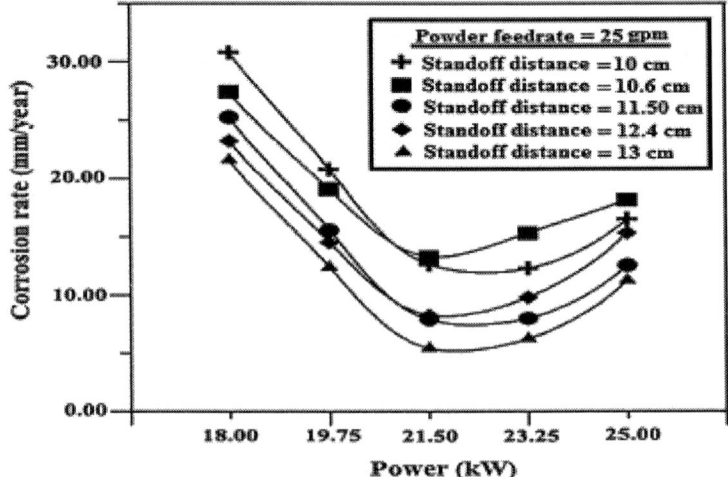

**Figure 10.** Effect of power and stand-off distance on corrosion rate.

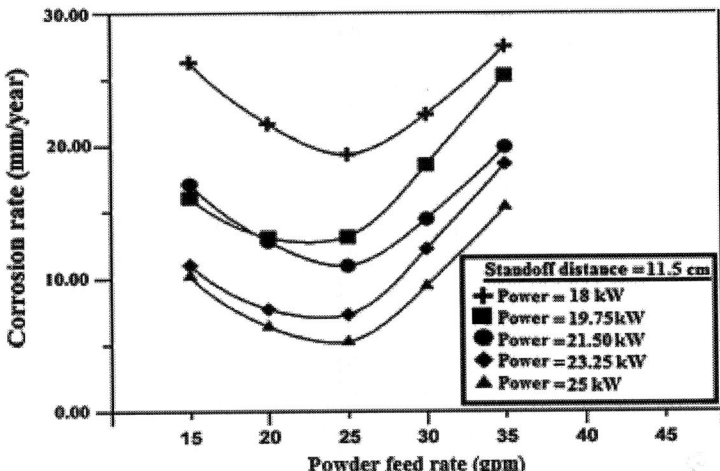

**Figure 11.** Effect of powder feed rate and power on corrosion rate.

## 4.1. Effect of Input Power on Corrosion Rate

The effects of input power on response of the coatings are illustrated in Fig. 10 and Fig. 11. The spraying power is an important parameter that affects the quality of the coating, since it can influence the temperature and velocity of the powder particles at the moment of impinging the substrate. More complete particle melting usually results in lower porosity content. It can be seen from Fig. 12a, lower power levels gave improper melting of the particles, which resulted in poor quality coatings. At the lowest spraying powers, the powder particles are poorly melted. When they impact on the substrate or the already formed coating, they are not able to spread out completely to form splats and therefore, could not conform to the surface [25]. In such a case, the interlamellar pores and cracks will be formed due to the solidification of the splats. Moreover, when the spraying power is relatively low, numerous unmelted and partially melted particles are existed in the coating. The SEM micrograph of the corroded specimens Fig. 13a reveals the at lower power levels, the alumina coated specimen which suffered a severe chemical dissolution in exposed area, the alumina coating flaked-off in a few regions. The coating was found to have been damaged at localized regions (Fig. 13a). Thus, the magnesium substrate underneath the coating was exposed to the electrolyte. The flake off of larger coating areas in NaCl solutions was caused possibly by the formation of corrosion products [26].

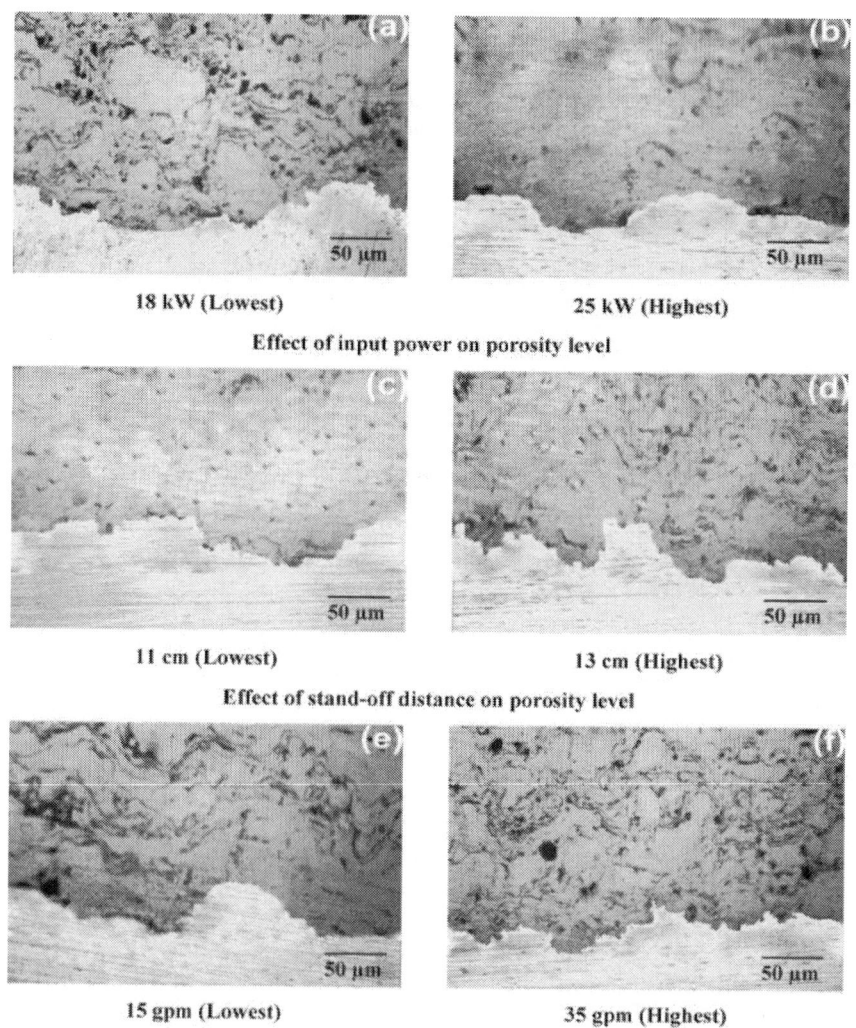

**Figure 12.** Effect of APS process parameters on porosity level.

As the arc current increases, the total and the net available energies in the plasma increase. This condition leads to a better in-flight particle molten state and higher velocities. The deposition yield reaches a plateau for the highest current levels because of the increase in plasma jet temperature, which in turn increases both the particle melting ratio and the plasma jet viscosity for particles to flatten better. Porosity decreases under high power levels (Fig. 12b) because the particles are more likely to melt at high plasma energy levels, thereby enhancing flow and compaction of the coating during its build up. If the velocity of the particles is increased and/or the viscosity is decreased, then particle spreading tends to increase [27]. The SEM micrograph (Fig. 13b) revealed that the coating surface did not undergo any discernible corrosion degradation. It is thus evident in this case that the alumina film at the interface was very stable and could resist the corrosion damage. When the power level was increased to 25, the alumina coated specimen exhibited a higher corrosion resistance and a more stable behavior than those in the lower power levels (Fig. 13a). It is clear from Fig. 13b, no obvious damage and coating degradation were observed on the surface of the material.

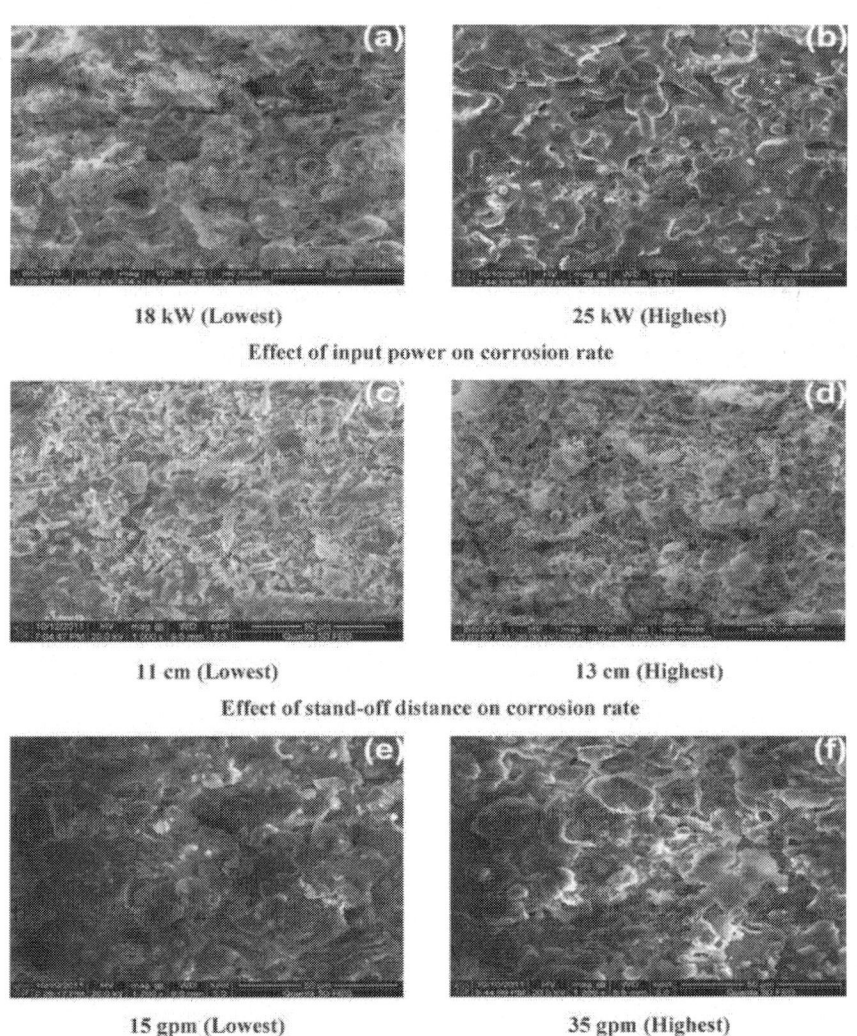

18 kW (Lowest)                25 kW (Highest)

Effect of input power on corrosion rate

11 cm (Lowest)                13 cm (Highest)

Effect of stand-off distance on corrosion rate

15 gpm (Lowest)                35 gpm (Highest)

Effect of powder feed rate on corrosion rate

**Figure 13.** Effect of APS process parameters on corrosion rate.

## 4.2. Effect of Stand-Off Distance on Corrosion Rate

The influence of the stand-off distance on the corrosion rate of the coatings is displayed in Fig. 10 and Fig. 11. In this case as the corrosion rate increases with the increase in the stand-off distance. The stand-off distance mainly controls the cohesion between splats because the temperature and velocity of particles in the plasma flame significantly change with stand-off distance. At short spraying distance, the droplets striking the substrate are semi or fully melted (Fig. 12c). With smaller stand-off distance, the splashing of material with possible fragmentation and quenching cracks may result. Therefore, better spreading and cohesion would be achieved with shorter spraying distances [28]. As shown in the SEM micrograph of Fig. 13c, it is also observed that at lower stand-off distance, coating has no pronounced deterioration in this condition. At this stage, the corrosion deterioration of coated specimens was dictated by the degradation of coatings especially in inner regions of the coating. Therefore, due to the denser and more compact inner layer in the alumina coating was superior and the corrosion deterioration was slower in lower stand-off distance (Fig. 13c).

With longer stand-off distances, the enthalpy of the molten ceramic particles is largely lost, and the particles are decelerated in a relatively longer flight path because of the interaction with the surrounding air. As a result, the particles striking on the substrate will not be adequately flattened to overlap the layers, resulting in porosity as shown in Fig. 12d. At higher spray distance, the sprayed powder has more time to react with the air entrained in the flame, which would result in an increase in oxide content with spray distance. The coatings deposited at a spraying distance of 130 mm were found to contain fewer unmelted particles and lower porosity than the trends would have suggested porosity level in the coating increased with the increase in spray distance (Fig. 12d). It has been reported that the longer spray distance increases the dwell time in the plume and allows more thorough heating/melting of the particles and the enthalpy of the molten ceramic particles is largely lost, and also the particles begin to slow down in a relatively longer flight path because of the interaction with the surrounding air [29].

At the stand-off distance not more than 100 mm, the coating was only deteriorated lightly on the edge of the samples. However, when the stand-off distance reaches 130 mm, chloride ions penetrate the coating and contact with the substrate, resulting in heavy corrosion reaction and a larger level of corrosion damage (Fig. 13d). Based on this investigation, it is concluded that the alumina coatings cannot provide a long term protection to the magnesium alloy substrate in higher stand-off distance.

## 4.3. Effect of Powder Feed Rate on Corrosion Rate

The effect of powder feed rate on the corrosion rate of the coatings is shown in Fig. 10 and Fig. 11. Varying the powder feed rate affects the number of particles having to share the kinetic and thermal energies of the flame, which in turn affects the particle velocity and the temperature. When the powder feed rate is extremely low (e.g., 15 gpm), most of the particles are fully melted. In such a case, the coating with dense microstructure and low porosity will be fabricated

(Fig. 12e). Too low a powder feed rate will result in vaporization, and over melting of the particles resulting in quench cracks[30], splashing, and high porosity levels, whereas too high a feed rate will end up in poor melting of the powder particles resulting in a decrease of the splat flattening ratio and an increase in the porosity. With a further increase in powder feed rate, the heat content in the plasma gas becomes insufficient for the melting of the powder particles. As a result, the boundary of the unmelted particle, the micro-cracks and pores can be found. Thus, it is clear that these micro-cracks and pores may be created due to the residual stress arisen from the material mismatch of unmelted particles and the splats in a molten state. The poorly melted (unmelted and partially melted) particles will be remained in the coating, resulting in a less-dense coating with high porosity as can be seen in Fig. 12f. This result indicates that, when the powder feed rate is high, the particles which obtain low thermal energy and kinetic energy cannot be fully melted.

The SEM micrograph of the coating exposure is shown in Fig. 13e, where no obvious indication of blistering of the coating can be observed. However, the coating seems to be locally attacked in some areas inside the coating (Fig. 13e). The SEM micrograph of alumina coating after 6 h of exposure showed that unlike the APS coating, localized attack can be detected (Fig. 13f) through the coating surface. In the case of APS coating, micro cracks and intersplat oxides are responsible for the electrolyte penetration into the coating. A localized corrosion can also take place at these defects which attacks the coating body.

## 5. CORROSION PRODUCTS CHARACTERISTICS

Fig. 14a and b displays the cross section and EDS analysis of as-sprayed alumina coating on AZ31B magnesium alloy after 6 h of immersion in NaCl solution. The cross section images of as-sprayed coatings revealed significant signs of degradation in the coating/substrate interface Fig. 14a evidences the extent of the corrosion process that occurs in the chloride medium, since the as-sprayed alumina coating was detached from the AZ31B substrate after 6 h of immersion. Examination of the coating/substrate interface showed the presence of corrosion products in this area, although only a part of them remained over the substrate or in the coating after the immersion tests[31] and [32]. This behavior is produced because the as-sprayed coating is highly porous, so that, there are a high number of pathways through this coating and the electrolyte rapidly reaches the magnesium alloy surface, giving rise to the substrate corrosion. Afterwards, the corrosion process progresses along the interface area, giving rise to the formation of corrosion products on the metal surface, which will finally cause the detachment of the coating. The growth of corrosion products would separate the coating from the substrate and their low mechanical properties would allow its detachment. According to EDX analysis (Fig. 14b), corrosion products rich in Mg and O were mainly detected in the interface area, along with a small amount of Al and of Cl. The main corrosion products responsible for the detachment of the coatings in immersion environment were identified as MgO (Fig. 14c).

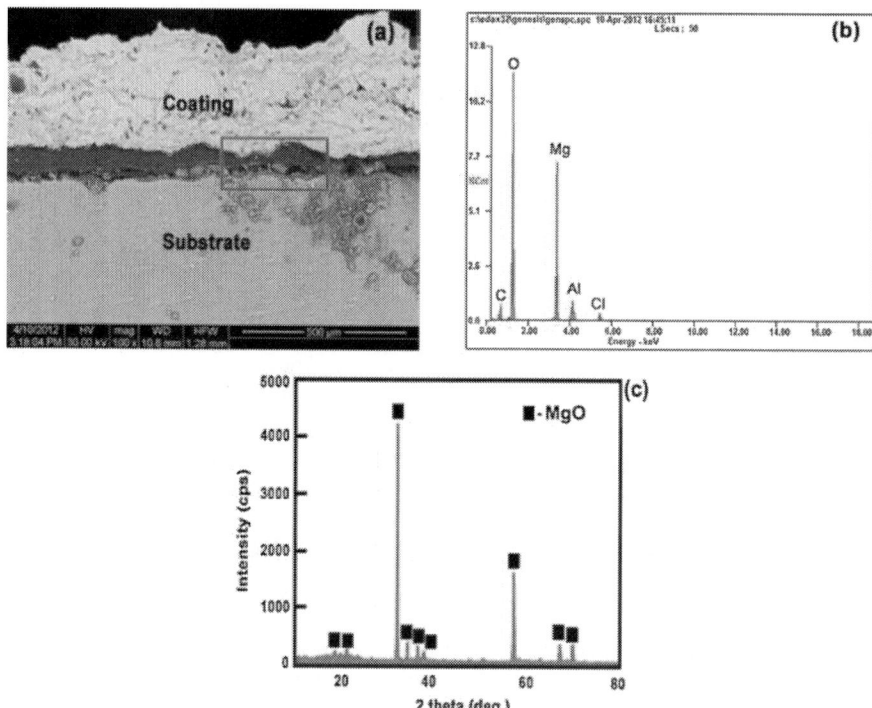

**Figure 14.** (a) Cross section of as-sprayed alumina coating on AZ31B magnesium alloy after immersion in NaCl solution for 6 h (b and c) EDX and XRD pattern analysis.

The corroded surfaces of the coated samples were examined using SEM and X-ray diffraction techniques immediately after the immersion test. The occurrence of uniform corrosion (Fig. 15a) can be observed. However, in the as-coated sample, an additional thicker top layer at discrete locations can be noted (Fig. 15a) indicative of higher corrosion rate. X-ray diffraction results obtained from the corroded surfaces of the samples are presented in Fig. 15b. The main corrosion products formed are bayerite $(Al(OH)_3)$ (JCPDS 33-0018) and aluminum oxide (AlO) (JCPDS 75-0278) as confirmed by EDX. Subsequent to the surface analysis, the coatings were cut in transverse direction and the cross section was immediately polished and observed under SEM. In the case of as-coated sample (Fig. 16a), presence of corrosion product can be observed as patches at discrete locations and an EDS analysis carried out on the patch (labeled P) indicates the presence of Cl and O besides Al as major constituents (Fig. 16b) as expected. The presence of corrosion product as discrete patches on the corroded sample cross section can be understood by noting that the intersplat cracks, through which electrolyte can penetrate the coating, follow a very tortuous path. Thus, the corrosion patch represents a situation where in the cut plane intersects with the tortuous crack path of intersplat cracks filled with electrolyte [33] and [34]. If the coating is nobler than the substrate, then pores/cracks only provide the path for electrolyte penetration into the pore since pore walls do not corrode being nobler than the substrate. In such cases, the secondary corrosion process can only be the corrosion at the coating–substrate interface. In contrast, in the present case, wherein the $Al_2O_3$ coating is less noble than the magnesium substrate, pore walls also undergo corrosion and the resulting corrosion product in the pores can impede further corrosion [35], [36] and [37]. To validate the proposal that the corrosion occurs only at the coating surface and on the intersplat pore/crack walls and not at the magnesium substrate–coating interface.

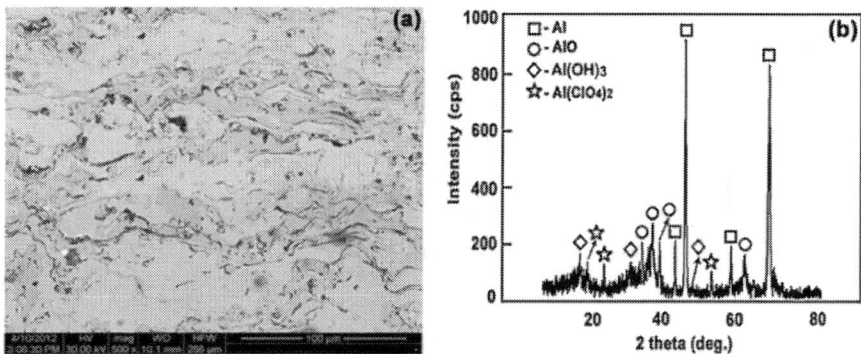

**Figure 15.** SEM micrographs (a) and X-ray diffraction analysis (b) of the corroded surface after 6 h exposure of coatings.

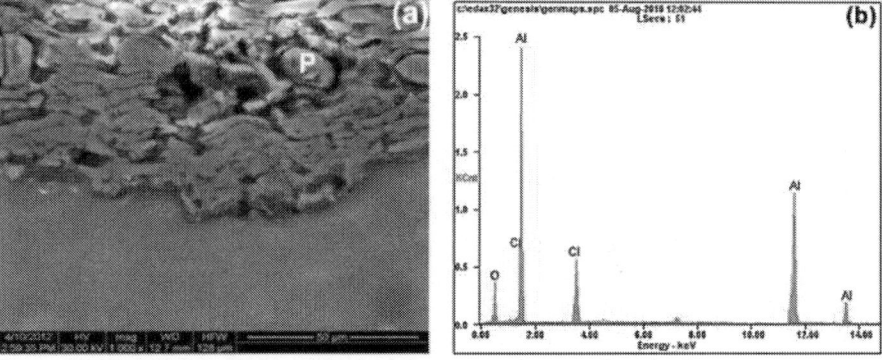

**Figure 16.** (a) SEM micrograph showing corrosion product in the cross section of as-coated coating exposed for 6 h and (b) EDS analysis.

# 6. RELATIONSHIP BETWEEN POROSITY AND CORROSION RATE OF ALUMINA COATINGS

The coating porosity and the corrosion rate obtained from the experimental results are related as shown in Fig. 17. The experimental data points are fitted by a straight line. The straight line is governed by the following regression equation:

$$CR\,(\text{mm/year}) = 1.721 + 24.27\,\langle\text{porosity in vol.}\,\%\rangle$$

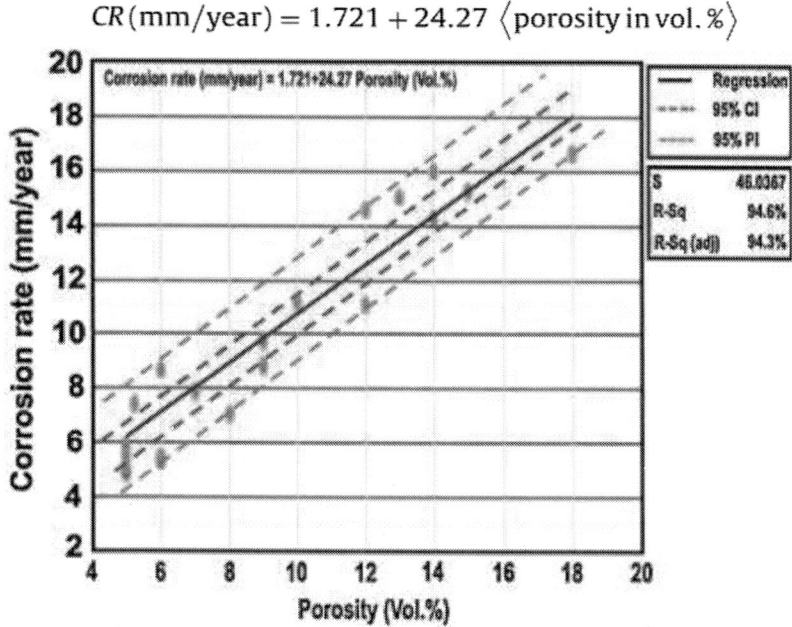

**Figure 17.** Relationship graph for porosity and corrosion rate.

The slope of the estimated regression equation (+24.27) is positive, implying that as porosity decreases, corrosion rate increases. The coefficient of determination is $R^2 = 94.6\%$. It can be interpreted as the percentage of the total sum of squares that can be explained by using the estimated regression equation. The coefficient of determination $R^2$ is a measure of the goodness of fit of the estimated regression equation[38].

The fitted regression line (Eq. (4)) may be used for two purposes:

a.    To estimate the mean value of corrosion rate for the given value of coating porosity.

b.    Predicting an individual value of corrosion rate for a given value of coating porosity level.

The confidence interval and prediction interval show the precision of the regression results. Narrower intervals provide a higher degree of precision (Fig. 17). Confidence interval (CI) is an interval estimate of the mean value of $y$ for a given value of $x$. Prediction interval (PI) is an interval estimate of an individual value of $y$ for a given value of $x$. The estimated regression equation provides a point estimate of the mean value of corrosion rate for a given value of porosity. The difference between CI and PI reflects the fact that it is possible to estimate the mean value of corrosion rate more precisely than an individual value of corrosion rate. The greater width of the PI is reflecting the added variability introduced by predicting a value of the random variable as opposed to estimating a mean value. From Fig. 17, it is also inferred that the closer the value to '$x$' (15.21 vol%) the narrower will be the interval.

# 7. CONCLUSIONS

The following important conclusions are obtained from this investigation:

- Empirical relationship was established using RSM to predict the corrosion rate of plasma sprayed alumina coatings, incorporating few important spray parameters. The developed relationship can be effectively used to predict the corrosion rate of alumina coatings on AZ31B magnesium alloy at 95% confidence level.
- The corrosion behavior of the coating was strongly affected by the power of the plasma. High power level of 25 kW improved the corrosion resistance of the coatings. Higher power also led to higher fractions of melted particles which consequently produce a higher probability of successful deposition.
- In higher stand-off distance, the alumina coatings were found to be highly susceptible to localized damage, and could not provide an effective corrosion protection to Mg alloy substrate. The level of the corrosion attack of alumina coated AZ31 alloy is much higher when stand-off distance is greater than 130 mm, which was validated by the surface micrographs.
- The coating fabricated at the long powder feed rate contained a considerable amount of pores and partially melted regions, thereby resulting in deterioration of the corrosion resistance. When the powder feed rate decreased, dense coatings containing reduced pores and partially melted regions could be fabricated. In these coatings, cracks or spalled-outs of oxides were hardly found on the smoothly worn surface.
- The input power was found to have greater influence on the corrosion rate of plasma sprayed alumina coatings followed by process parameters such as stand-off distance and powder feed rate.

# REFERENCES

1. G.L. Song, J. Adv. Eng. Mater., 7, 563–586 (2005).
2. D. Thirumalaikumarasamy, K. Shanmugam and V. Balasubramanian, Trans. Indian Inst. Met., 67, 19–32 (2014).
3. K.A. Khor and Y.W. Gu, Thin Solid Films, 372, 104–113 (2000).
4. R. McPherson, Surf. Coat. Technol., 173, 39–40 (1989).
5. T. Lampke, D. Meyer, G. Alisch, B. Wielage, H. Pokhmurska, M. Klapkiv and M. Student, J. Mater. Sci. (Ukrainian Original), 46, 591–598 (2011).
6. E. Celik, I. Ozdemir, E. Avci and Y. Tsunekawa, Surf. Coat. Technol., 193, 297–302 (2005).
7. T. Lampke, D. Meyer, G. Alisch, D. Nickel, I. Scharf, L. Wagner and U. Raab, Surf. Coat. Technol., 206, 2012–2016 (2011).
8. Z. Yin, S. Tao and X. Zhou, Mater. Charact., 62, 90–93 (2011).
9. S. Costil, C. Verdy, R. Bolot and C. Coddet, J. Therm. Spray. Technol., 16, 839–843 (2007).

10. D. Thirumalaikumarasamy, K. Shanmugam and V. Balasubramanian, Prog. Nat. Sci. Mater. Int., 22, 468–479 (2012).

11. E. Celik, A.S. Demirkiran and E. Avci, Surf. Coat. Technol., 116, 1061–1064 (1999).

12. S.T. Aruna, N. Balaji, J. Shedthi and V.K. William Grips, Surf. Coat. Technol., 208, 92–100 (2012).

13. W. Zhongshan, L. Liufa and D. Wenjiang, J. Mater. Sci. Forum, 488, 685–688 (2005).

14. L. Chiu, H. Lin, C. Chen, C. Yang, C. Chang and J. Wu, J. Mater. Sci. Forum, 419, 909–914 (2003).

15. R. Arrabal, A. Pardo, M.C. Merino, M. Mohedano, P. Casajus and S. Merino, Surf. Coat. Technol., 204, 2767–2774 (2010).

16. M. Carboneras, M.D. Lopez, P. Rodrigo, M. Campo, B. Torres, E. Otero and J. Rams, Corros. Sci., 52, 761–768 (2010).

17. C.S. Ramachandran, V. Balasubramanian and P.V. Ananthapadmanabhan, Surf. Eng., 27, 217–229 (2011).

18. ASTM B276-05, Standard Test Method for Apparent Porosity in Cemented Carbides (2010).

19. B. Roge, A. Fahr, J.S.R. Giguere and K.I. McRae, J. Therm. Spray. Technol., 12, 530–535 (2003).

20. J. Zhang and V. Desai, Surf. Coat. Technol., 190, 98–109 (2005).

21. G. Antou, G. Montavon, F. Hlawka, A. Cornet and C. Coddet, Mater. Charact., 53, 361–372 (2004).

22. N. Ahmed, M.S. Bakare, D.G. McCartney and K.T. Voisey, Surf. Coat. Technol., 204, 2294–2301 (2010).

23. ASTM G31-72, Standard Practice for Laboratory Immersion Corrosion Testing of Metals (2002).

24. S. Kumar, P. Kumar and H.S. Shan, J. Mater. Process. Technol., 182, 615–623 (2007).

25. H. Wu, H.-j. Li, Q. Lei, Q.-g. Fu, C. Ma, D.-j. Yao, Y.-j. Wang, C. Sun, J.-f. Wei and Z.-h. Han, Appl. Surf. Sci., 257, 5566–5570 (2011).

26. Y. Wang, W. Tian, T. Zhang and Y. Yang, Corros. Sci., 51, 2924–2931 (2009).

27. R. Venkataraman, G. Dasa, S.R. Singh, L.C. Pathak, R.N. Ghosha, B. Venkataraman and R. Krishnamurthy, J. Mater. Sci. Eng., 445, 269–274 (2007).

28. A. Portinha, V. Teixeira, J. Carneiro, J. Martins, M.F. Costa, R. Vassen and D. Stoever, Surf. Coat. Technol., 195, 245–251 (2005).

29. A. Kucuk, C.C. Berndt, U. Senturk, R.S. Lima and C.R.C. Lima, J. Mater. Sci. Eng. A, 284, 29–40 (2000).

30. S. Kuroda, T. Fukushima and S. Kitahara, J. Therm. Spray. Technol., 1, 325–332 (1992).

31. W.-m. Zhao, C. Liu, L.-x. Dong and Y. Wang, J. Therm. Spray. Technol., 518, 702–707 (2009).
32. C. Blawert, T.V. Heitmann, W. Dietzel, H.M. Nykyforchyn and M.D. Klapkiv, Surf. Coat. Technol., 200, 68–72 (2005).
33. L. Gil and M.H. Staia, Thin Solid Films, 420, 446–454 (2002).
34. K. Spencer, D.M. Fabijanic and M.X. Zhang, J. Therm. Spray. Technol., 204, 336–344 (2009).
35. D. Dzhurinskiy, E. Maeva, E.V. Leshchinsky and R.Gr. Maev, J. Therm. Spray Technol., 21, 304–313 (2012).
36. F.L. Toma, C.C. Stahr, L.M. Berger, S. Saaro, M. Herrmann, D. Deska and G. Michael, J. Therm. Spray. Technol., 19, 137–147 (2009).
37. J. Zhang, W. Zhang, C. Yan, K. Du and F. Wang, Electrochim. Acta, 55, 560–571 (2009).
38. S. Karthikeyan, V. Balasubramanian and R. Rajendran, J. Ceram. Int., 40, 3171–3183 (2014).

CHAPTER 7

# Formation of Grooved and Porous Coatings on Titaniumby Plasma Electrolytic Oxidationin H₂SO₄/H₃PO₄ Electrolytes and Effects of Coating Morphology on Adhesive Bonding

*O.A. Galvisa, D. Quinteroa, J.G. Castañoa, H. Liub, G.E. Thompsonb, P. Skeldonb, F. Echeverríaa*

a Centro de Investigación, Innovación y Desarrollo de Materiales – CIDEMAT, Universidad de Antioquia UdeA, Calle 70 No. 52-21, Medellín, Colombia
b Corrosion and Protection Centre, School of Materials, The University of Manchester, Manchester, M13 9PL, UK

## ABSTRACT

The paper reports a change in the morphology of coatings formed galvanostatically on titanium by plasma electrolytic oxidation in phosphoric/sulfuric acid mixtures, and investigated using scanning electron microscopy, X-ray diffraction, and glow discharge optical emission spectroscopy. An initial grooved morphology, containing anatase, is transformed to a more usual porous morphology, which may also contain rutile. The coatings also contain phosphorus species, but comparatively small amounts of sulfur species. The morphological change occurs over a range of cell charge that is strongly dependent on the molar ratio of the acids but weakly dependent on the applied current. With the change in the coating morphology, the efficiency of coating formation reduces and the sparking becomes more localized and intense. Lap shear tests show that the grooved morphology provides a ~ 60% increase in the strength of adhesively bonded joints compared with a porous morphology.

**Keywords:** Titanium; Plasma electrolytic oxidation; Sulfuric acid; Coating morphology

## 1. INTRODUCTION

Titanium and its alloys have been adopted as engineering materials for a broad range of aerospace, petrochemical, catalysis, cryogenic and biomaterial applications, due to their attractive mechanical properties, low density, biocompatibility and corrosion resistance. In order to further enhance their surface performance, a range of surface modification techniques have been employed, including sol–gel coating [1] and [2], chemical vapor deposition (CVD) [3], hydrothermal treatment [4] and [5] and anodic oxidation [6], [7],[8] and [9]. Of relevance to the present study, plasma electrolytic oxidation (PEO), a development from anodizing, has attracted much recent interest due to its potential for formation of relatively thick, ceramic coatings on light metals [10], [11], [12] and [13]. The coatings often display good corrosion protection, high hardness, good wear resistance and excellent substrate adhesion [14], [15], [16] and [17]. DC, AC and pulsed electrical regimes may be used, leading to discharges at the treated surface where the coating material is formed under high anodic potentials [10] and [18]. The coatings usually contain constituents originating from both the substrate and the electrolyte and owing to the high temperatures and pressures at the locations of discharges may contain phases that are not generated under conditions of conventional anodizing [19]. They also contain significant porosity, which may be created by the discharges and gas generation that accompanies the process [20]. The coating morphology is affected by the nature of the discharges, which may change in characteristics, such as optical emission, lifetime, and acoustic emission, during the PEO treatment [21]. However, there is no detailed understanding of the relationship between the discharges and the morphology, composition and structure of the resultant coating that allows a prediction of the coating properties under particular growth conditions. Further, the mechanism of coating formation is poorly understood, possibly involving anodic oxidation, thermal oxidation, thermolysis and plasma chemical reactions.

PEO studies of titanium have examined the influences of process parameters on the composition and microstructure of coatings, as well as their mechanical, corrosion-resistant, bioactive, catalytic, wear-resistant and antifriction properties [15], [22], [23],[24], [25], [26] and [27]. The coatings can be up to tens of microns in thickness and generally contain anatase and/or rutile. They exhibit numerous pores of up to few microns in diameter on their surfaces. The pores are formed by the discharges and gas evolution, and are typical of PEO processing of light metals. Grooved surfaces have also been occasionally reported [28], [29], [30] and [31]. In the case of titanium, grooves were observed in a coating formed under high frequency voltage pulses [32], which promoted deposition of ZnO in a following treatment designed to improve biocompatibility. However, the development of the grooved morphology and its possible transformation to a porous morphology have not been investigated in detail for any substrate.

In the present work, coating formation is investigated for galvanostatic PEO of titanium in two electrolytes consisting of mixtures of phosphoric and sulfuric acids. It is shown that a transition in the coating morphology, from formation of grooves to formation of pores, is mainly dependent on the charge passed in the cell for a particular electrolyte composition, with little influence of the current density. Further, the critical charge is shown to be reduced by increasing the concentration of phosphoric acid in the electrolyte. The work is potentially of practical relevance to improving both the biocompatibility and adhesive bonding of titanium, which are dependent upon the details of the surface composition and topography. For such applications, it is well established that micron- and sub-micron scale features can play an important role in the biocompatibility of surfaces [33] and [34] and the strength of adhesive bonds [35]. The use of adhesively bonded structures is widespread in various engineering fields, as they provide many advantages over conventional mechanical joints [36]. In the present work, lap shear tests are employed to compare the strength of adhesively joined titanium pre-treated with either porous or grooved morphologies, the results revealing a significant benefit from the latter morphology.

# 2. EXPERIMENTAL

## 2.1. Pre-Treatment of Samples

Cylindrical specimens (1.2 mm thick) of titanium (grade 2, purity 99.6%, ASTM F-67-13[37]) were cut from a rod of 20 mm of diameter. The specimens were ground with silicon carbide paper to a 600 grit finish, then degreased with acetone in an ultrasonic bath for 15 min. Subsequently, the specimens were soaked in a mixture of HF (3%) and $HNO_3$(30%) for 15 s according to ASTM B600-11 standard [38], in order to remove the naturally formed oxide layer, and finally washed with distilled water and dried with cool air.

## 2.2. Plasma Electrolytic Oxidation

Specimens were processed at current densities of 10, 30 and 50 mA cm$^{-2}$ in 0.1 M $H_3PO_4$/1.5 M $H_2SO_4$ and 0.9 M $H_3PO_4$/1.5 M $H_2SO_4$ electrolytes, without stirring the electrolytes. During the treatments, the temperature of the electrolytes rose above room temperature. The final temperature was ~ 35, 75 and 90 °C following treatments at 10, 30 and 50 mA cm$^{-2}$, respectively. In order to determine whether the rise in temperature had a significant effect on the formation of the coatings, additional experiments were carried out at 50 mA cm$^{-2}$, using a stirred electrolyte and an ice bath to maintain the temperature of the electrolyte at 20 °C. The ionic conductivities of the electrolytes were $3.65 \pm 0.01$ and $3.11 \pm 0.01$ mS cm respectively. A two-electrode electrochemical cell was employed, with a titanium specimen of area 7 cm$^2$ as the anode and a platinum mesh as the cathode. The current was provided by a DC power supply (Kepco BHK 500-0.4 MG). The cell voltage was registered electronically by Labview 8.1 Software (National Instruments) interfaced with a personal computer. The maximum time of treatment in the 0.1 M $H_3PO_4$/1.5 M $H_2SO_4$ electrolyte was 2500 s for all current densities. In contrast, the maximum time in the 0.9 M $H_3PO_4$/1.5 M $H_2SO_4$ electrolyte was 950 s at 30 and 50 mA cm$^{-2}$ and 2500 s at 10 mA cm$^{-2}$. The various times were selected to avoid a large voltage drop that occurred if treatments were carried on for longer periods. Before PEO, all specimens were measured and weighed. After anodizing, the specimens were removed immediately from the electrolyte, rinsed with deionized water in ultrasonic bath for 15 min, dried in a cool air, and then re-weighed. For weight measurements, a Metter Toledo UMX5 ($\pm$ 0.1 μg) microbalance was used. Images of specimens were acquired every 1 s during PEO, using USB microscopes controlled by in-house software.

## 2.3. Characterization of Coatings

Coatings were observed by scanning electron microscopy (SEM), using a JEOL JSM 6940 LV instrument. The thicknesses of coatings were determined using cross-sections prepared using successive grades of SiC paper, followed by polishing to 0.25 μm diamond paste, and examined by SEM employing backscattered electrons. Phase compositions were determined by X-ray diffraction (XRD), using a Philips XiPERT-MPD (PW3050) instrument with Cu K$\alpha$ radiation ($\lambda$ = 0.15405 nm) at an accelerating voltage of 45 kV and current of 40 mA, with an incident angle of 3°, a scanning speed of 0.00625° s$^{-1}$, a step size of 0.05° and a scan range from 15 to 85° (in 2θ). Depth profiling was carried out by glow discharge optical emission spectroscopy (GDOES) using a Horiba-Jobin-Yvon GD Profiler 2 in an argon atmosphere of 635 Pa by applying rf of 13.56 MHz and power of 35 W. Light emissions at 130.223, 178.291, 365.355, and 180.738 nm for oxygen, phosphorus, titanium and sulfur respectively were monitored during the analysis with a sampling time of 0.1 s. The area of analysis was of ~ 4 mm diameter.

The influence of porous and grooved morphologies on the shear strengths of adhesively bonded titanium substrates was compared using single lap joint tests that were carried out using a tensile machine, adhesive type and surface preparation in accordance with ASTM standard D1002-10. However, a modified size of the specimens was necessary due to the dimensions of the round titanium bar used for the PEO treatments [39]. The tests employed a universal tensile machine (model MX-5000) with a maximum load capacity of 5 kN, and with a constant crosshead speed of 5 mm min$^{-1}$. The specimens were assembled with an overlap of 7 mm and were held together after adhesive application with similar load blocks and cured according to the adhesive manufacturer's instructions, i.e. 24 h at room temperature. Two-component epoxy-resin (SinteSolda, Sinteco) was used as the adhesive, which is commonly used for joining materials, in protective coatings, as a structural material and for preparing moulds. Coatings employed in the tensile tests were produced at the intermediate current density of 30 mA cm$^{-2}$ in 0.9 M $H_3PO_4$/1.5 M $H_2SO_4$ electrolyte. They were formed to cell charge densities of either 6.8 or 28.5 C cm$^{-2}$, which correspond to stages of coating growth where grooved and porous morphologies respectively completely cover the specimen surfaces (Fig. 7). After the PEO process, the specimens were rinsed with deionized water in an ultrasonic bath for 15 min, dried in a cool air, and then tested in the tensile machine. The tests were replicated seven times for the porous morphology and eight times for the grooved morphology.

## 3. RESULTS AND DISCUSSION

Voltage–time responses during PEO at 10, 30 and 50 mA cm$^{-2}$ are shown in Fig. 1(a) and (b) for the 0.1 M H$_3$PO$_4$/1.5 M H$_2$SO$_4$ and 0.9 M H$_3$PO$_4$/1.5 M H$_2$SO$_4$ electrolytes respectively. The average rate of voltage increase over the interval up to 100 V was 1.2, 3.2, and 7.0 V s$^{-1}$ at current densities of 10, 30 and 50 mA cm$^{-2}$ respectively for 0.1 M H$_3$PO$_4$/1.5 M H$_2$SO$_4$ (Fig. 1(a)) and at 1.8, 5.4 and 9.5 Vs$^{-1}$ respectively for 0.9 M H$_3$PO$_4$/1.5 M H$_2$SO$_4$ (Fig. 1(b)). The rates increase approximately linearly with increase of the current density and also with increase of the concentration of H$_3$PO$_4$ at a particular current density. A decrease in rate was reported for anodizing in H$_2$SO$_4$ solution with increase of the electrolyte concentration [6], unlike the present trend, suggesting that H$_3$PO$_4$ may reduce the coating solubility and/or oxygen generation [40].

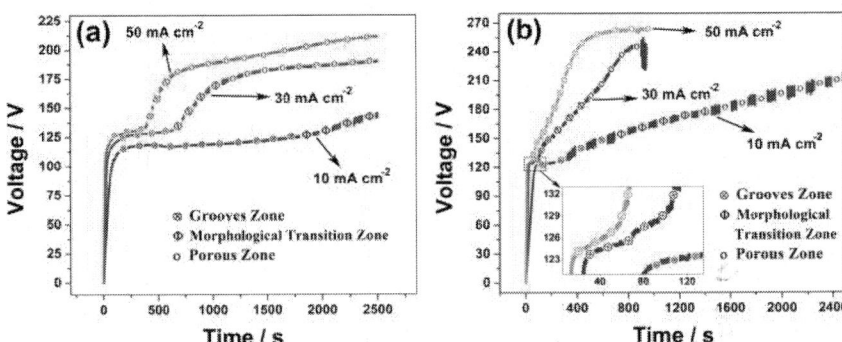

**Figure 1.** Voltage responses during PEO of titanium at different current densities in (a) 0.1 M H$_3$PO$_4$/1.5 M H$_2$SO$_4$and (b) 0.9 M H$_3$PO$_4$/1.5 M H$_2$SO$_4$ electrolytes.

The end of the initial linear regions coincided with the appearance of sparks on the specimen surface that persisted until the end of the process. All of the curves then displayed an approximate plateau followed by a voltage increase. For treatments carried out at 30 and 50 mA cm$^2$, the rate at which the voltage increased diminished with time and appeared to approach a relatively constant value. The potential ranges of each stage depend on the electrolyte and the current density employed. The durations of each stage reduce with increase of the current density and the concentration of H$_3$PO$_4$. Ferdjani et al. [41] found similar trends in the voltage for treatments carried out in H$_3$PO$_4$ solutions, proposing that differences in slopes are related to the phase composition of the coatings. However, it is later shown that the changes of the present curves coincide also with a transition from a grooved morphology to a porous one and an increase in size of the discharges.

Fig. 2 shows the voltage–time responses recorded at 50 mA cm$^{-2}$ in 0.9 M H$_3$PO$_4$/1.5 M H$_2$SO$_4$, with the temperature of the electrolyte maintained at

20 °C. The results of several experiments are presented, which were terminated at different charge densities. The response is also shown for a specimen that was treated under similar conditions except that the electrolyte temperature was allowed to rise due to the heat generated by the PEO process. The responses were similar for all specimens in the first ~ 120 s for all specimens. After this time, the voltage increased more rapidly for the specimen treated under a rising electrolyte temperature. The increase was particularly noticeable from ~ 175 V, when charge of 10 C cm$^{-2}$ had been passed through the cell.

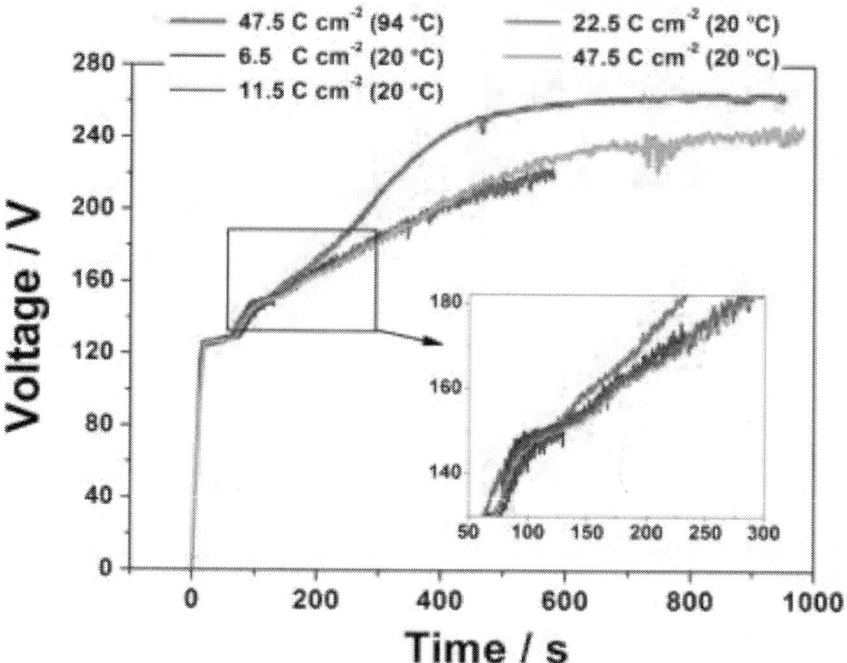

**Figure 2.** Voltage responses during PEO of titanium at 50 mA cm$^{-2}$ in 0.9 M H$_3$PO$_4$/1.5 M H$_2$SO$_4$ electrolyte.

Voltage–cell charge density curves are shown in Fig. 3(a) and (b) for 0.1 M H$_3$PO$_4$/1.5 M H$_2$SO$_4$ and 0.9 M H$_3$PO$_4$/1.5 M H$_2$SO$_4$ electrolytes respectively. The curves for the former electrolyte reveal a relatively large increase in the gradient at a charge density of ~20 C cm$^{-2}$ at all current densities, and a subsequent decrease in the gradient at a charge density of ~28 C cm$^{-2}$ for current densities of 30 and 50 mA cm$^{-2}$. In the case of the latter electrolyte, an increase in the gradient occurred at a charge density of ~ 5 C cm$^{-2}$ electrolyte and a subsequent decrease at a charge density of ~ 25 C cm$^{-2}$ for current densities of 30 and 50 mA cm$^{-2}$. For both electrolytes, a reduction in the gradient was not observed at a current density of 10 mA cm$^{-2}$.

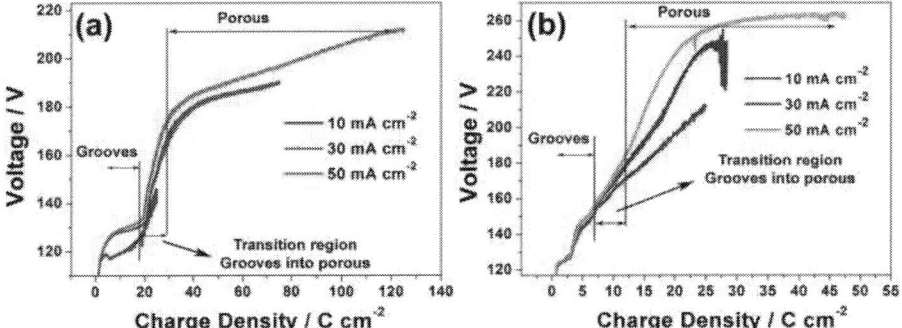

**Figure 3.** Dependence of voltage on cell charge density during PEO of titanium at different current densities in (a) 0.1 M $H_3PO_4$/1.5 M $H_2SO_4$ and (b) 0.9 M $H_3PO_4$/1.5 M $H_2SO_4$ electrolytes.

Fig. 4 shows the dependence of the weight gain of the specimens on the cell charge density. The results for the 0.1 M $H_3PO_4$/1.5 M $H_2SO_4$ electrolyte reveal an initial linear region, with a slope of $0.012 \pm 0.001$ mg $C^{-1}$ for all current densities, that extends up to a charge of ~ 28 C $cm^{-2}$. The slope then reduces to $0.005 \pm 0.001$ mg $C^{-1}$ at 30 and 50 mA $cm^{-2}$. The weight gains at the transition are in the range 0.30 to 0.35 mg $cm^{-2}$. No transition is evident at 10 mA $cm^{-2}$, probably due to the relatively low charge passed and the low weight gain achieved. In the case of the 0.9 M $H_3PO_4$/1.5 M $H_2SO_4$ electrolyte, the first region shows an average weight gain of $0.026 \pm 0.004$ mg $C^{-1}$ for all current densities. The weight gain is non-linear, decreasing with increase of charge. The transition occurs at a charge density of ~ 7 C $cm^{-2}$, when the weight gain is ~ 0.20 to 0.25 mg $cm^{-2}$. In contrast to the first region, the slope in the second region depends significantly on the current density, with values of $0.001 \pm 0.001$, $0.006 \pm 0.001$ and $0.0085 \pm 0.0005$ mg $C^{-1}$ at 10, 30 and 50 mA $cm^{-2}$ respectively. The conversion of titanium into $TiO_2$ at 100% Faradaic efficiency results in a weight gain of 0.083 mg $C^{-1}$. Assuming that the weight gain is mainly due to formation of $TiO_2$, the slopes of the regions of Fig. 4 allow an efficiency of oxide formation of ~ 15% to be estimated up to the transition at ~ 28 C $cm^{-2}$ for the 0.1 M $H_3PO_4$/1.5 M $H_2SO_4$ electrolyte, and ~ 6% after the transition. An efficiency of ~ 36% is estimated up to the transition for the 0.9 M $H_3PO_4$/1.5 M $H_2SO_4$ electrolyte, thereafter reducing to ~ 1, 7 and 10% at current densities of 10, 30 and 50 mA $cm^{-2}$ respectively. It is clear that the efficiency for both electrolytes is highest at the initial stages of the PEO and then decreases significantly between the stages of groove formation and pore formation, with the coatings formed in 0.9 M $H_3PO_4$/1.5 M $H_2SO_4$ showing a higher efficiency than those formed in 0.1 M $H_3PO_4$/1.5 M $H_2SO_4$. The calculations of the efficiency neglect the incorporation of phosphorus and sulfur species into the coating and possible dissolution of titanium. Thus, they represent upper limits of the true efficiencies. Later results of GDOES indicate that sulfur is a negligible constituent of the coatings. Work of others revealed 3 to 5 at.% phosphorus within coatings formed in an electrolyte similar to those employed in the present study [41]. Delplancke and Winand also found low oxide yields for PEO of titanium in 1 M $H_2SO_4$ electrolyte, in the range of 11 to 21% [42].

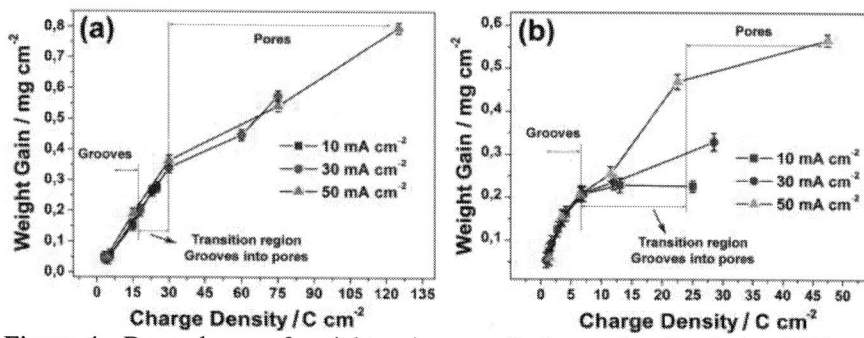

**Figure 4.** Dependence of weight gain on cell charge density during PEO of titanium at different current densities in (a) 0.1 M $H_3PO_4$/1.5 M $H_2SO_4$ and (b) 0.9 M $H_3PO_4$/1.5 M $H_2SO_4$ electrolytes.

The appearances of specimens during treatment at 30 mA cm$^2$ in 0.9 M $H_3PO_4$/1.5 M $H_2SO_4$ are shown in the optical images of Fig. 5. Other anodizing conditions produced similar results. Evolution of oxygen was observed from the first few seconds of the process, followed by a sequence of interference colors due to formation of an oxide of increasing thickness [43] and [44]. The oxygen evolution probably accounts for the low efficiencies of coating growth. Numerous, blue sparks of small size were observed after ~ 28 s at a voltage of ~ 123 V. The sparks gradually covered the whole coating surface, causing a color change from a faint pink to gray. At about 70 s, the population density of the sparks decreased and their size increased. At the same time their color changed to white. Similar observations have been reported in the literature [45], [46] and [47]. Between 225 and 900 s, a relatively small number of large, bright sparks were present at the coating surface. The charge densities that had passed through the cell at 225 and 900 s were 6.8 and 27 C cm$^{-2}$. Later results of SEM indicate that the appearance of the larger, brighter sparks coincided with a transition in the morphology of the coating surface.

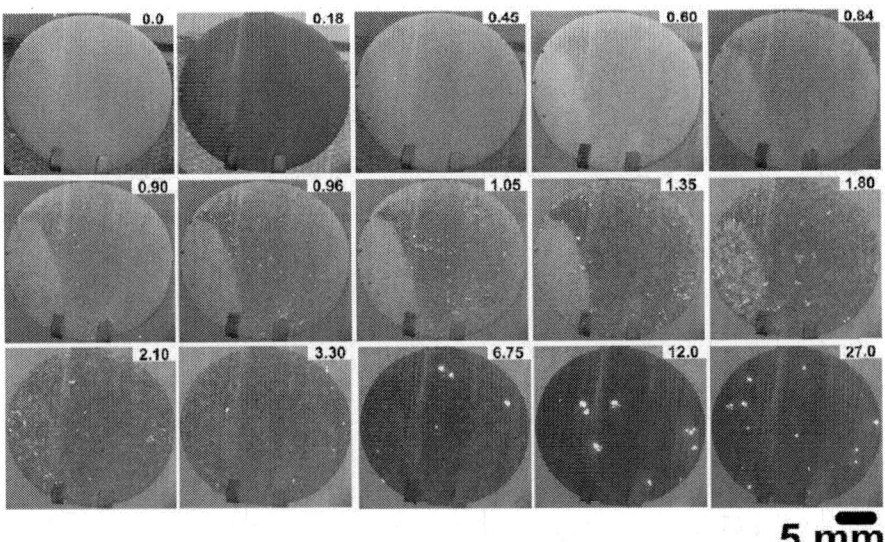

**Figure 5.** Discharge appearances during PEO of titanium at 30 mA cm$^{-2}$ in 0.9 M H$_3$PO$_4$/1.5 M H$_2$SO$_4$ electrolyte. The cell charges (C cm$^{-2}$) are indicated for each specimen.

Fig. 6 and Fig. 7 show scanning electron micrographs of the surfaces of specimens treated at 10, 30 and 50 mA cm$^{-2}$ in 0.1 M H$_3$PO$_4$/1.5 M H$_2$SO$_4$, and 0.9 M H$_3$PO$_4$/1.5 M H$_2$SO$_4$ respectively. Fig. 6 reveals that a major change in the morphology of the coating surfaces occurred at a charge density between ~ 18 and 30 C cm$^{-2}$ at 30 and 50 mA cm$^{-2}$. During growth of coatings to charge densities of 3.1 and 5.5 C cm$^{-2}$, small craters with short, linear pores were formed across the coating surfaces. Clusters of small, approximately circular pores were also evident in occasional locations. Following passing of a charge density of ~ 18 C cm$^{-2}$, a grooved morphology, which has developed from the short linear pores, extended across most of the specimen surfaces. However, patches of approximately circular pores occurred in places, and some locations showed a nodular surface, which appeared to be unaffected by dielectric breakdown, revealing neither pores nor grooves. In contrast, at a charge density of ~ 30 C cm$^{-2}$, the coatings exhibited a mainly porous morphology. Comparison with the results of Fig. 3, which shows the dependence of the voltage on the charge density during formation of the coatings, indicates that the change in slope at a charge density of ~ 20 C cm$^{-2}$ coincides closely with the extension of the grooved morphology to most regions of the coating surfaces, while the second change in slope, at a charge density of ~ 28 C cm$^{-2}$, coincides closely with the attainment of almost fully porous surfaces. The intermediate regions of charge, of increased slope compared with the preceding and succeeding regions, represent the transition between the two morphologies. Further, the weight gain data of Fig. 4 reveal a significant change in slope at a charge density of ~ 28 C cm$^{-2}$, which correlates with the formation of the porous morphology according to the observations of SEM. The arrows in Fig. 3 and Fig. 4 indicate the dominant morphologies of the breakdown regions that are developed at the various stages of coating growth. Lin et al. [48] have previously observed an oxide film with barrier layer features and small areas of pores, similar to those observed in the early stages of present PEO process. In contrast, at higher charge densities, the coating displayed a porous morphology, with pores of diameter $1.1 \pm 0.2$ and $1.5 \pm 0.3$ μm for coatings formed at 30 and 50 mA cm$^{-2}$ respectively.

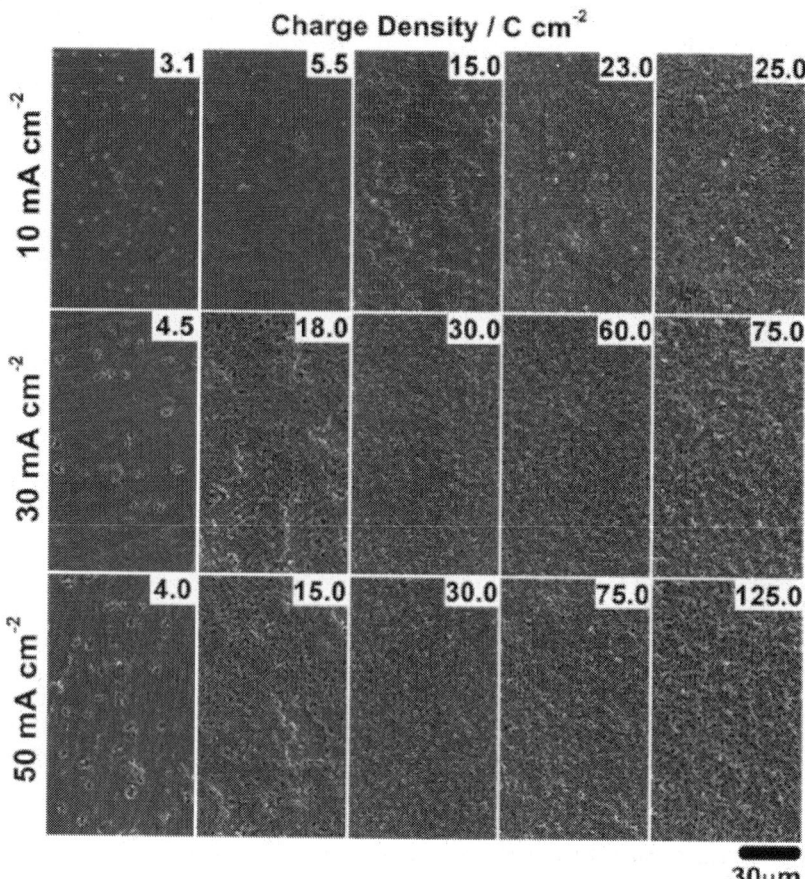

**Figure 6.** Scanning electron micrographs of coatings formed by PEO of titanium at different current densities in 0.1 M $H_3PO_4$/1.5 M $H_2SO_4$ electrolyte.

**Charge Density / C cm$^{-2}$**

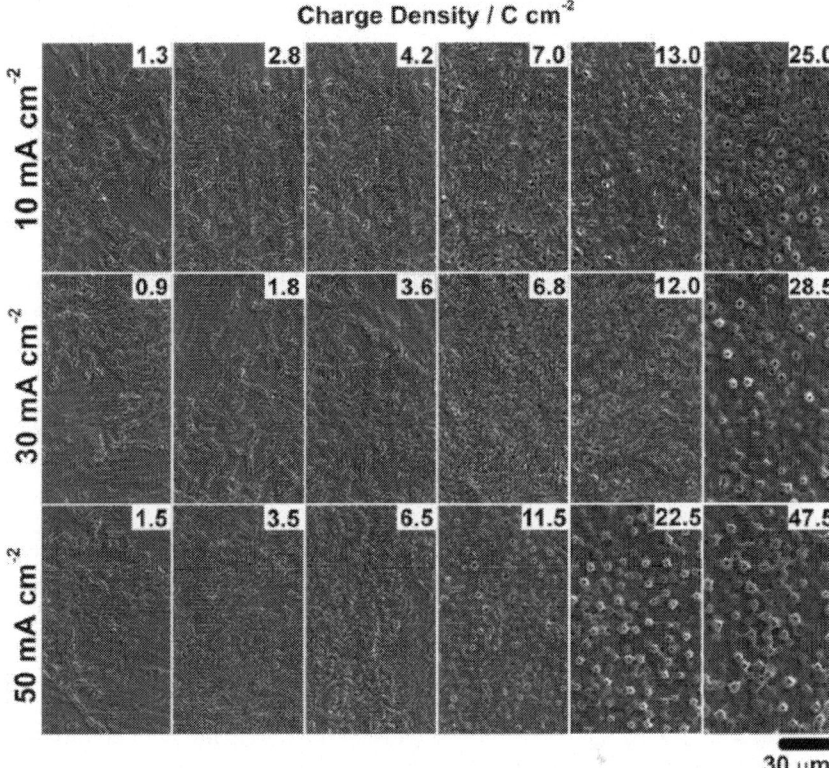

**Figure 7.** Scanning electron micrographs of coatings formed by PEO of titanium at different current densities in 0.9 M H$_3$PO$_4$/1.5 M H$_2$SO$_4$ electrolyte.

The morphology of the coating formed at 10 mA cm$^{-2}$ evolved more slowly than at higher current densities, and a uniform porous morphology had not been fully formed by the termination of the treatment. The coating formed to the lowest charge density of 3.1 C cm$^{-2}$ displayed formation of incipient grooves, occasional patches of small, approximately circular pores and relatively more numerous nodules that appeared to consist of cracked oxide. The appearance of the nodules resembled the morphology produced by field-crystallization of anodic oxides [49]. Following formation to a charge density of 15.0 C cm$^{-2}$, the grooves dominated the regions of dielectric breakdown on the coating surface, with small pores, cracked nodules and regions unaffected by breakdown also being present. At higher charge densities, the surface displayed a mixture of pores and short grooves.

Fig. 7 shows a similar transition from a grooved to porous morphology for the coatings formed in the 0.9 M H$_3$PO$_4$/1.5 M H$_2$SO$_4$ electrolyte. Unlike the coatings formed in the previous electrolyte, small pores, pre-cursor grooves and nodules were not observed at the lowest charge densities for which observations were made, i.e. in the range 0.9 to 1.5 C cm$^{-2}$. The formation of grooves appeared to take place from the start of dielectric breakdown. At a charge density of $\sim 7$ C cm$^{-2}$, the surfaces of the coatings are mainly grooved at all current densities. In contrast, at a charge density of $\sim 25$ C cm$^{-2}$, the surfaces are mainly of the porous type, with a few remnants of grooves in the coatings formed at 10 and 30 mA cm$^{-2}$. An intermediate morphology, consisting of pores and remnants of the grooved morphology, is evident at $\sim 12$ C cm$^{-2}$. Fig. 3 and Fig. 4, depicting the dependence of voltage and weight gain on the charge density, reveal significant changes in slope at a charge density of $\sim 5$ to 7 C cm$^{-2}$ at all current densities, coinciding closely with the charge at which the grooved morphology has been developed at most coating regions. Thereafter, the weight gains for the different current densities diverge in association with the transition to the porous morphology. Fig. 5 also shows that the appearances of the sparks of a specimen treated at 30 mA cm$^{-2}$ had increased in size and brightness at a charge density of 6.8 C cm$^{-2}$, with the appearance remaining relatively similar up to the final time illustrated when the charge density had reached 27 C cm$^{-2}$. In this range of charge, the morphology of the coating surface was changing from grooved to porous. The pore diameters were $1.4 \pm 0.3$ and $1.7 \pm 0.5$ µm for coatings obtained at 30 and 50 mA cm$^2$ respectively. At 10 mA cm$^{-2}$, the transformation is incomplete, with circular and elongated pores being evident. The pores appear to arise from fusion of the earlier formed grooves. Formation of grooves takes place from the onset of breakdown and continues at voltages up to $\sim 150$ V in 0.1 M H$_3$PO$_4$/1.5 M H$_2$SO$_4$ and $\sim 175$ V in 0.9 M H$_3$PO$_4$/1.5 M H$_2$SO$_4$, with more prominent grooves forming in the latter electrolyte.

Fig. 8 compares the morphologies of coating surfaces following treatments at 50 mA cm$^{-2}$ in the 0.9 M H$_3$PO$_4$/1.5 M H$_2$SO$_4$ without and with control of the electrolyte temperature. For both temperature conditions, a transition is from a grooved morphology to a porous morphology. Treatments at 6.5 C cm$^{-2}$ resulted in similar grooved morphologies for both electrolyte conditions. An increased charge density of 11.5 C cm$^{-2}$ led to partial transformation of the morphology, as pores replaced grooves. However, the transformation of the coating surface had progressed further when the electrolyte temperature was not controlled and by a charge density of 22.5 C cm$^{-2}$ pores occupied the whole of the coating surface. In contrast, grooves were still prevalent for the coating formed at 20 °C and remnants of grooves could still be observed at a charge density of 47.5 C cm$^{-2}$. Furthermore, under the latter condition, relatively numerous large craters were formed in the coating.

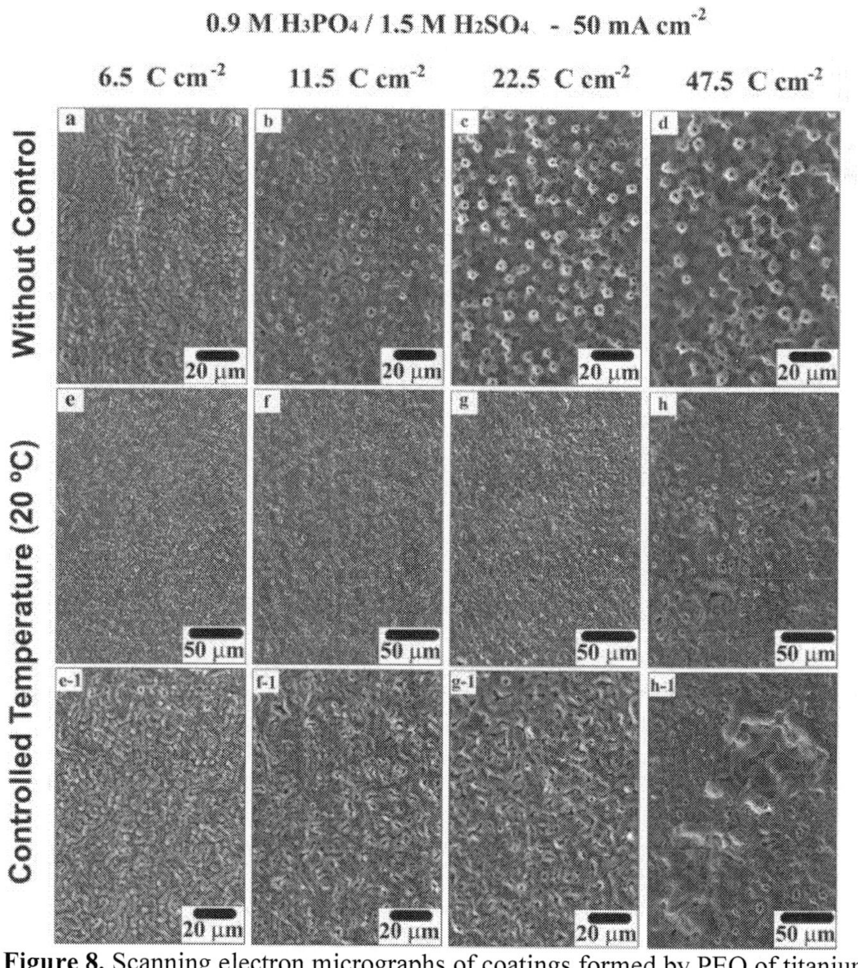

**Figure 8.** Scanning electron micrographs of coatings formed by PEO of titanium at 50 mA cm$^{-2}$ in 0.9 M H$_3$PO$_4$/1.5 M H$_2$SO$_4$ electrolyte.

It is clear from Fig. 6 and Fig. 7 that the morphological transition requires less time to complete when a higher current is employed. These results and those from weight gain measurements and observations of sparks (Fig. 4 and Fig. 5 respectively) indicate that the coating morphologies depend on the charge density supplied to the specimen. Sparks initially form elongated pores and then the grooves. It appears at this stage that the sparks travel on the coating surface. The grooves may develop due to local heating and pressure at the initial spark site, which causes breakdown of adjacent oxide. Thicker oxide is formed as a consequence of the spark, which causes subsequent breakdown events to occur preferentially on unaffected regions of the coating surface, where the oxide is thinner. The process continues until the whole surface is covered by grooves and thickened oxide. The sparks then intensify, accompanied by a voltage rise (see Fig. 1), leading to the formation of the pores, which appear to be distributed randomly in the coating. The sites of breakdown of the coating are presumed to be locations of defects, such as oxygen bubbles or cracks within the coating. It is widely reported that the decrease of the population density and increase of the size and intensity of sparks are associated with larger pores [14], [26], [46], [50] and [51]. Several researchers[9] and [52] have studied anodizing of titanium in mixtures of sulfuric and phosphoric acids, reporting only porous morphologies. Previously, the grooved morphology was speculatively attributed to fusion of pores under increased heating at the higher frequency [32].

The transition to a porous morphology during a treatment at 50 mA cm$^{-2}$ is promoted in an electrolyte in which the temperature of the electrolyte is allowed to rise during the treatment and results in a more uniform surface. A similar effect of the electrolyte temperature may be anticipated for PEO at 30 mA cm$^{-2}$, which also causes a large increase in temperature. However, the temperature rise at 10 mA cm$^{-2}$ is relatively modest and is likely to have a much smaller influence on the coating morphology. The lower electrolyte temperature at 10 mA cm$^{-2}$ results from both the lower current density and the lower cell voltage. The PEO process generates heat from the electrochemical and chemical reactions that form the coating material and from the Joule heating due to the resistance of the coating to passage of the ionic and electronic current. The increase in temperature of the coating under a particular condition of PEO is dependent on the rate of heat flow from the coating to the titanium substrate and to the electrolyte. The temperature of the electrolyte at the coating surface will exceed the temperature of the bulk solution due to the temperature gradient across the boundary layer at the coating surface. However, the turbulence created by the generation of gas bubbles at the coating surface will assist the heat transfer. The large craters formed at the later stage of coating formation at 50 mA cm$^{-2}$ in the 0.9 M H$_3$PO$_4$/1.5 M H$_2$SO$_4$ when the electrolyte was controlled suggest that current becomes concentrated at these sites, possibly due to a self-accelerating process associated with an initial local temperature rise of the coating. In contrast, in the higher temperature electrolyte, a more uniform distribution of the current appears to be maintained despite a higher voltage being achieved under this electrolyte condition.

Fig. 9 presents backscattered SEM micrographs of polished cross-sections of the coatings formed in each electrolyte to the final time of PEO. The coatings increase in thickness with increase of current density, with average values of 3.0, 7.4 and 11.2 μm for the coatings formed in the 0.1 M H$_3$PO$_4$/1.5 M H$_2$SO$_4$ electrolyte and 3.4, 4.0 and 6.1 μm for coatings formed in the 0.9 M H$_3$PO$_4$/1.5 M H$_2$SO$_4$ electrolyte for current densities of 10, 30 and 50 mA cm$^{-2}$ respectively. The coatings reveal porous morphologies typical of PEO coatings, the pores resulting from the formation of discharge channels and oxygen evolution. The coatings formed the 0.9 M H$_3$PO$_4$/1.5 M H$_2$SO$_4$ electrolyte reveal relatively large pores that tend to be distributed non-uniformly, often being present in the inner region of the coating. In comparison, the coatings formed in the 0.1 M H$_3$PO$_4$/1.5 M H$_2$SO$_4$ electrolyte appear to have more numerous, finer pores distributed relatively uniformly through the coating thickness. Detailed studies of porosity in PEO coatings have revealed a wide distribution of pore size, including significant porosity of nanometer size that cannot be revealed by conventional SEM [53].

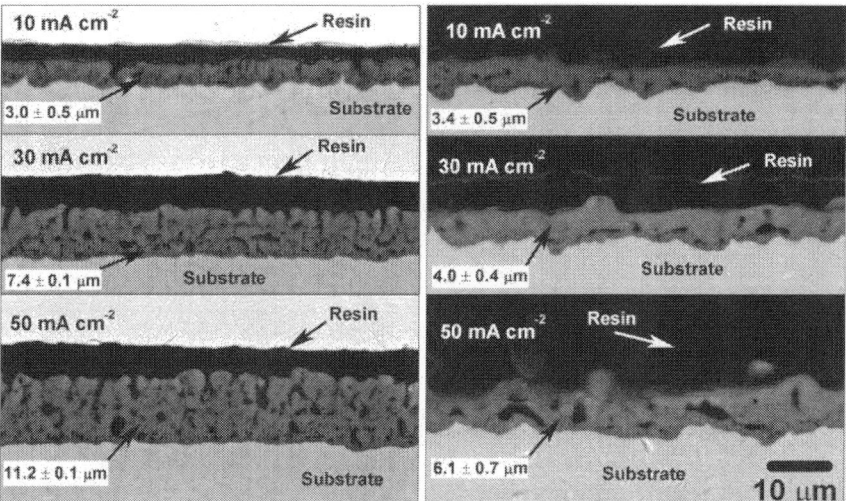

**Figure 9.** Scanning electron micrographs (backscattered electrons) showing cross-sections of coatings formed by PEO of titanium at different current densities in 0.1 M $H_3PO_4$/1.5 M $H_2SO_4$ and 0.9 M $H_3PO_4$/1.5 M $H_2SO_4$ electrolytes. The coatings were formed to the final times of Fig. 1.

XRD patterns of specimens anodized to the final times in the two electrolytes are shown in Fig. 10. Characteristic peaks of the substrate and the coatings are evident. Coatings formed in both electrolytes at 10 mA cm$^{-2}$, and in 0.9 M $H_3PO_4$/1.5 M $H_2SO_4$ at 30 mA cm$^{-2}$ reveal only anatase. Other coatings, formed in 0.1 M $H_3PO_4$/1.5 M $H_2SO_4$ at 30 to 50 mA cm$^{-2}$ and in 0.9 M $H_3PO_4$/1.5 M $H_2SO_4$ at 50 mA cm$^{-2}$, revealed both anatase and rutile. The coatings formed in the latter electrolyte also contained amorphous material, as evident from the broad peak centered at ~ 25°. In contrast, XRD patterns of Fig. 11 for specimens oxidized to the stage of the grooved morphology reveal anatase as the only crystalline phase. The occurrence of anatase at potentials near 150 V and the appearance of rutile at higher potentials and current densities is consistent with previous reports [6], [48], [50] and [54], suggesting that anatase transforms to rutile during the PEO treatment. Further, changes in the slopes of voltage responses in the first seconds of anodizing have been related to transformation of the initial amorphous phase into anatase or rutile [55]. It is known that anatase is stable at low temperature, but may be converted into rutile when heated above 900 °C [56]. It has been suggested that the temperature and pressure in the discharge zones may reach 3700–37,000 °C and $10^2$–$10^3$ MPa respectively, thus being sufficient to develop and transform the coating structure, with an increased intensity of discharges at higher voltages [57].

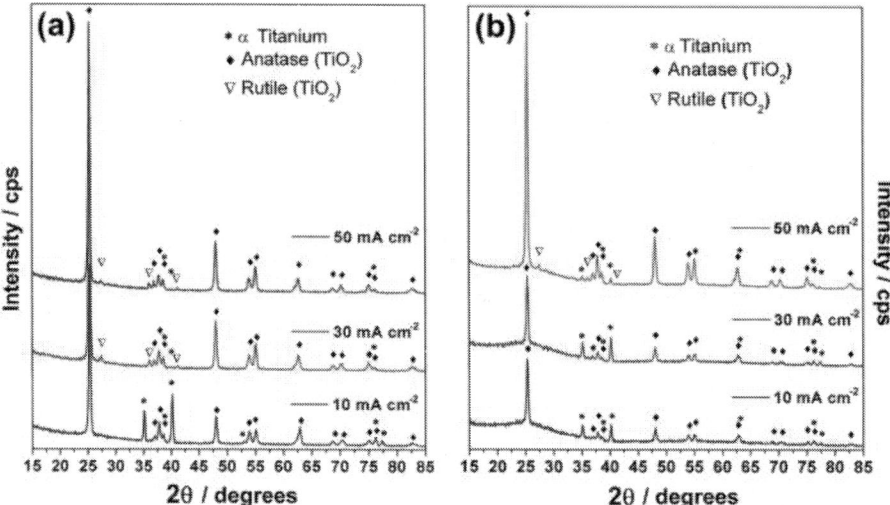

**Figure 10.** XRD analysis of porous PEO coatings formed on titanium in (a) 0.1 M $H_3PO_4$/1.5 M $H_2SO_4$ at 2500 s for all current density values and (b) 0.9 M $H_3PO_4$/1.5 M $H_2SO_4$ at 2500, 950 and 950 s for 10, 30 and 50 mA cm$^{-2}$, respectively.

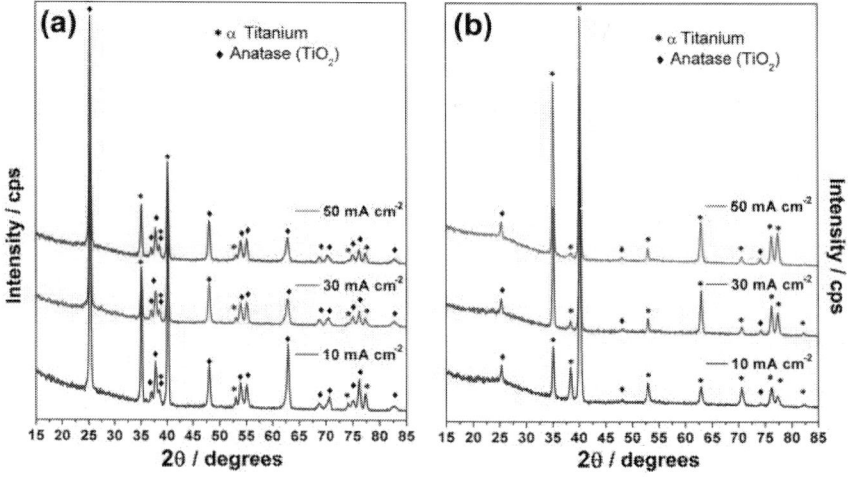

**Figure 11.** XRD analysis of grooved PEO coatings formed on titanium at different current densities in (a) 0.1 M $H_3PO_4$/1.5 M $H_2SO_4$ and (b) 0.9 M $H_3PO_4$/1.5 M $H_2SO_4$ electrolytes.

GDOES depth profiles are displayed in Fig. 12(a) and (b) for the coatings formed in 0.1 M $H_3PO_4$/1.5 M $H_2SO_4$ and 0.9 M $H_3PO_4$/1.5 M $H_2SO_4$ respectively. The profiles show that the coatings contain oxygen, phosphorus and sulfur, with similar distributions in the layer regardless of the anodizing conditions. The amount of phosphorus incorporated into the coating increases with the concentration of $H_3PO_4$ in the electrolyte, as expected [41], while the concentrations of sulfur in the coatings were low for both electrolytes, with signals in most regions appearing to be close to the background level, apart from a possible increase near the coating base. The distribution of phosphorus indicates the access of the electrolyte to the inner regions of the coating through breakdown channels. Others have suggested that precipitation and re-dissolution of phosphate may occur in the discharge channel, with migration of phosphate into the inner parts of the coating under the high electric field [24]. Consequently, the concentration of phosphorus species is increased toward the film/ substrate interface. This could explain the increased phosphorus and sulfur signals near the base of the coatings in the depth profiles ofFig. 12. Lee et al. [52] formed PEO coatings by PEO of titanium at a constant voltage of 180 V in 1.5 M $H_2SO_4$/0.3 M $H_3PO_4$, 1.5 M $H_2SO_4$/0.6 M $H_3PO_4$, and 1.5 M $H_2SO_4$/0.9 M $H_3PO_4$ electrolytes, finding mainly phosphates ($H_2PO_4$, $PO_4^-$, $HPO_4^-$, $PO_3^-$) by X-ray photoelectron spectroscopy. Marino et al. [58] reported the incorporation of phosphates and phosphides into the coating during anodizing in phosphate buffers at pH 1 and 5. Others have suggested the presence of $P_2O_5$[9]. Future work of the authors will explore the composition and morphology of the present coatings in more detail, using high resolution analytical electron microscopy. The presence of oxygen and titanium in the elemental distributions obtained by GDOES for the present coatings is consistent with the occurrence of rutile and anatase as detected by XRD. No crystalline phosphorus-containing phase was detected by XRD, which may be due to the low concentration of such phases or to the presence of the phosphorus in amorphous material.

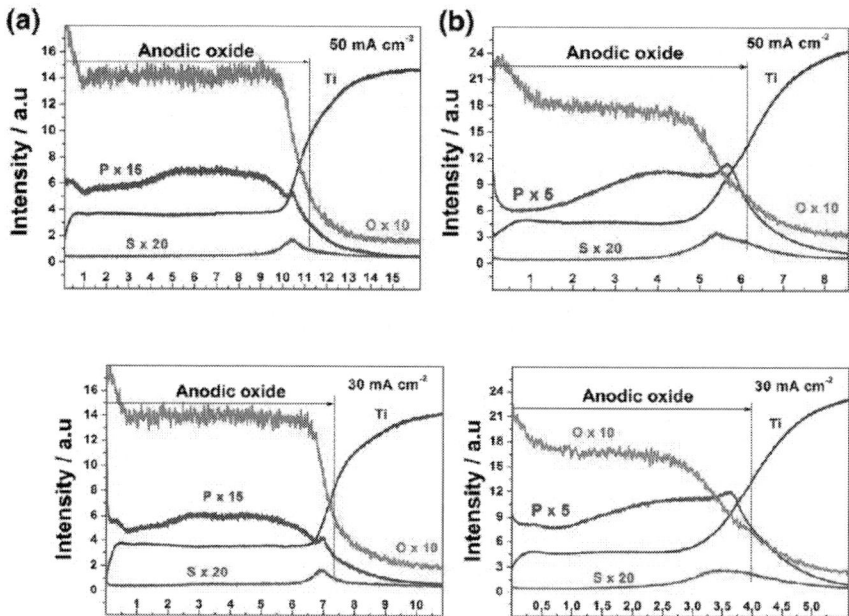

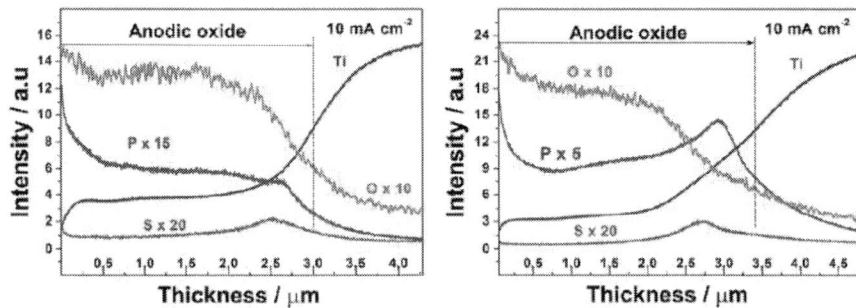

**Figure 12.** GDOES depth profiles of PEO coatings formed on titanium at different current densities in (a) 0.1 M $H_3PO_4$/1.5 M $H_2SO_4$ and (b) 0.9 M $H_3PO_4$/1.5 M $H_2SO_4$ electrolytes.

Fig. 13 shows that the values of the shear strength obtained from tensile tests of lap-shear joints prepared with porous and grooved morphologies. In order to prepare the surface morphologies, PEO was carried out at the intermediate current density of 30 mA $cm^{-2}$ in 0.9 M $H_3PO_4$/1.5 M $H_2SO_4$ electrolyte. The treatments were terminated at charge densities of 6.8 and 28.5 C $cm^{-2}$ in order to generate grooves and pores respectively. Under these conditions, a relatively complete coverage of the surface by either grooves or pores can be achieved as revealed in Fig. 7. The porous morphology resulted in shear strengths in the range 489 to 892 N $mm^{-2}$, with an average value and standard deviation of $713 \pm 196$ N $mm^{-2}$. In comparison, the shear strengths with the grooved morphologies ranged from 747 to 1566 N $mm^{-2}$, with an average value and standard deviation of $1155 \pm 295$ N $mm^2$. Thus, the average shear strength was ~ 62% increase greater for the grooved morphology compared with the porous morphology. Only one test of the grooved morphology gave a shear strength that was significantly less than those of all tests with the porous morphology. SEM and visual examinations showed that the joint failed close to the coating–resin interface. Fig. 14 shows the difference in the amounts of adhesive on the porous and grooved surfaces. Some regions of the coating surface appeared to be resin-free. In other places, the resin appeared to have penetrated the grooves and pores (indicated by arrows), but with more resin being evident within the grooves than within the pores. The results suggest that the grooves allow greater penetration of the epoxy-resin adhesive into the coating than the pores, thereby improving the keying of the resin to the coating. There may also be differences in the details of the composition and nanoscale topography of the surfaces that affect the bonding. However, any such differences could not be resolved by the present analyses. The coatings were formed in the same electrolyte and anatase was the only crystalline phase in both coatings, such that differences in chemical composition are possibly relatively small. At the available resolution of the SEM examination, the failure appeared to be of a mixed adhesive-cohesive mode. Further, the stronger bonding was obtained with the thinner coating, such that coating thickness, which might allow deeper penetration of adhesive, does not appear to be a significant factor in the strength of the joints. Heida et al. [59] fabricated a nanostructured anodic oxide on a titanium alloy in 1.0 M

$H_3PO_4$ solution containing 0.5% NaF in order to improve the adhesive strength of a medical polymer. They found that the adhesive strength of the polymer coating on the nanostructured surface was 144% greater than on an untreated surface, concluding that the nanostructured surface provided anchoring of the coating, with failure occurring within the anodic oxide.

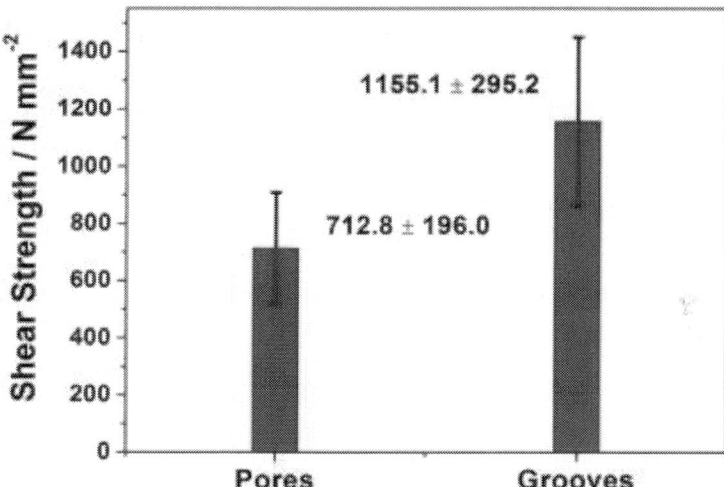

**Figure 13.** Influence of (a) porous and (b) grooved morphologies on the shear strength of adhesively joined titanium.

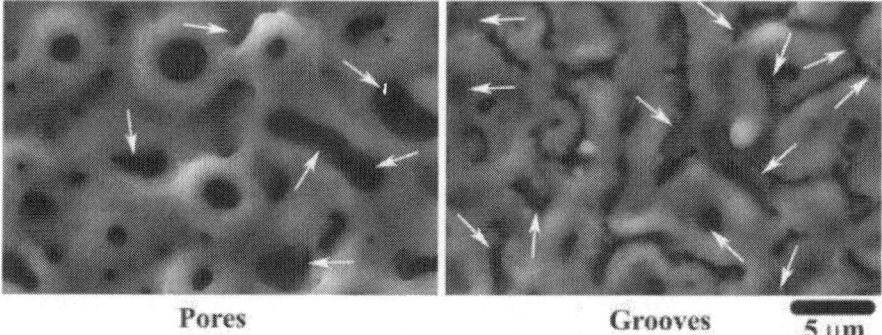

**Figure 14.** Scanning electron micrographs (secondary electrons) of the failed surfaces showing the difference in the amounts of adhesive on the porous and grooved surfaces.

## 4. CONCLUSIONS

Coatings formed galvanostatically on titanium, at current densities from 10 to 50 mA cm$^{-2}$ in 0.1 M H$_3$PO$_4$/1.5 M H$_2$SO$_4$ and 0.9 M H$_3$PO$_4$/1.5 M H$_2$SO$_4$ electrolytes reveal initially a grooved surface morphology, which precedes the formation of a more usual porous morphology. The morphological transition occurs in a range of cell charge that depends strongly on the composition of the electrolyte, but weakly on the current density, occurring over a significantly higher range of charge for the 0.1 M H$_3$PO$_4$/1.5 M H$_2$SO$_4$electrolyte compared with the 0.9 M H$_3$PO$_4$/1.5 M H$_2$SO$_4$ electrolyte.

The formation of the grooved coating occurs at a higher efficiency than the formation of the porous coating, with the weight gain per unit charge passed in the cell for the grooved morphology being $\sim 2$ times greater in 0.1 M H$_3$PO$_4$/1.5 M H$_2$SO$_4$, and at least 3 times greater in 0.9 M H$_3$PO$_4$/1.5 M H$_2$SO$_4$, than for the porous morphology.

The morphological transition coincides with increases in the cell voltage and changes in the population density, color and intensity of sparks. Anatase is present in grooved coatings, while rutile may also occur in porous coatings. GDOES also revealed the presence of phosphorus species, and relatively negligible amounts of sulfur species, the concentration of the former species being dependent on the concentration of H$_3$PO$_4$ in the electrolyte.

The grooved morphology resulted in a $\sim 62\%$ higher shear strength than a porous morphology in adhesive bonding tests. The increased strength is suggested to be due to improved penetration and keying of the adhesive into the grooved surface compared with the porous surface.

## ACKNOWLEDGMENTS

The authors express their gratitude to "Departamento Administrativo de Ciencia, Tecnología e Innovación – COLCIENCIAS", Universidad de Antioquia for financial assistance (project 111545221209 and Estrategia de Sostenibilidad 2013–2014 de la Universidad de Antioquia) and the Engineering and Physical Sciences Research Council (U.K.) for the support of the LATEST2 Programme Grant.

## REFERENCES

1.  A. Ochsenbein, F. Chai, S. Winter, M. Traisnel, J. Breme, H.F. Hildebrand, Acta Biomater. 4(2008) 1506.

2.  F.T. Cheng, P. Shi, H.C. Man, Scr. Mater. 51 (2004) 1041.

3.  H. Lee, M.Y. Song, J. Jurng, Y.-K. Park, Powder Technol. 214 (2011) 64.

4.  R. Yoshida, Y. Suzuki, S. Yoshikawa, J. Solid State Chem. 178 (2005) 2179.

5.  J. Jitputti, Y. Suzuki, S. Yoshikawa, Catal. Commun. 9 (2008) 1265.

6.  M.V. Diamanti, M.P. Pedeferri, Corros. Sci. 49 (2007) 939.

7.  S.A. Fadl-Allah, R.M. El-Sherief, W.A. Badawy, J. Appl. Electrochem. 38 (2008) 1459.

8.  E. Santos, N.K. Kuromoto, G.A. Soares, Mater. Chem. Phys. 102 (2007) 92.

9.  H.-J. Oh, J.-H. Lee, Y. Jeong, Y.-J. Kim, C.-S. Chi, Surf. Coat. Technol. 198 (2005) 247.

10. R.O. Hussein, P. Zhang, X. Nie, Y. Xia, D.O. Northwood, Surf. Coat. Technol. 206 (2011)1990.

11. A. Ghasemi, V.S. Raja, C. Blawert, W. Dietzel, K.U. Kainer, Surf. Coat. Technol. 202 (2008)3513.

12. A.L. Yerokhin, X. Nie, A. Leyland, A. Matthews, S.J. Dowey, Surf. Coat. Technol. 122 (1999)73.

13. A.L. Yerokhin, L.O. Snizhko, N.L. Gurevina, J. Phys. D. Appl. Phys. 36 (2003) 2110.

14. S. Stojadinović, R. Vasilić, M. Petković, L. Zeković, Surf. Coat. Technol. 206 (2011) 575.

15. M. Shokouhfar, C. Dehghanian, M. Montazeri, A. Baradaran, Appl. Surf. Sci. 258 (2012)2416.

16. H. Habazaki, T. Onodera, K. Fushimi, H. Konno, K. Toyotake, Surf. Coat. Technol. 201 (2007)8730.

17. Y. Cheng, E. Matykina, P. Skeldon, G. Thompson, Electrochim. Acta 56 (2011) 8467.

18. J. Martin, A. Melhem, I. Shchedrina, T. Duchanoy, A. Nominé, G. Henrion, T. Czerwiec, T.Belmonte, Surf. Coat. Technol. 221 (2013) 70.

19. J.A. Curran, H. Kalkancı, Y. Magurova, T.W. Clyne, Surf. Coat. Technol. 201 (2007) 8683.

20. L.O. Snizhko, A.L. Yerokhin, A. Pilkington, N.L. Gurevina, D.O. Misnyankin, A. Leyland, A.Matthews, Electrochim. Acta 49 (2004) 2085.

21. F. Jaspard-Mécuson, T. Czerwiec, G. Henrion, T. Belmonte, L. Dujardin, A. Viola, J. Beauvir,Surf. Coat. Technol. 201 (2007) 8677.

22. M.V. Diamanti, M. Ormellese, M. Pedeferri, Corros. Sci. 52 (2010) 1824.

23. M.V. Diamanti, M. Ormellese, E. Marin, A. Lanzutti, A. Mele,M.P. Pedeferri, J. Hazard. Mater.186 (2011) 2103.

24. M. Nakajima, Y. Miura, K. Fushimi, H. Habazaki, Corros. Sci. 51 (2009) 1534.

25. Y.M. Wang, B.L. Jiang, L.X. Guo, T.Q. Lei, Appl. Surf. Sci. 252 (2006) 2989.

26. Z. Yao, Y. Jiang, F. Jia, Z. Jiang, F. Wang, Appl. Surf. Sci. 254 (2008) 4084.

27. I. Han, J.H. Choi, B.H. Zhao, H.K. Baik, I.-S. Lee, Curr. Appl. Phys. 7 (2007) e23.

28. F. Monfort, A. Berkani, E. Matykina, P. Skeldon, G.E. Thompson, H. Habazaki, K. Shimizu,Corros. Sci. 49 (2007) 672.

29. D. Shen, G. Li, C. Guo, J. Zou, J. Cai, D. He, H. Ma, F. Liu, Appl. Surf. Sci. 287 (2013) 451.

30. L. Zhao, C. Cui, Q. Wang, S. Bu, Corros. Sci. 52 (2010) 2228.

31. H.B. Jiang, Y.K. Kim, J.H. Ji, I.S. Park, T.S. Bae, M.H. Lee, Surf. Coat. Technol. 259 (2014)310–317.

32. C. Chu, Z. Liu, X. Rao, Q. Sun, P. Lin, F. Chen, P. Chu, Surf. Coat. Technol. 232 (2013) 68.

33. G. Zhao, A. Raines, M. Wieland, Z. Schwartz, B. Boyan, Biomaterials 28 (2007) 2821.

34. J. Lincks, B. Boyan, C. Blanchard, C. Lohmann, Y. Liu, D. Cochran, D. Dean, Z. Schwartz,Biomaterials 19 (1998) 2219.
35. G. Critchlow, D. Brewis, Int. J. Adhes. Adhes. 15 (1995) 173.
36. G. Stamoulis, N. Carrere, J.Y. Cognard, P. Davies, C. Badulescu, Int. J. Adhes. Adhes. 51(2014) 148.
37. ASTM, Standard F67-13, Standard Specification for Unalloyed Titanium, for SurgicalImplant Application, ASTM International, West Conshohocken, PA, 2013.
38. ASTM, Standard B600-11, Standard Guide for Descaling and Cleaning Titanium andTitanium Alloy Surfaces, ASTM International, West Conshohocken, PA, 2011.
39. ASTM, Standard D1002-10, Standard Test Method for Apparent Shear Strength of Single-Lap-Joint Adhesively Bonded Metal Specimens by Tension Loading (Metal-to-Metal),ASTM International, West Conshohocken, PA, 2010.
40. Y.T. Sul, C.B. Johansson, Y. Jeong, T. Albrektsson, Med. Eng. Phys. 23 (2001) 329.
41. S. Ferdjani, D. David, G. Beranger, J. Alloys Compd. 200 (1993) 191.
42. J.L. Delplancke, R. Winand, Electrochim. Acta 33 (1988) 1551.
43. M.V. Diamanti, B. Del Curto, V. Masconale, C. Passaro, M.P. Pedeferri, Color. Res. Appl. 37(2012) 384.
44. M.V. Diamanti, B. Del Curto, M. Pedeferri, Color. Res. Appl. 33 (2008) 221.
45. Y.-l. Cheng, X.-Q. Wu, Z.-g. Xue, E. Matykina, P. Skeldon, G.E. Thompson, Surf. Coat.Technol. 217 (2013) 129.
46. M. Petković, S. Stojadinović, R. Vasilić, L. Zeković, Appl. Surf. Sci. 257 (2011) 10590.
47. S. Stojadinović, J. Jovović, M. Petković, R. Vasilić, N. Konjević, Surf. Coat. Technol. 205(2011) 5406.
48. C.S. Lin, M.T. Chen, J.H. Liu, J. Biomed. Mater. Res. A 85 (2008) 378.
49. H. Habazaki, T. Ogasawara, K. Fushimi, K. Shimizu, S. Nagata, T. Izumi, P. Skeldon, G.Thompson, Electrochim. Acta 53 (2008) 8203.
50. Y. Wang, T. Lei, B. Jiang, L. Guo, Appl. Surf. Sci. 233 (2004) 258.
51. Y. Cheng, F. Wu, E. Matykina, P. Skeldon, G.E. Thompson, Corros. Sci. 59 (2012) 307.
52. J.-H. Lee, S.-E. Kim, Y.-J. Kim, C.-S. Chi, H.-J. Oh, Mater. Chem. Phys. 98 (2006) 39.
53. J.A. Curran, T.W. Clyne, Acta Mater. 54 (2006) 1985.
54. A.F. Yetim, Surf. Coat. Technol. 205 (2010) 1757.
55. J.S. Choi, R.B. Wehrspohn, J. Lee, U. Gosele, Electrochim. Acta 49 (2004) 2645.
56. D.J. Won, C.H. Wang, H.K. Jang, D.J. Choi, Appl. Phys. A 73 (2001) 595.

57. L. Xie, G. Yin, D. Yan, X. Liao, Z. Huang, Y. Yao, Y. Kang, Y. Liu, J. Mater. Sci. Mater. Med. 21(2010) 259.

58. C.E.B. Marino, P.A.P. Nascente, S.R. Biaggio, R.C. Rocha, N. Bocchi, Thin Solid Films 468(2004) 109.

59. J. Hieda, M. Niinomi, M. Nakai, K. Cho, T. Mohri, T. Hanawa, Mater. Sci. Eng. C 36(2014) 244.

# CHAPTER 8

# Washcoat Deposition of Ni- and Co-ZrO$_2$ Low Surface Area Powders onto Ceramic Open-Cell Foams: Influence of Slurry Formulation and Rheology

*Riccardo Balzarotti \*, Mirko Ciurlia †, Cinzia Cristiani † and Fabio Paparella †*

Politecnico di Milano, Dipartimento di Chimica, Materiali e Ingegneria Chimica "G. Natta", Piazza Leonardo da Vinci 32, 20133 Milano, Italy

## ABSTRACT

The effect of formulations and procedures to deposit thin active layers based on low surface area powders on complex geometry substrates (open-cell foams) was experimentally assessed. An acid-free liquid medium based on water, glycerol, and polyvinyl alcohol was used for powder dispersion, while a dip-coating technique was chosen for washcoat deposition on 30 PPI ceramic open-cell foams. The rheological behavior was explained on the bases of both porosity and actual powder density. It was proved that the use of multiple dippings fulfills flexibility requirements for washcoat load management. Multiple depositions with intermediate flash drying steps at 350 °C were carried out. Washcoat loads in the 2.5 to 22 wt. % range were obtained. Pore clogging was seldom observed in a limited extent in samples with high loading (>20 wt. %). Adhesion, evaluated by means of accelerated stress test in ultrasound bath, pointed out good results of all the deposited layers.

**Keywords:** ceramic open-cell foams; washcoat; catalyst deposition; rheology; structured support

# 1. INTRODUCTION

Structured catalysts and reactors for process intensification are receiving a large interest from the modern chemical engineering community [1]. A structured catalyst is intended as a continuum metallic or ceramic geometrical matrix (support) where voids (*i.e.*, channels) are present. The catalytically active phases are properly dispersed onto the support surface by direct incorporation or coating deposition. The latter is the most used technique, due to its simplicity and versatility [2]. Commonly, the deposited washcoat consists of a high surface area carrier, generally ceramic, where the metal active phase is properly dispersed. A variety of materials have been investigated as high surface area ceramic supports, such as alumina, silica, titania, and ceria [3,4,5,6]. In some process applications, low surface area catalysts have been investigated instead of high surface area ones. As an example, cerium oxide has been proposed as an active phase carrier, due to its oxygen storage properties [7]. Unfortunately, cerium oxide undergoes a fast surface area decrease when it is treated at high temperature [8]. Moreover, in several cases, this transition to lower surface area values occurs at temperatures that are lower than process operative conditions. Thus, in many cases, low surface area cerium-based catalysts need to be deposited onto structured supports [9].

The active phase usually consists of metal ions, and it is the core of the catalysis process. A variety of metals have been proposed in view of the different chemical processes. Among others, cobalt [10,11] and nickel [12,13] have been chosen because of their catalytic activity in many different reactions. Moreover, they have been proposed as a low cost and effective alternative to noble metals.

A variety of geometrical supports are now commercially available for catalytic purposes. They mainly differ for the structural and physical properties, such as chemical nature, surface area per unit volume or mechanical properties. Three way catalysts (TWCs) are one of the most diffused and investigated applications of structured supports to catalysis [14] because they are extensively used for gas pollution control in vehicle exhaust [15].

Among the different supports, solid open-cell foams are highly promising. Open-cell foams are characterized by a high interface area and high porosity, which could result in lower pressure drops and higher energy efficiency [16,17,18]. Due to their geometrical structure, high performance in terms of fluid/solid mass transfer are guaranteed.

As already reported, structured catalyst preparation is usually based on coating deposition onto the geometrical support surface. Depending on the structured substrate geometry, several deposition techniques for the catalytic thin layer have been made up [19,20]. Among them, dip-coating technique is the simplest, most versatile and cheapest one to be used in industrial practice. Moreover, dip-coating deposition can be easily applied to both metallic and ceramic supports of a variety of shapes [21].

The first step in dip-coating technique consists of the preparation of a slurry that is composed of a liquid phase in which the final powder to be deposited has to be suspended or better, dispersed. Then, the structured support is dipped in the liquid medium in order to fill the voids with a catalytic slurry precursor. Finally, the geometrical support is withdrawn from the slurry at a controlled rate. Depending on viscosity and support geometry, excess liquid removal is guaranteed by the opposite forces acting on the liquid during the withdrawal step. As a result, coating thickness depends on the balance between gravitational force, which promotes the removal of the liquid phase, and the viscous forces acting in the slurry, which involve the sliding resistance [22,23]. Therefore, the control of the coated layer properties, *i.e.*, thickness and adhesion, are mainly ruled by the slurry rheological behavior and the withdrawal velocity. As far as rheological behavior is concerned, the suspension formulation (e.g., water/powder ratio, acid/powder ration, and surfactant content) can be easily tuned in order to achieve the desired viscosity at the typically applied dip coating shear rates [21]. Indeed, at low viscosity values, which promote good adhesion, lower coating loads are obtained. On the other hand, when a higher viscosity is applied, the higher coating load obtained is counterbalanced by poor adhesion, and sometimes a few difficulties are faced in applying the method [24].

Different methods are available for both powder suspension (or dispersion) and slurry stabilization. A well-known method is based on the generation of electrostatic repulsions among the powder particles, promoted by the surface charging of the material to be dispersed in an acidic environment [24,25,26]. This procedure has been applied with success to high surface area powders, as their surface can be easily charged simply by managing the pH of the suspension. Unfortunately, this route is not easily applicable in the case of powders with low surface area, those characterized by chemical un-reactivity, or in the case of a possible dissolution of components. In all of these cases, an acid-free dispersion method needs to be used, and the selection of an alternative dispersant agent is required. Generally speaking, in the acid-free method organic molecules, more typically macromolecules, are used as dispersants. They are dissolved into the liquid phase so they can closely interact with the powder surface, allowing particle dispersion and slurry stabilization [27,28,29]. Many papers are reported in the literature regarding formulation and its effect on resulting slurries [30,31,32]. However, to our knowledge, fewer works have been reported on the production of slurries for coating deposition, via dipping, of low surface area powders onto open cell foams. In particular, scarce information is reported on the effect of the different dispersants and the composition of the final slurry properties. This study is of fundamental importance to correlate easily tunable experimental parameters, such as ratios between the components, with the rheological behavior of the slurry. Indeed, rheology is the main operative parameter to drive the final coating load, thickness, and adhesion, *i.e.*, which are the parameters of interest for industrial application [33].

Accordingly, the aim of this work is to clarify these aspects by studying the formulations and the procedures to deposit thin active layers of low surface area model catalyst powder, such as Ni- or Co-supported, onto low surface area $ZrO_2$. The catalytically active powders were produced by means of the incipient wetness impregnation technique using a commercial support. Different acid-free formulations based on water (H), glycerol (G), and polyvinyl alcohol (PVA) were studied, and the effect of the components that determine the slurry rheology was assessed. The different slurries were tested to coat 30 PPI (Pores Per Inch) ceramic open-cell foams via the dip-coating technique. A correlation among the final coating load, thickness, and adhesion after thermal treatments with the slurry composition and rheology was proposed. An attempt to rationalize the rheological behavior at the light of composition was also made.

## 2. RESULTS AND DISCUSSION

### 2.1. Catalyst Characterization

Catalytic powders characterization is reported in Figure 1 and Table 1. For the sake of comparison, the characterization of the pristine morphological carrier ($ZrO_2$) is also reported.

XRD patterns (Figure 1) clearly showed the reflections of the monoclinic $ZrO_2$ support. Additional reflections at 37.2°, 43.2° and 63° 2θ can be clearly seen in case of the Ni-containing sample; the latter were attributed to NiO, while the one that was detected at 36.9° in the Co-based sample was attributed to $Co_3O_4$.

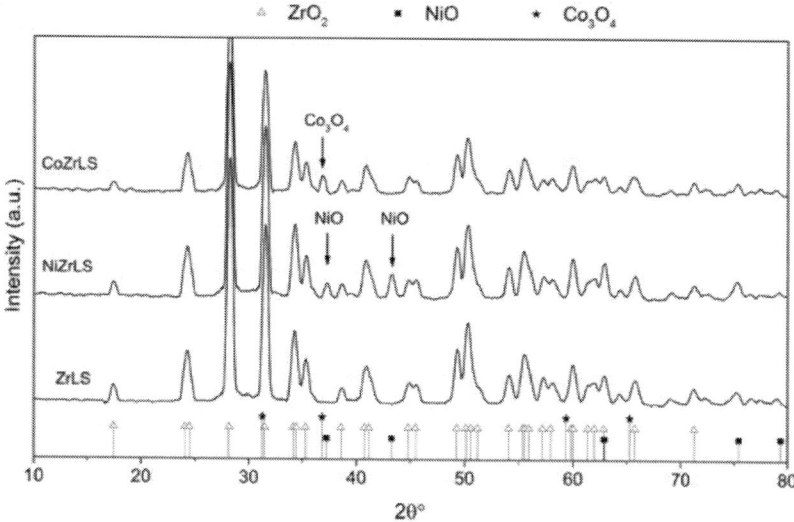

**Figure 1.** XRD spectra of the impregnated powders.

**Table 1.** Morphological characterization of the powders.

| Sample | Active Phase Crystal Size (nm) (by XRD) | Surface Area (m²·g⁻¹) | Pore Volume (cm³·g⁻¹) |
|---|---|---|---|
| ZrLS | - | 27 | 0.2 |
| NiZrLS | 28 | 14 | 0.1 |
| CoZrLS | 25 | 19 | 0.1 |

The crystallite dimensions of the carrier and the active phase, which were calculated according to Scherrer equation, were in the range of 25–28 nm, thus, almost comparable.

Regarding morphology, very close surface areas and pores volumes were measured for the active materials: a decrease of the surface area was observed upon impregnation that was accompanied by a decrease of the pore volumes. This effect is clearly explained by the partial occupation of pores due to the presence of the active phase.

According to the procedure reported in the experimental section, all the powders, carrier and active materials were dispersed in the HGP liquid medium, and their rheological behaviors were compared. Results are reported in Figure 2.

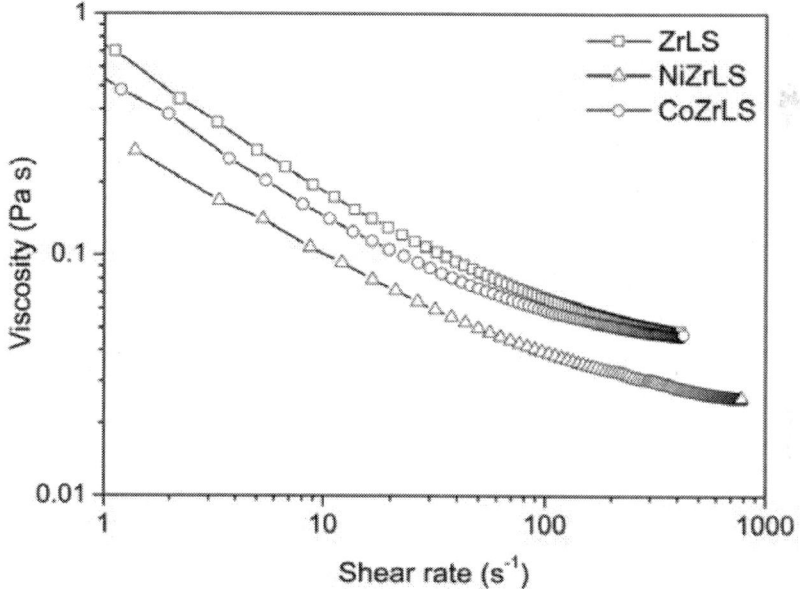

**Figure 2.** Rheological behavior of zirconia-based slurries.

Apparently, active phase presence did not affect rheological properties. Indeed, shear thinning behaviors were found for all the samples. In order to get information on the non-Newtonianity degree of the slurries, the flow curves slope was determined in the $10–100$ s$^{-1}$ shear rate range. Generally speaking, higher slope values (in modulus) correspond to a more marked shear thinning behavior. For all samples, slopes in the range of 0.6–0.4 were calculated; this highlighted a very close rheological behavior for the three slurries. The powder composition exerted a different effect, since the presence of the active phase induced lower viscosity values in all the shear rate range (Figure 2). Such differences cannot be directly related neither to the surface area nor to the porosity due to the close values of these parameters detected for the three samples; similar considerations can be done for the active phase content, too. On the contrary, the absolute viscosity appeared to be much more specific for the metal cations present at the carrier surface. The effect of the presence of an active phase onto the carrier surface was already reported in the literature [34]. Accordingly, the modification of the surface nature was reported as responsible for a different powder-dispersant interaction, which influenced the rheological behavior. However, in this case, this effect should be quite peculiar considering the strong chemical similarity between Ni and Co both in terms of atomic weight and atomic number.

Accordingly, an attempt to rationalize the viscosity behavior was done, considering properties other than those mentioned above or a combination of properties reported so far.

A possible explanation for the rheological behavior could be found by considering the actual volume fraction of the powders. To evaluate this aspect, a representative model to describe the powders has to be built. In the model, three components were considered: (1) powder nature (mainly related to the molar weight); (2) powder density (mainly related to porosity); and (3) powder size and size distribution.

The first consideration regarded the composition of the different powders, since slurries formulation was obtained by considering the powders content on mass base. The presence of the active phase in its oxidized form should be taken into account, since it affects the powder molar weight and consequently, the powder concentration in the slurry. In order to quantify this effect, the actual amount of the active phase (in the oxidized form ($m_{Ox.Act.Ph}$)) was determined according to Equation (1)

$$m_{Ox.Act.Ph} = \frac{m_{Met.Act.Ph} \ (g)}{M_{MetalIon} \ \left(\dfrac{g}{mol}\right)} \cdot M_{Ox}\left(\frac{g}{mol}\right)$$

where $m_{Met.Act.Ph}$ (g) is the mass of the active phase (metallic form), $M_{MetalIon}$ is the atomic mass of the metal ion and $M_{Ox}$ the molecular weight of the active phase (oxidized form). Starting from Equation (1), support weight fraction was recalculated according to Equation (2).

$$K = \frac{m_{sup\,port}}{m_{sup\,port} + m_{Ox.Act.Ph}}$$

$K$ equals to one for the bare support, while it is lower than one for the active powders (Table 2). However, $K$ cannot completely describe the powders because no powder density was taken into account in Equation (1). Powder density is directly correlated to the morphological properties and particularly to porosity that can be evaluated by means of BET analysis. Therefore, in order to properly evaluate the actual density of the powders, crystallographic density and sample porosity were introduced in the calculation. In the case of active powders, crystallographic density was assumed as the average of the densities of all the phases, weighted by composition as reported in Equation (3):

$$\rho_{bulk} = \sum_{i=1}^{N.C.} \%wt._i \cdot \rho_{bulk,i}, \quad \text{with} \quad \%wt._i = \frac{m_i}{m_{sup\,port} + m_{m_{Ox.Act.Ph}}}$$

where "$i$" is the any phase present in the powder in the oxidized form. The "real" density was thus calculated by taking into account the evaluated material porosity. Porosity values were used to determine the powder void fraction ($\varphi$) and, thus, the real powder density ($\rho_{real}$) calculated according to Equation (4):

$$\rho_{real} = \rho_{bulk} \cdot (1-\varphi), \quad \text{with} \quad \varphi = \frac{V_p}{V_m} = \frac{\text{pore volume}}{\text{specific volume}}$$

Moreover, to give a complete description of the powders and to evaluate their actual concentration in the slurry, the particle size and size distribution, evaluated by granulometric analysis, were introduced in the model. Accordingly, the weighted volumetric fraction was calculated by Equation (5):

$$V_{Fract} = \sum_{i=1}^{Dp} V_i \cdot \text{Percentage}_i$$

where $V_{fract}$ is the volume of the particle ($V_i$) and Percentage$_i$ is the relative amount of particles with the $i$-th diameter. For volume calculation, a spherical shape was assumed for particles while the specific volume was assumed as equal to the reciprocal of density. Finally, the number of particles per unit volume was obtained by using Equation (6)

$$\text{Particles concentration} = \frac{m_{powder}}{\rho_{real} \cdot V_{Fract} \cdot V_{liq}}$$

where $\rho_{real}$ is the real density of powder and $V_{liq}$ is the volume of dispersing HGP liquid medium. Results are reported in Table 2.

**Table 2.** Powder physical properties and slurry viscosity values.

| Powder | K | Bulk Density (g·cm⁻³) | Real Density (g·cm⁻³) | Particles Concentration (Particles cm⁻³) | Viscosity (Pa·s) at Shear Rate: 1 s⁻¹ | 10 s⁻¹ | 100 s⁻¹ |
|---|---|---|---|---|---|---|---|
| ZrLS | 1 | 5.68 | 1.13 | $1.50 \times 10^8$ | 0.743 | 0.185 | 0.067 |
| NiZrLS | 0.91 | 5.77 | 2.43 | $4.24 \times 10^7$ | 0.309 | 0.1 | 0.04 |
| CoZrLS | 0.76 | 5.78 | 2.16 | $4.10 \times 10^7$ | 0.532 | 0.14 | 0.059 |

On these bases, viscosity at three selected shear rates was plotted as a function of powder density (Figure 3).

Viscosity was found to decrease linearly as powder density increased. Once the mass of powder to be dispersed is fixed, a larger material density results in a lower number of particles per volume unit. The number of particles per volume unit directly influences the rheological behavior: a higher number of particles leads to higher viscosity values [35,36]. Thus, the rheological slurry behavior can be explained on these bases. Pure $ZrO_2$ (the carrier), which is a less dense material than the impregnated one, showed the highest viscosity because the same powder amount (weight base) had a larger number of particles present. When powder density increases, such as in the case of the active powders, the number of particles decreases; therefore, powder concentration and slurry viscosity also decreases.

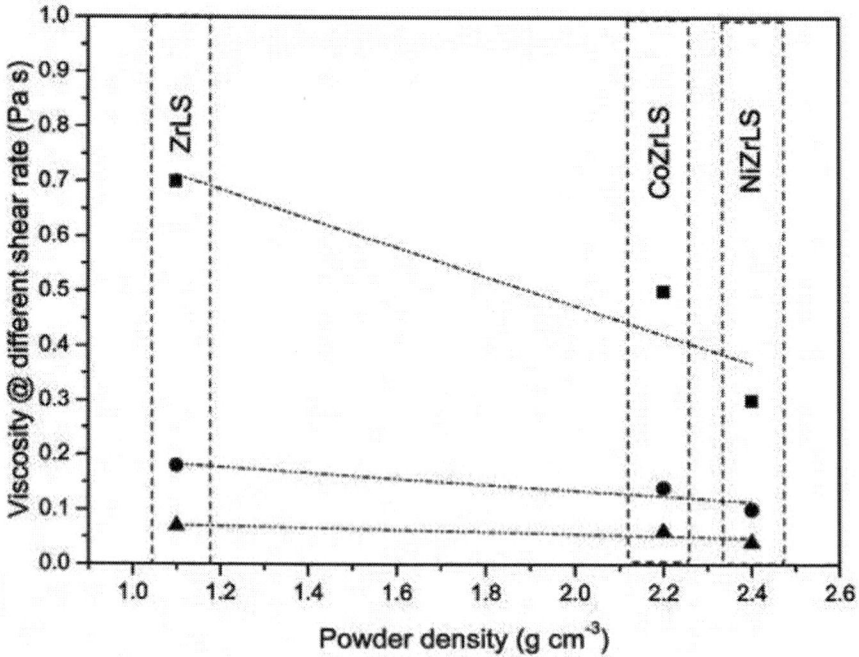

**Figure 3.** Viscosity at three selected shear rates as the function of the powder density (triangles: SR 100 $s^{-1}$; circles: SR 10 $s^{-1}$; squares: SR 1 $s^{-1}$).

This picture was confirmed by plotting the viscosity as a function of the particle concentration, which was calculated as reported above (Figure 4). At any shear rate, the viscosity increases with the particles concentration. Results of Figure 4 are in line with the conventional behavior of slurry viscosity. It is well known that the viscosity of concentrated dispersions is higher than that of diluted ones. This effect may be related with an increase of particle-particle interaction. Moreover, the fact that parallel lines were found suggested the presence of the same interaction mechanism at any shear rates.

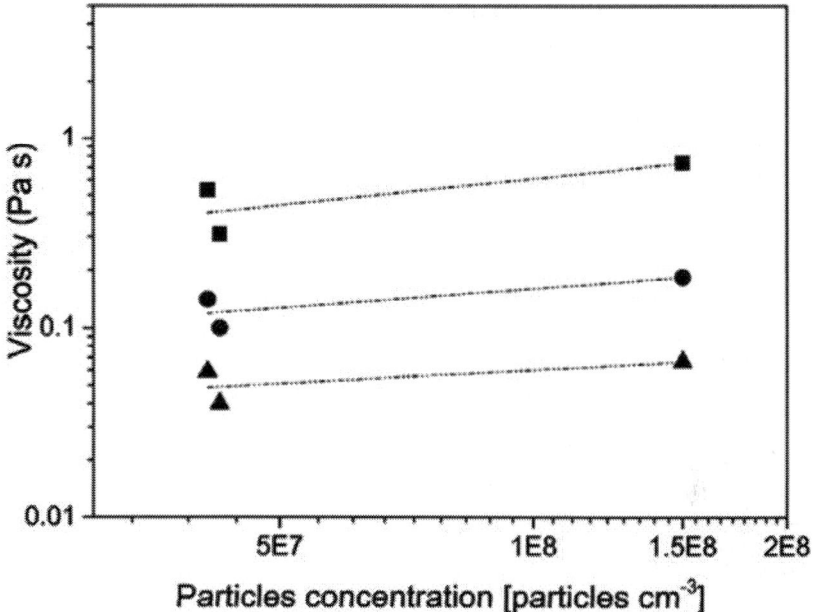

**Figure 4.** Viscosity as a function of particles concentration (triangles: SR 100 $s^{-1}$; circles: SR 10 $s^{-1}$; squares: SR 1 $s^{-1}$).

## 2.2. Washcoat Deposition

HGP-based slurries were deposited onto the 30 PPI ceramic foams via dip-coating process and then, thermally treated. Up to three multiple depositions were performed. It is well known that washcoat load, the parameter of interest, is correlated with viscosity, *i.e.*, the main operative variable of the process [37]. Accordingly, the coating load after calcination was plotted as a function of the viscosity values at shear rate $10 \cdot s^{-1}$, the one of interest for dip-coating application (Figure 5). Regardless slurry formulation, a quite linear load-viscosity correlation was found.

As expected, the washcoat load increased with viscosity and was observed at any dipping number. Due to the highest viscosity values, the higher loads were obtained with the pristine $ZrO_2$ slurries which were found to reach approximately 22 wt. % after three dippings.

When more than one dipping was applied, the deposition seemed not to be affected by the presence of the previous washcoat layer. This effect is better evidenced when the washcoat load after calcination is plotted as a function of the dipping number (Figure 6).

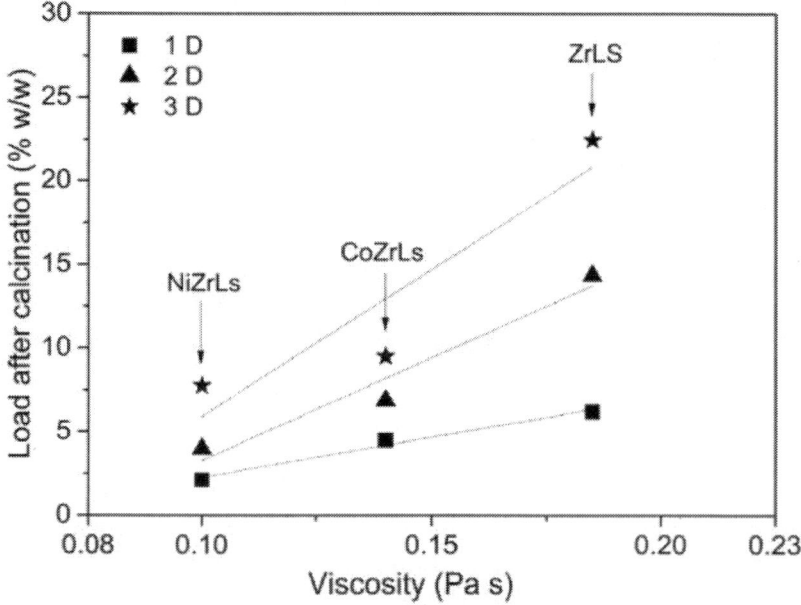

**Figure 5.** Coating Load as a function of viscosity and of the number of subsequent depositions.

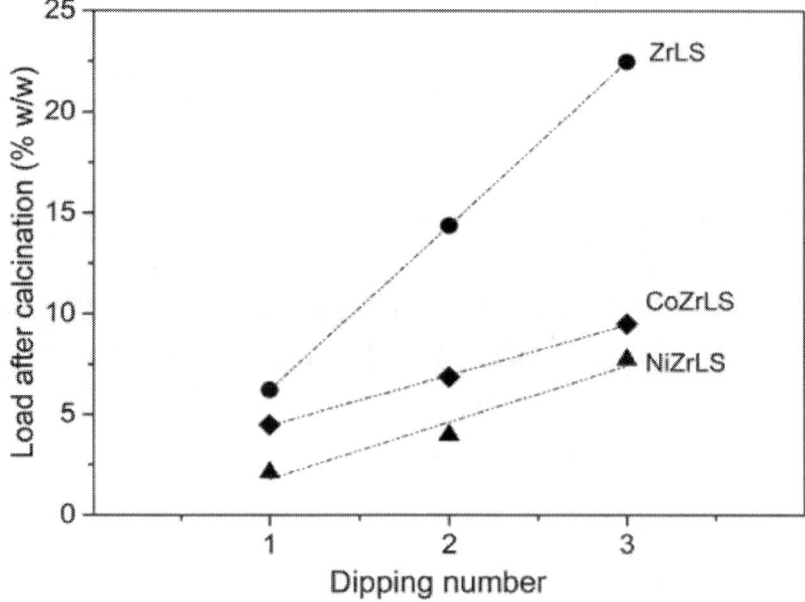

**Figure 6.** Coating load as a function of the dipping numbers.

A linear trend between load and dipping number was found, suggesting dipping number as a useful tool to manage washcoat load.

Results on adhesion tests—by means of an ultrasound stress test in petroleum ether—are shown in Figure 7.

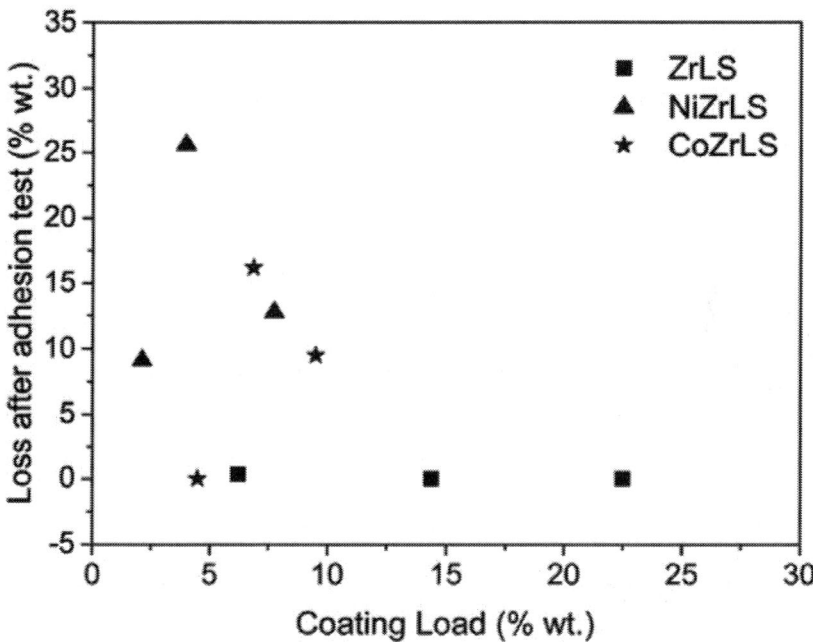

**Figure 7.** Weight loss after adhesion test for Zr-based samples as a function of load after calcination.

From the analysis of points of Figure 7, adhesion seemed strongly related to the coating composition; in the case of pure $ZrO_2$, negligible or no weight losses were present while the presence of active phase led to higher weight losses, with a relative maximum in correspondence of the intermediate washcoat load values. Generally speaking, higher washcoat load should determine higher losses after adhesion test, due to layer thickness. On the contrary, in these samples an overall increase of adhesion was found with load increase. For the considered samples, the effect of the different surface composition should also be taken into account even though this effect cannot be easily evaluated. Anyway, due to the large complexity of the involved phenomena and to the brittleness of the support foam, this point deserves more study to be clarified.

A qualitative washcoat analysis as a function of dipping number was performed on the active powders by using an optical microscope (Figure 8).

Figure 8 shows the images after flash drying. As described in the experimental section, this step is performed at 350 °C for 6 min: these operative parameters are suitable for solvent removal, but they are not enough for the total decomposition of the organic components which is still incomplete [38]. To analyze the washcoat at this step can be highly useful: the partial decomposition of the organic compound will result in a darkening of the surface that should allow for a better vision of coating coverage and homogeneity. Both samples clearly showed a surface darkening that was qualitatively interpreted as a homogeneous distribution of the washcoat load. As a matter of fact, white areas, which correspond to the bare substrate, were still evident only after one dipping; then, they gradually decreased and tended to disappear upon multiple depositions. After three dippings, a good coverage was reached. Almost no pore clogging occurred: only limited pore clogging was found for the CoZrLS (left side in the picture).

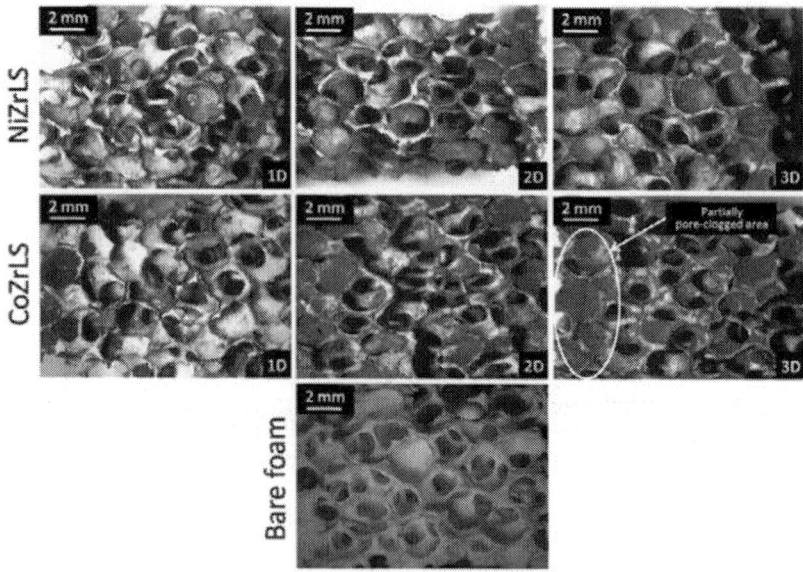

**Figure 8.** Optical images of NiZrLS (1D to 3D) and CoZrLS (1D to 3D) deposited on 30 PPI foams: effect of multiple dippings after flash drying. For the sake of comparison, an image of the bare foam has been added at the bottom of the image.

Figure 8 shows the images after flash drying. As described in the experimental section, this step is performed at 350 °C for 6 min: these operative parameters are suitable for solvent removal, but they are not enough for the total decomposition of the organic components which is still incomplete [38]. To analyze the washcoat at this step can be highly useful: the partial decomposition of the organic compound will result in a darkening of the surface that should allow for a better vision of coating coverage and homogeneity. Both samples clearly showed a surface darkening that was qualitatively interpreted as a homogeneous distribution of the washcoat load. As a matter of fact, white areas, which correspond to the bare substrate, were still evident only after one dipping; then, they gradually decreased and tended to disappear upon multiple depositions. After three dippings, a good coverage was reached. Almost no pore clogging occurred: only limited pore clogging was found for the CoZrLS (left side in the picture).

In order to evaluate the deposited layers at higher magnification, washcoat after calcination was analyzed by SEM measurements (Figure 9). The results after three dippings are reported for the active powders.

Acquisitions were performed in back scattering, and they once more demonstrated the good coverage homogeneity of the surface (Figure 9). Few defects were present, but they were of limited extent and localized. Coating surface, analyzed at higher magnification (Figure 9e,f), showed evidence of the presence of cracks of limited depth; another coating layer (not the bare support) was seen underneath.

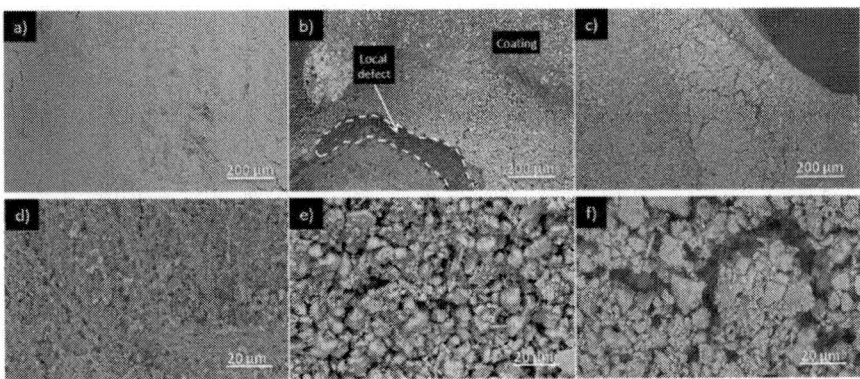

**Figure 9.** Back scattering SEM analysis of ZrLS, (**a,d**), NiZrLS (**b,e**) and CoZrLS (**c,f**) coated samples after three dippings at different magnifications (100X and 1000X).

# 3. EXPERIMENTAL SECTION

## 3.1. Catalytic Powders Preparation and Characterization

A commercially available low surface area carrier was used for catalyst production, namely zirconium oxide supplied by Melcat (in the following, ZrLS).

Catalysts were produced by using the incipient wetness impregnation method [19]. Nickel nitrate hexahydrate (98.5%, Sigma-Aldrich, St. Louis, MO, USA) and cobalt nitrate hexahydrate (98%, Sigma-Aldrich, St. Louis, MO, USA) were used as nickel and cobalt active phases precursors, respectively. These metal salts were used to obtain a final load equal to 7 wt. %, on a metal base.

The precursors solution was wisely dropped onto the carrier. Impregnated powders were dried overnight at 120 °C, and then, they were calcined in order to decompose the nitrate precursors and to obtain the final oxide. In particular, Co-based samples were calcined at 400 °C while Ni-based samples were treated at 800 °C; in both cases, dwell was set at 10 h, with heating and cooling rate of 2 °C·min$^{-1}$. The two different calcination temperatures were chosen in accordance with possible catalytic applications (*i.e.*, Oxy-Steam Reforming for Ni-based catalysts [39] and Fisher Tropsch synthesis for Co-bases samples [11]. A final metal load about 10 wt. % was measured for all powders.

Impregnated powders were characterized by means of X-ray diffraction. A D8 Advance diffractometer (Bruker, Billerica, MA, USA) and a Cu-$K_\alpha$ radiation were used (10–80° 2θ range, 40 kV and 40 mA, step scan 0.02° 2θ, time 1 s·step$^{-1}$). Crystallite dimensions were evaluated from the reflection line broadening (FWHM, calculated by Topas) using the Scherrer equation [40].

The powders particle size was evaluated by using a CILAS 1180 laser granulometer (Compagnie Industrielle des Lasers, Orleans, France).

BET surface area and pore volume were determined by $N_2$ adsorption and Hg intrusion; in the first case, a Tristar 3000 device was used (Micromeritics, Norcross, GA, USA). $N_2$ physisorprion measurements were carried out after heating at 150 °C overnight, under vacuum. An Autopore IV instrument (Micromeritics, Norcross, GA, USA) was used for Hg intrusion.

## 3.2. Washcoating

The powders dispersion formulation is based on a dispersant, glycerol (G) (87% *w/w* water solution, Sigma-Aldrich, St. Louis, MO, USA), a solvent/diluent, distilled water (H), and a rheology modifier, polyvinyl alcohol (PVA) (Mowiol, Sigma-Aldrich, St. Louis, MO, USA). Weight ratios among the three components were calculated with respect to the powders (PW), and they were respectively set at: G/PW = 1.9, H/PW = 1.8 and PVA/PW = 0.07. In the following, the liquid medium based on this formulation will be labeled as HGP.

The slurry was obtained by means of a procedure reported elsewhere [24]. Briefly, in a typical experiment, PVA was dissolved in distilled water at 85 °C; then glycerol was added, always under magnetic stirring. The obtained HGP liquid medium was used to disperse catalyst powders. The powder was added to the HGP solution, and the resulting slurry was ball-milled for 24 h (50 rpm of rotation rate) in a polyethylene jar using $ZrO_2$ spheres as grinding bodies. After the milling process, a sonication pre-treatment was performed for 30 min on the slurries in order to reduce foaming.

The slurries rheological behavior was evaluated in the $1–10^3$ s$^{-1}$ shear rate range by means of a DSR 200 instrument (Rheometrics, New Castle, DE, USA) by using the parallel plates geometry and plates of 40 mm of diameter.

Before coating deposition, supports were cleaned with acetone for 30 min in an ultrasound bath.

Slurries were deposited on Yttria-stabilized Zirconia Alumina (YZA) open-cell foams (Selee Company, Hendersonville, NC, USA), with a nominal pore density of 30 PPI (pore per inch). Structured supports were cut in parallelepiped shape with squared section; dimensions were set at 1.5 cm and 1 cm for length and section, respectively.

Dip-coating was used as the deposition technique. Both the dipping and withdrawal rate were set at 13 (cm·min$^{-1}$). After the dipping step, coated samples were flash dried [38] for 6 min at 350 °C in a sealed oven. Then, a final calcination thermal treatment was performed for 10 h at 400 °C and 800 °C for Co- and Ni-based materials respectively; in both cases, dwell was set at 10 h with a heating and cooling rate of 2 °C·min⁻¹. When necessary, multiple depositions were performed; the dipping procedure was repeated, and flash drying was performed between two subsequent dippings.

Washcoat load was evaluated by the weight difference of bare and coated foam.

Coated layers homogeneity and morphology were evaluated by means of optical (SZ-CTV microscope, Olympus, Tokio, Japan) and scanning electronic microscopy (Stereoscan 360, Cambridge Instruments microscope, Somerville, MA, USA).

Coating adhesion was determined by coated samples sonication for 30 min in a petroleum ether bath, according to literature [34]. In Figure 10, a schematic representation of a typical procedure for structured catalyst production is reported.

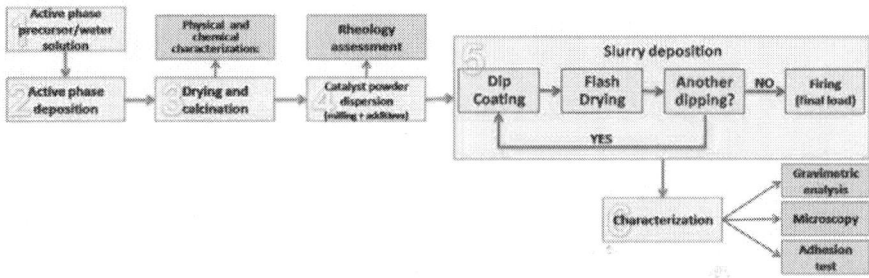

**Figure 10.** Scheme for catalyst production and washcoat deposition (slurry composition and details of the procedures as reported above).

## 4. CONCLUSIONS

1) The use of an acid-free water-based formulation proved to be effective for the dispersion of model catalytic powders characterized by low surface areas. The obtained slurries are suitable for deposition on ceramic open cell foams via dip-coating.

2) A dependence between the powder properties and the final rheology was evidenced. In particular, the rheological behavior, being directly related to both powder particles porosity and density, could be managed by taking into account these properties during formulation. Active phase presence influences particles properties, although present in a limited amount.

3) From a practical point of view, chemical composition (*i.e.*, molar weight) and porosity were found to be simple and detectable parameters to be used in slurry formulation to control the rheological behavior.

4) The use of multiple dippings proved to fulfill the flexibility requirements for washcoat load management. All the formulations obtained good results in terms of washcoat load and adhesion. Local and limited pore clogging occurred only in the washcoat at very high load degree (higher than 20 wt. %).

5) Depending on the powder to be deposited and on the operative condition, promising results have been obtained in terms of adhesion. Most samples displayed losses lower than 10 wt. %, which was reported in literature as satisfactory for washcoat adhesion on open-cell foams [38].

6) Although, up to now an "*a priori*" formulation of the slurry is hardly to be obtained without any experimental base, the results here reported can help to reduce experimental tests in a trial-and-error approach.

## ACKNOWLEDGMENTS

This work was funded by the Ministry of Education, University and Research, Italy (MIUR, Progetti di Ricerca Scientifica di Rilevante Interesse Nazionale 2010–2011) within the project IFOAMS ("Intensification of catalytic processes for clean energy, low-emission transport and sustainable chemistry using open cell FOAMS as novel advanced structured materials", protocol no. 2010XFT2BB).

## AUTHOR CONTRIBUTIONS

R.B. and C.C. conceived and designed the experiments. M.C. and F.P. performed the experiments; R.B., M.C., F.P. and C.C. analyzed the data; C.C. contributed reagents/materials/analysis tools; R.B. and C.C. wrote the paper.

## REFERENCES

1. Tronconi, E.; Groppi, G. Preface. *Catal. Today* **2009**, *147*, S1.

2. Montebelli, A.; Visconti, C.G.; Groppi, G.; Tronconi, E.; Cristiani, C.; Ferreira, C.; Kohler, S. Methods for the catalytic activation of metallic structured substrates. *Catal. Sci. Tech.* **2014**, *4*, 2846–2870.

3. Bagheri, S.; Muhd Julkapli, N.; Bee Abd Hamid, S. Titanium dioxide as a catalyst support in heterogeneous catalysis.*Sci. World J.* **2014**, *2014*, 727496.

4. Huirache-Acuna, R.; Nava, R.; Peza-Ledesma, C.L.; Lara-Romero, J.; Alonso-Nunez, G.; Pawelec, B.; Rivera-Munoz, E.M. SBA-15 mesoporous silica as catalytic support for hydrodesulfurization catalysts—Review. *Materials* **2013**, *6*, 4139–4167.

5. Trueba, M.; Trasatti, S.P. γ-Alumina as a support for catalysts: A review of fundamental aspects. *Eur. J. Inorg. Chem.***2005**, *17*, 3393–3403

6. Maciel, C.G.; Silva, T.D.; Hirooka, M.I.; Belgacem, M.N.; Assaf, J.M. Effect of nature of ceria support in CuO/CeO$_2$catalyst for PrO$_x$-CO reaction. *Fuel* **2012**, *97*, 245–252.

7. Maupin, I.; Mijoin, J.; Barbier, J.; Bion, N.; Belin, T.; Magnoux, P. Improved oxygen storage capacity on CeO$_2$/zeolite hybrid catalysts. Application to VOCs catalytic combustion. *Catal. Today* **2011**, *176*, 103–109.

8. Periyat, P.; Laffir, F.; Tofail, S.A.M.; Magner, E. A facile aqueous sol-gel method for high surface area nanocrystalline $CeO_2$. *RSC Adv.* **2011**, *1*, 1794–1798.

9. Pino, L.; Vita, A.; Laganà, M.; Recupero, V. Hydrogen from biogas: Catalytic *tri*-reforming process with Ni/La-Ce-O mixed oxides. *Appl. Catal. B* **2014**, *148–149*, 91–105.

10. Iglesia, E. Design, synthesis, and use of cobalt-based Fischer-Tropsch synthesis catalysts. *Appl. Catal. A* **1997**, *161*, 59–78.

11. Khodakov, A.Y.; Chu, W.; Fongarland, P. Advances in the development of novel cobalt Fischer-Tropsch catalysts for synthesis of long-chain hydrocarbons and clean fuels. *Chem. Rev.* **2007**, *107*, 1692–1744.

12. Wang, C.M.; Friedrich, S.; Younkin, T.R.; Li, R.T.; Grubbs, R.H.; Bansleben, D.A.; Day, M.W. Neutral nickel(II)-based catalysts for ethylene polymerization. *Organometallics* **1998**, *17*, 3149–3151.

13. Li, Y.D.; Li, D.X.; Wang, G.W. Methane decomposition to $CO_x$-free hydrogen and nano-carbon material on group 8–10 base metal catalysts: A review. *Catal. Today* **2011**, *162*, 1–48.

14. Matsumoto, S.I. Recent advances in automobile exhaust catalysts. *Catal. Today* **2004**, *90*, 183–190.

15. Koltsakis, G.C.; Stamatelos, A.M. Catalytic automotive exhaust aftertreatment. *Prog. Energy Combust. Sci.* **1997**, *23*, 1–39

16. Ghosh, I. Heat transfer correlation for high-porosity open-cell foam. *Int. J. Heat Mass Transf.* **2009**, *52*, 1488–1494.

17. Giani, L.; Groppi, G.; Tronconi, E. Mass-transfer characterization of metallic foams as supports for structured catalysts. *Ind. Eng. Chem. Res.* **2005**, *44*, 4993–5002.

18. Giani, L.; Groppi, G.; Tronconi, E. Heat transfer characterization of metallic foams. *Ind. Eng. Chem. Res.* **2005**, *44*, 9078–9085

19. Campanati, M.; Fornasari, G.; Vaccari, A. Fundamentals in the preparation of heterogeneous catalysts. *Catal. Today* **2003**, *77*, 299–314.

20. Meille, V. Review on methods to deposit catalysts on structured surfaces. *Appl. Catal. A* **2006**, *315*, 1–17.

21. Avila, P.; Montes, M.; Miro, E.E. Monolithic reactors for environmental applications—A review on preparation technologies. *Chem. Eng. J.* **2005**, *109*, 11–36.

22. Brinker, C.J.; Scherer, G.W. *Sol-Gel Science*; Academic Press: San Diego, CA, USA, 1990.

23. Middleman, S. *Fundamentals of Polymers Processing*; McGraw-Hill Companies: New York, NY, USA, 1977; p. 448.

24. Valentini, M.; Groppi, G.; Cristiani, C.; Levi, M.; Tronconi, E.; Forzatti, P. The deposition of $\gamma$-$Al_2O_3$ layers on ceramic and metallic supports for the preparation of structured catalysts. *Catal. Today* **2001**, *69*, 307–314.

25. Visconti, C.G.; Tronconi, E.; Lietti, L.; Groppi, G.; Forzatti, P.; Cristiani, C.; Zennaro, R.; Rossini, S. An experimental investigation of Fischer-Tropsch synthesis over washcoated metallic structured supports. *Appl. Catal. A* **2009**, *370*, 93–101.

26. Cristiani, C.; Visconti, C.G.; Finocchio, E.; Stampino, P.G.; Forzatti, P. Towards the rationalization of the washcoating process conditions. *Catal. Today* **2009**, *147*, S24–S29.

27. Phan, X.K.; Bakhtiary-Davijany, H.; Myrstad, R.; Pfeifer, P.; Venvik, H.J.; Holmen, A. Preparation and performance of Cu-based monoliths for methanol synthesis. *Appl. Catal. A* **2011**, *405*, 1–7.

28. Won, J.Y.; Jun, H.K.; Jeon, M.K.; Woo, S.I. Performance of microchannel reactor combined with combustor for methanol steam reforming. *Catal. Today* **2006**, *111*, 158–163.

29. Germani, G.; Stefanescu, A.; Schuurman, Y.; van Veen, A.C. Preparation and characterization of porous alumina-based catalyst coatings in microchannels. *Chem. Eng. Sci.* **2007**, *62*, 5084–5091.

30. Lin, K.S.; Pan, C.Y.; Chowdhury, S.; Lu, W.; Yeh, C.T. Synthesis and characterization of $CuO/ZnO-Al_2O_3$ catalyst washcoat thin films with $ZrO_2$ sols for steam reforming of methanol in a microreactor. *Thin Solid Films* **2011**, *519*, 4681–4686.

31. Barth, N.; Zimmermann, M.; Becker, A.E.; Graumann, T.; Garnweitner, G.; Kwade, A. Influence of $TiO_2$ nanoparticle synthesis on the properties of thin coatings. *Thin Solid Films* **2015**, *574*, 20–27.

32. Truyen, D.; Courty, M.; Alphonse, P.; Ansart, F. Catalytic coatings on stainless steel prepared by sol-gel route. *Thin Solid Films* **2006**, *495*, 257–261.

33. Agrafiotis, C.; Tsetsekou, A.; Leon, I. Effect of slurry rheological properties on the coating of ceramic honeycombs with yttria-stabilized-zirconia washcoats. *J. Am. Ceram. Soc.* **2000**, *83*, 1033–1038.

34. Cristiani, C.; Finocchio, E.; Latorrata, S.; Visconti, C.G.; Bianchi, E.; Tronconi, E.; Groppi, G.; Pollesel, P. Activation of metallic open-cell foams via washcoat deposition of $Ni/MgAl_2O_4$ catalysts for steam reforming reaction. *Catal. Today* **2012**, *197*, 256–264.

35. Olhero, S.M.; Ferreira, J.M.F. Influence of particle size distribution on rheology and particle packing of silica-based suspensions. *Powder Technol.* **2004**, *139*, 69–75.

36. Zupancic, A.; Lapasin, R.; Kristoffersson, A. Influence of particle concentration on rheological properties of aqueous α-$Al_2O_3$ suspensions. *J. Eur. Ceram. Soc.* **1998**, *18*, 467–477.

37. Agrafiotis, C.; Tsetsekou, A. The effect of processing parameters on the properties of γ-alumina washcoats deposited on ceramic honeycombs. *J Mater. Sci.* **2000**, *35*, 951–960.

38. Balzarotti, R.; Cristiani, C.; Latorrata, S.; Migliavacca, A. Washcoating of low surface area cerium oxide on complex geometry substrates. *Part. Sci. Technol.* **2015**. in press.

39. Sutton, D.; Kelleher, B.; Ross, J.R.H. Review of literature on catalysts for biomass gasification. *Fuel Process. Technol.* **2001**, *73*, 155–173

40. Cullity, B.D.; Stock, S.R. *Elements of X-ray Diffraction*, 3rd ed.; Prentice Hall: Upper Saddle River, NJ, USA, 2001.

# CHAPTER 9

# Thermal Barrier Ceramic Coatings — A Review

*Sumana Ghosh[1]*

[1] Bio-ceramics and Coating Division, CSIR-Central Glass and Ceramic Research Institute, Kolkata, India

## ABSTRACT

Thermal barrier coatings (TBCs) provide effective thermal barrier to the components of gas turbine engines by allowing higher operating temperatures and reduced cooling requirements. Plasma spraying, electron-beam physical vapor deposition, and solution precursor plasma spray techniques are generally used to apply the TBCs on the metallic substrates. The present article addresses the TBCs formed by different processing techniques, as well as the possibility of new ceramic, glass-ceramic, and composite materials as TBCs. Promising bond coat materials for a TBC system have been also stated.

**Keywords:** Processing techniques, new TBC materials, engine applications

## 1. INTRODUCTION

Thermal barrier coatings (TBCs) enable the engines to operate at higher temperatures without raising the base metal temperatures using cooling systems inside the hot section components and thus, enhance the operating efficiency of the engines [1]. Therefore, continued development of TBCs is essential to increase the inlet gas temperature further for improving the performance of gas turbines. Hence, TBCs with low thermal conductivity, phase stability, and high resistance to sintering have ever increasing demands [2]. Generally, TBCs consist of a ceramic (e.g., yttria partially stabilized zirconia) top coat and a NiCoCrAlY/PtAl-based metallic bond coat. A bond coat is deposited between the metallic substrate and the top coat to protect the metal substrate from oxidation and high temperature corrosion and assist the coupling of the ceramic top coat and the metallic substrate [3]. Two methods are generally used to deposit the ceramic top coat. These are the electron beam physical vapor deposition (EB-PVD) and the atmospheric plasma sprayed (APS) methods. TBCs with EB-PVD top coats generally provide longer thermal cycle lifetimes because of its more strain-tolerant columnar structure than those observed with APS TBCs. However, APS TBCs are widely applied due to lower thermal conductivity and lower processing costs [3]. Recently, various processing techniques have been developed to deposit the ceramic coatings.

The objective of this article is to present an overview of the TBC requirement, application of TBCs, degradation mechanisms for TBCs, different processing techniques used for preparation of TBCs, and their thermal properties. Recent developments in TBC material have been also described. The prospect of innovative materials as bond coat in a TBC system has been elucidated.

## 2. MAIN REQUIREMENTS FOR TBCS

TBCs must have low weight and low thermal conductivity and they should withstand large stress variations due to heating and cooling, as well as thermal shock. They must be chemically compatible with the underlying metal and the thermally grown oxide (TGO) and should operate in an oxidizing environment. TBCs must provide thermal insulation to the underlying superalloy engine parts. They must have strain compliance in order to minimize the thermal expansion mismatch stresses with the superalloy parts. Additionally, they must reflect much of the radiant heat from the hot gas and thereby, preventing it from reaching the superalloy substrate. Further, TBCs must provide thermal protection to the substrate for prolonged service times and thermal cycles without failure [4].

## 3. APPLICATIONS OF TBCS

TBCs provide thermal insulation to superalloy engine parts such as the combustor, rotating blades, stationary guide vanes, blade outer air-seals, shrouds in the high-pressure section behind the combustor, and afterburners in the tail section of jet engines. Significant gas-temperature increase can be achieved by using TBCs in association with innovative air-cooling approaches than that obtained by earlier materials including single-crystal Ni-based superalloys [4].

## 4. DEGRADATION MECHANISMS

During service, several kinetic processes occur in parallel. Interdiffusion occurs between the bond coat and the superalloy. Consequently, Al diffuses from the bond coat to form the TGO. Microstructural, chemical, and phase changes occur in all the materials including the ceramic top coat. The rates of these thermally activated processes are expected to increase exponentially with temperature. These processes generally lead to degradation and failure of the coating [5]. During service, failure of the TBC system occurs depending upon the following three factors:

### 4.1. Bond Coat Degradation

Bond coat plays an important role in promoting the durability of the TBC system. But the role of the bond coat is very complex and poorly understood. In most practical cases, oxidation of the bond coat becomes the predominant coating failure mechanism. During high temperature exposure, NiCrAlY bond coat is oxidized resulting in a TGO layer on the bond coat. After reaching a critical thickness, the TGO becomes prone to spallation, which in turn results to the failure of the TBC system. It is very difficult to establish the exact mechanisms of bond coat-induced TBC failure for various coating types. However, all researchers agree on the significance of spallation or cracking of the TGO for the failure of air plasma-sprayed TBC system. Detailed research is being conducted to find out a solution to the problem regarding bond coat degradation in the TBC system [5]. Nanostructured NiCrAlY bond coat may improve the life expectancy of thermal barrier coatings. Daroonparvar et al. [6] investigated the microstructural evolution of TGO layer on the conventional and nanostructured atmospheric plasma sprayed (APS) NiCrAlY coatings in TBC systems during oxidation. It was observed that the growth of $Ni(Cr,Al)_2O_4$ (as spinel) and NiO on the surface of the $Al_2O_3$ layer (as pure TGO) in nano TBC systems was much lower compared to that of normal TBC systems during thermal exposure at 1,150°C. These two oxides play a detrimental role in causing crack nucleation and growth, reducing the life of the TBC in air. This microstructure optimization of the TGO layer is primarily associated with the formation of a continuous, dense, uniform $Al_2O_3$ layer over the nanostructured NiCrAlY coating [6]. The interfacial failure mechanism of the TBC system was numerically investigated by Xu et al. [7], considering the role of mixed oxides (MO), which was induced by the discontinuous $\alpha$-$Al_2O_3$ at the top coat-bond coat interface. High growth rate of MO will stimulate the initiation and propagation of interface cracks resulting in debonding of the top coat. The high coverage ratio of MO at the interface will accelerate the propagation of an interface crack. Therefore, the durability and performance of TBCs can be improved by suppressing the formation of MO [7]. The prediction for spallation of thermal barrier coatings has proven to be an intricate problem [8]. The spallation usually occurs through buckling that is driven by strain energy release within the ceramic top coat. If the delamination interface is at the bond-coat/TGO interface, then spallation occurs within the underlying TGO layer. Prior to spallation, substantial sub-critical damage must develop at one or both of the TGO interfaces. Evans [8] stated that the strain energy within the TGO produced during cooling contributes significantly to this damage development and not that within the top coat. Critical strain energy within the TGO layer is assumed to be a possible pragmatic method of predicting the spallation. Several factors such as phase changes in the bond coat, mechanical constraint imposed by the top coat on the mechanical stability of the bond coat interface, TGO growth on a non-planar interface on stress development, and localized Al depletion in nucleating fast-growing non-protective TGO influence the TBC failure [8].

## 4.2. Generation of High Residual Stress

Residual stress has a vital effect on the performance of a TBC system. The role of residual stress is very complex and varies with the difference in system configurations. Thermal expansion mismatch between the three layers generates residual stress resulting in degradation of a TBC system. Although extensive research has been initiated to study the effect of residual stress on TBC life, there is still ample scope to carry out this study on novel TBC systems involving novel compositions [9]. High residual stresses are induced in the TBC due to thermal expansion mismatch and bond coat (BC) oxidation leading to failure by spalling and delamination. An analysis of the stress distributions in TBC systems, which is a prerequisite for the understanding of failure mechanisms, was performed by Sfar et al. [10] using the finite element method (FEM). Cracks in the interface region were considered in the FE models in order to determine the loading conditions for their propagation and thus, the failure criteria of the TBCs as cracking usually occurs at or near the interfaces between BC/TGO and TBC/TGO depending on the processing mode of the TBC. The modified crack closure integral (MCCI) method combined with an FE analysis led to highly accurate energy release rate values. Moreover, this method enables the determination of mode-dependent energy release rates. TBC failure models could be developed and verified using this tool and appropriate crack propagation criteria [10]. Yang et al. [11] investigated the residual stress evolution in air plasma-sprayed yttria-stabilized zirconia (YSZ) TBCs after thermal treatments at 1,150°C. The residual stress in the YSZ layer was measured using Raman spectroscopy and the curvature method. Generally, as-deposited YSZ layer was under compressive stress and subsequently after thermal treatment for 30 h it was under tensile stress partly due to the monoclinic to tetragonal phase transformation in the YSZ layer. Sintering of the YSZ layer occurred with prolonged thermal treatment resulting in the gradual transformation of the residual stress, from tensile to compressive stress. Further, β-NiAl to γ/γ′-Ni₃Al phase transformation in the bond coat also plays an important role on the stress development in the top coat [11].

## 4.3. Top Coat Degradation

Top coat degradation is another parameter that governs TBC failure. The ceramic top coat has a tendency to crack due to stress generated from thermal expansion mismatch between the three layers of the TBC system. When the top coat cracks, oxygen easily diffuses to the bond coat leading to the catastrophic failure of the TBC system. Significant research is being carried out to improve the microstructure, mechanical properties, and stability of the ceramic top coat [12]. TBCs are subject to many kinds of degradation, e.g., erosion, foreign object damage (FOD), oxidation, etc., which deteriorate the integrity and mechanical properties of the whole system. Moreover, a new type of damage has been highlighted, i.e., corrosion by molten Calcium-Magnesium-Alumino Silicates, known as CMAS with the aim to increase the turbine inlet temperature. Basu et al. studied interactions between YSZ materials synthesized via the sol-gel process and synthetic CMAS powder via a step-by-step methodology. However, CMAS can cause faster sintering of the ceramic and thereby, leading to loss of strain tolerance in the protective coating. Further, a dissolution/re-precipitation mechanism between YSZ and CMAS resulted in the transformation of the initial tetragonal YSZ into globular particles of monoclinic zirconia. In addition, CMAS infiltrated both EB-PVD and sol-gel YSZ coatings at 1,250°C for 1 h [12]. Thompson and Clyne [13] deposited a vacuum plasma spray (VPS) MCrAlY bond coat and atmospheric plasma spray (APS) zirconia top coat onto a nickel superalloy substrate. They measured the stiffness of detached top coats by cantilever bending and also by nanoindentation technique. Measurements were made on as-sprayed specimens and after various heat treatments. Significant changes were detected in the Young's modulus of the heat treated top coat. The rate of sintering was found to be a function of temperature and weather. The coating was detached with the substrate during heat treatment. During high temperature exposure the effects of stiffening of the top coat on the stress development within the TBC system was included by using a well-known, modified numerical model. Sintering of the top coat enhanced debonding at the top coat-bond coat interface resulting in top coat spallation under service conditions [13]. It has been found by Abubakar et al. [14] that the use of low grade fuels in land-based turbines in Saudi Arabia results in hot corrosion due to the diffusion of a molten salt ($V_2O_5$) into the top coat of the TBCs. Consequently, volumetric expansion of the coating occurs due to the tetragonal-to-monoclinic transformation of zirconia in the planar reaction zone near the surface of the coating. They used a phase field model for estimating the kinetics of microstructure evolution during the corrosion process at 900 °C and close agreement between numerical and experimental results was achieved. The transformation-induced stresses were predicted by coupling the phase transformation with elasticity. The result showed that the coating spallation occurred due to very high compressive stress development within the coating cross section [14].

# 5. PROCESSING TECHNIQUES FOR TBCS

## 5.1. EB-PVD Process

In the EB-PVD process, the source material is heated with an electron beam, vapors are produced, and the evaporated atoms condense on the substrate. Crystal nuclei form on favored sites and grow laterally and in thickness to form individual columns that provide in-plane compliance [15]. A TGO layer often forms on the bond coat in these TBC systems and increases the residual stress. Further, brittleness of the top coat increases with the sintering of the coating. Consequently, the adhesion of the bond coat to the top coat becomes weak at high temperatures. Therefore, the TGO layer is very detrimental for TBC performance [16]. Movchan and Yakovchuk [17] described the design of a new generation of electron beam units for the deposition of the TBCs and cost-effectiveness of the one-step deposition process. They produced variants of graded TBC, which consist of bond coats of NiAl or MCrAlY+NiAl and YSZ-based outer ceramic layer in a one-step cycle by evaporation of a composite ingot. The composition and structure of the bond coats, outer ceramic layer, and the transition barrier zones of the substrate/bond coat and bond coat/outer ceramic layers was controlled in a broad range. They have shown distributions of chemical elements in the coating/substrate system and microstructure after deposition and after heat treatment. Various types of graded TBCs were subjected to thermal cycling tests at 1,150°C and their thermal cyclic resistance was monitored [17]. Current numerical approaches in modeling the intrinsic failure of TBC relies largely on the fact that spallation occurs when the accumulating strain energy stored in the coating exceeds a fixed critical value resembling interfacial adhesion. If this is to be entirely correct, one would expect that this critical value of interfacial adhesion varies with different materials, but stays independent of their thermal exposure history. Wu et al. [18] characterized the adhesion of oxide-bond coat interface among five systematically prepared material systems using a unique cross-sectional indentation technique. The results re-confirmed that interfacial adhesion is a material-specific property and the adhesion is dynamic, particularly with time and temperature. Certain parameters such as the oxide growth rate, rumpling of the oxide-bond coat interface, and phase transformation of bond coat were studied as a function of thermal exposure to understand the dynamics. They clearly indicated that the oxide-bond coat interfacial adhesion depends strongly on the phase distribution of the bond coats and TGO growth rate while having little effect from TGO rumpling and residual stress [18].

## 5.2. APS Process

In the APS process, ceramic powders are introduced into a high temperature plasma plume, melted inside the plume, and accelerated towards the substrate wherein molten droplets spread and form splats that are rapidly quenched. In one pass, several successive splats are deposited on the substrate and the coating thickness is increased by means of several passes [19]. A typical fractured cross-section of the plasma sprayed ceramic coating show layers of splats along with interlamellar pores, cracks, and globular pores [15]. Coating compliance is increased by the presence of the cracks and thereby, extending their lifetimes [19]. The resulting coating microstructure is strongly dependent on processing conditions such as spray parameters (e.g., torch current, plasma gas flow rate, carrier gas flow rate, torch traverse velocity, and stand off distance) and feedstock materials (e.g., size, temperature, and velocity). Splat morphologies are changed with the angle of impact of impinging particle [15]. Higher substrate temperatures lead to lower porosity and improved inter-splat contact resulting in enhanced coating properties [20]. During service operations at high temperatures, a TGO layer, mainly an $Al_2O_3$ layer, is developed between the bond coat and the top coat due to the oxidation of the bond coat. This is the most important factor that determines the lifetime of the TBC system. The thickness of this layer increases with increasing operation time. High stresses are present at the bond coat and TGO interface because of oxide layer growth, thermal expansion misfit, and applied loads. As a result, crack initiates and propagates resulting in spallation of the ceramic layer, and finally, system degradation [3]. During thermal exposure at $\geq 1,000$°C, $Ni(Cr, Al)_2O_4$ (spinels) and NiO clusters are also formed at the interface of the $Al_2O_3$ layer and the ceramic coating in the TBC system with MCrAlY (M=Ni, Co) bond coat. Cracks were nucleated on these oxide clusters and grew into the ceramic coating leading to premature TBC separation. A heat treatment in a low pressure oxygen environment was found to promote the formation of a uniform, thin protective layer of $Al_2O_3$ at the ceramic-bond coat interface and can reduce these detrimental oxides [21].

Thermo-mechanical properties of TBCs have been studied to improve TBC performance. The Young's modulus of the ceramic top coat is an important factor that affects the thermal stress distribution in TBCs and thus, thermal fatigue behavior. Apparent Young's modulus ($E_{ap}$) indicates the macro-elastic properties of the coatings. $E_{ap}$ of the top coat is usually much lower than the value for dense YSZ due to the porous microstructure. The extremely low $E_{ap}$ values are also attributed to the weak bonding between the particles because of the extremely high cooling rate. Tang and Schoenung [22] conducted bending tests of the TBC specimens exposed to thermal cycling to determine their $E_{ap}$. The $E_{ap}$ decreased with increasing thermal cycles, up to certain thermal cycles, and then remained unchanged for increased thermal cycles. The breaking of the bonds at the splat boundaries or the formation of new cracks caused by thermal strain is the reason for the decrease in $E_{ap}$ with increasing thermal cycles. Effect of heat treatment on the elastic properties of the separated porous plasma sprayed zirconia TBCs was investigated by D. Basu et al. [23]. The depth-sensitive indentation technique was employed to determine the elastic moduli of the coatings. The characteristic moduli were dependent on the indentation load. The increase of moduli with decreased indentation load was attributed to the presence of small pores and micro-cracks at the subsurface. Heat treatment of the coatings at $1,100°C$ increased the elastic moduli appreciably due to the formation of sintering necks and the elimination of the micro-pores within the lamellae.

Functionally graded $Al_2O_3$–$ZrO_2$ TBC was prepared by plasma spraying technique and reported elsewhere [24]. Functionally-graded TBC was found to reduce the oxidation rate of the TBC system. Thus, large residual stress associated with the formation of TGO was minimized. The $Al_2O_3$ interlayer should be very thin to increase the adhesion of the layers. However, low fracture toughness of $Al_2O_3$ might lead to TBC failure. In addition, phase transformation of $γ$-$Al_2O_3$ to $α$-$Al_2O_3$ could induce additional residual stress, which should be minimized to get reliable TBC systems. Thick thermal barrier coatings (thickness >1 mm) have been developed for increased thermal protection by using the APS method [25]. However, low thermal shock resistance is the problem with the thick coating. Certain degrees of porosity and micro-cracks, preferably segmentation cracks, in TBCs favor to achieve high thermal shock resistance. Chen et al. [26] prepared a new functionally-graded thermal barrier coating based on $LaMgAl_{11}O_{19}$ (LaMA)/YSZ by using air plasma spraying technique. The coefficient of thermal expansion (CTE) of the functionally-graded coating varied gradually from the YSZ bottom layer to the LaMA top layer, resulting in the decrease in residual stress level than that of the LaMA/YSZ double ceramic layered TBC system. Excellent thermal cycling lifetime ($\sim$11,749 cycles at$\sim$ $1,372°C$) of the functionally graded TBC proved the potential of these TBCs for advanced applications [26].

## 5.3. Plasma-Enhanced Chemical Vapor Deposition (PECVD) Method

Thick, partially yttria-stabilized zirconia coatings have been deposited by plasma-enhanced chemical vapor deposition (PECVD) method. The morphology and phase composition of the coatings was studied after annealing treatments at the temperature range of 1,100 to 1,400°C up to 1,000 h. The as-deposited columnar morphology of the coating was similar to that observed in the coating prepared by the EBPVD technique. The PECVD method is suitable for developing TBCs as it provides thermally stable coating at elevated temperatures [27].

## 5.4. Electrostatic Spray-Assisted Vapor Deposition Method (Esavd)

TBCs, such as 8 wt. % $Y_2O_3$–$ZrO_2$ (YSZ), provide effective thermal barrier to the gas turbine blades and are able to protect them, leading to further increase in the operating temperature. A novel and cost-effective electrostatic spray-assisted vapor deposition (ESAVD) technique was utilized to prepare YSZ coatings, which involves spraying atomized zirconium and yttrium alkoxide precursor droplets within an electric field wherein they are subjected to decomposition and/or chemical reactions in the vapor phase near the heated substrate. The coatings were characterized by scanning electron microscopy, X-ray diffraction, and Raman spectroscopy. Vyas and Choy [28] produced thick and uniform YSZ films using the ESAVD method. Raman spectroscopy identified carbon to be present in the as-deposited coatings. When heat treatment of the YSZ coating was conducted at 1,000°C for 2 h, carbon was removed and the adhesion of the TBC coating to the bond coat improved [28].

## 5.5. Solution-Precursor Plasma Spray (SPPS) Process

In this process, an aqueous chemical precursor feedstock is injected into the plasma jet where the droplets undergo a series of physical and chemical reactions and then deposited on the substrate as coating. Microstructural observations of this type of TBC show fine splats and vertical cracks in a porous matrix. TBCs deposited by the optimized solution-precursor plasma spray (SPPS) process exhibit superior durability relative to TBCs formed by the APS and EB-PVD processes. Thick and durable TBCs can be deposited by this process. Failure of these TBCs occurs by large scale buckling of the ceramic top coat [29]. The efficiency of TBCs used to protect and insulate metal components in engines increases with the thickness of the TBCs. However, the durability of thick TBCs deposited using conventional deposition methods has not been adequate. Jadhav et al. [30] deposited highly durable, 4 mm-thick $ZrO_2$–7 wt% $Y_2O_3$ (7YSZ) TBCs on bond-coated superalloy substrates using the SPPS method. The average thermal cycling life of the SPPS TBCs was 820 cycles, while most of the conventional air plasma-sprayed coatings of the same composition and thickness deposited on similar bond-coated superalloy substrates were observed to be detached partially from the substrates in the as-sprayed condition. Only the APS TBC failed after 40 thermal cycles.

Significantly higher in-plane indentation fracture toughness and high degree of strain tolerance due to the presence of the vertical cracks in the SPPS TBCs led to the dramatic improvement in the thermal cycling life of the SPPS TBCs over APS TBCs [30].

## 5.6. Sol-Gel Process

Recently, a new, attractive sol-gel route has been successfully developed to synthesize and deposit the TBCs [31–34]. Non-directional deposition and formation of thin or thick coating by dip or spray technique or the combined method of both techniques can be performed by this technique. Sol-gel TBCs show an isotropic microstructure having randomly distributed porosities leading to a good compromise between thermal conductivity and mechanical strength. The degradation of sol-gel TBCs is initiated by the formation of a regular crack network either during the post-deposition thermal treatment required to sinter the deposit or during the first cycles of oxidation. In both cases, this regular surface crack network forms on account of the in-plane stress release due to the sinter-induced shrinkage of the zirconia scale. Subsequently, enlargement and coalescence of the cracks occur under cumulative oxidation cycles promoting the detachment of individual TBC layers and finally, the complete spallation of the TBC. To improve the cyclic oxidation resistance of the TBCs, the sintering efficiency after the TBC deposition needs to be improved or the crack network needs to be stabilized by filling crack grooves by supplementary dip or spray coating passes [33]. In addition, the feasibility of consolidating sol-gel TBCs by additional fillings of zirconia into the sinter-induced cracks was investigated by adjusting different process parameters such as the choice of either dip-coating or spray-coating and the modification of the slurry viscosity [34]. Basically, the optimization of both the sintering heat treatment and the procedure for filling the initial crack network promotes a significant improvement of the sol-gel TBC durability during cyclic oxidation at 1,100°C. Typically, a sol-gel TBC that is properly sintered and adequately reinforced can be cycled for 1 h at 1,100°C one thousand and five hundred times without spalling, which is nearly equivalent to the performance of EB-PVD TBCs [33, 34].

## 5.7. Composite Sol-Gel Method

Composite sol-gel method and pressure filtration microwave sintering (PFMS) technologies were utilized to form novel YSZ ($ZrO_2$–6 wt% $Y_2O_3$)–($Al_2O_3$/YAG) (alumina–yttrium aluminum garnet, $Y_3Al_5O_{12}$) double-layer ceramic coatings. The thin $Al_2O_3$/YAG layer showed good adhesion with the substrate. Cyclic oxidation tests were carried out at 1,000°C, which indicated that double-layer ceramic coatings can prevent the oxidation of alloy and improve the spallation resistance. The 250 μm coating had better thermal barrier effect than that of the 150 μm coating during thermal stability tests at 1,000°C and 1,100°C at different cooling gas rates. These beneficial effects are mainly attributed to the decrease of the rate of TGO scale development and the reduced thermal stresses by means of nano/micro-composite structure. This double-layer coating can be considered as a promising TBC [35].

## 5.8. Spark Plasma Sintering (SPS) Method

Pt-modified Ni aluminides and MCrAlY coatings (where M=Ni and/or Co) are widely used on turbine blades and vanes for protection against oxidation and corrosion and as bond coat in TBC systems. The SPS method can be used by Monceau et al. [36] to develop rapidly new coating compositions and microstructures. This technique allows the formation of multi-layered coatings on a superalloy substrate. They have shown the possibility of fabricating MCrAlY overlays with local Pt and/or Al enrichment and coatings made of $\zeta$-$PtAl_2$, $\varepsilon$-PtAl, $\alpha$-AlNiPt$_2$, martensitic $\beta$-(Ni,Pt)Al, or Pt-rich $\gamma/\gamma'$ phases. Further, they have demonstrated the prospect of achievement of a complete TBC system with a porous and adherent YSZ layer on a $\gamma/\gamma'$ low mass bond coating. Additionally, they have discussed the difficulties of fabrication such as Y segregation, risks of carburization, local overheating, or difficulty to coat complex shape parts [36]. Recently, Boidot et al. [37] prepared complete TBC systems on single crystal Ni-based superalloy substrate in a one-step SPS process. A proto-TGO layer in situ was formed during the fabrication of the TBC systems. Formation of a dense, continuous, slow-growing alumina layer (TGO) between a ceramic top coat and an underlying bond coat during service influences the lifetime of the TBC systems. During thermal treatment at 1,100°C in air, the amorphous oxide layer transforms to $\alpha$-$Al_2O_3$ in the as-deposited samples. Oxidation kinetics during annealing was in good agreement with the protective $\alpha$-$Al_2O_3$ layer formation [37]. In the last decade, an increasing interest was given to Pt-rich $\gamma-\gamma'$ alloys and coatings as they have shown good oxidation and corrosion properties. SPS has been proved to be a fast and efficient tool to fabricate coatings on superalloys including entire TBC systems. Selezneff et al. [38] used the SPS technique to fabricate doped Pt-rich $\gamma-\gamma'$ bond coatings on the superalloy substrate, whereas the doping elements were reactive elements (e.g., Hf, Y or Zr, Si) and metallic additions of Ag. These samples were then coated with Y-PSZ TBC through the EBPVD method. The performance of such TBC system was compared to a conventional TBC system consisting of a $\beta$-(Ni,Pt)Al-based bond coat. Thermal cycling tests were performed in air and spallation was observed during this test. It was noted that most of the Pt-rich $\gamma-\gamma'$ samples showed better adherence of the ceramic coating than that of the $\beta$-samples. Cross-sectional scanning electron microscopy was used to characterize the thickness and the composition of the oxide scales after cyclic oxidation test. It was proved that the doping elements have significant influence on the oxide scale formation, metal/oxide roughness, Al and Pt content under the oxide scale, and TBC adhesion. It was established that RE-doping can not improve the oxidation kinetics of Pt-rich $\gamma-\gamma'$ bond coat. Moreover, $\gamma-\gamma'$-based systems were superior to $\beta$-(Ni,Pt)Al bond coat with respect to ceramic top coat adherence and better oxide scale adherence [38].

## 5.9. Low-Pressure Plasma Spraying Process

The TBC must exhibit high thickness (100–300 μm), vertical cracks should be present in the TBC in order to be a strain tolerant layer, and it must have high porosity to decrease the thermal conductivity. Rousseau et al. [39] prepared a Y-PSZ layer using low-pressure plasma spraying technique by introducing a solution of nitrate salt into a low-pressure plasma discharge. The characteristics and stability of the Y-PSZ layers were analyzed by several techniques. Optical emission spectroscopy indicated that the oxidant chemistry of the plasma caused oxide formation and the nitrate elimination at low temperature (T<300°C). Effects of the several parameters such as power of the plasma discharge, post-treatment and heat treatment on structure, morphology, and stability of the Y-PSZ coatings was studied by X-ray diffraction (XRD), scanning electron microscopy (SEM), water porosimetry, and thermal diffusivity measurement. It was observed that Y-PSZ coating (porosity-50%) had good thermal barrier property at high temperatures [39].

## 5.10. Thermal Plasma Process

Superior properties such as high-melting point, high phase stability, low sintering ability, low thermal conductivity, and low oxygen permeability of lanthanum zirconate (LZ) have made it one of the most promising TBC materials for high-temperature applications. However, the production methods used to synthesize lanthanum zirconate are highly time-consuming and the powder is not commercially available. Hence, the thermal plasma process was utilized to synthesize, spheroidize, and spray deposits of lanthanum zirconate material by Ramachandran et al. [40]. They demonstrated the effectiveness of thermal plasma as a major materials processing technique. Suitable characterization techniques were used to study the material modifications after respective plasma processing exposures [40].

## 5.11. Cathodic Plasma Electrolytic Deposition (CPED) Method

Inconel alloys (IN738) have a wide range of applications in industries as high temperature structural materials. Further, different surface treatments and coatings have been developed for the improvement of the properties of Inconel alloys. Bahadori et al. [41] deposited $Al_2O_3$ ceramic coating on MCrAlY bond-coated Ni-based superalloy using the CPED method in an ethanol solution of Al $(NO_3)_3.9H_2O$ (18 g/l). Several samples were prepared under different deposition conditions and characterized by XRD, SEM, and energy dispersive X-ray spectrometer (EDS). The XRD analysis confirmed the presence of $Al_2O_3$ and $Ni_3Al$ phases. The results were in good agreement with the composition of the MCrAlY bond coat based on the thermal expansion data. SEM micrograph showed changes in the microstructure of the specimen by varying the pH of the solution [41].

## 5.12. Detonation Gun Spray Technique

Kim et al. [42] had taken a new approach and fabricated an excellent functionally-graded thermal barrier coating (FGM TBC) by using the detonation gun spray process in association with a newly-proposed shot-control method. FGM TBCs were sprayed in the form of multi-layered coatings having a compositional gradient across the thickness. FGM TBCs consisted of a finely mixed microstructure of metals and ceramics with no interfaces between the layers. The gradient ranged from 100% NiCrAlY metal on the substrate to a 100% $ZrO_2$–8 wt% $Y_2O_3$ ceramic for the topcoat. In the FGM layer of the FGM TBCs, the ceramics and metals maintained their individual properties without any phase transformation during the spraying process. They investigated the thermal shock properties of FGM TBCs and compared the data obtained with those for traditional duplex TBCs [42].

## 5.13. Plasma Laser Hybrid Spraying Technique

Post-treatments of sprayed coatings and simultaneous spraying processes by a plasma laser hybrid technique have been tried by Chwa and Akira [43] to improve the lifetime of TBC coatings. An analytic technique using a low-viscosity resin with a fluorescent dye under a high vacuum has been investigated for the accurate observation of the microstructure of TBCs prepared by a post-laser treatment and a laser hybrid spraying process. Coatings formed by post-laser treatments and laser hybrid spraying processes showed significantly improved thermal shock resistance compared to as-sprayed coatings as a consequence of water quenching tests. The relationship of the microstructure of TBCs modified by laser treatment and thermal shock resistance has been evaluated by the careful observation of samples. They suggested the optimum process conditions for improving the thermal shock resistance of TBCs [43].

## 5.14. Electrophoretic Deposition Method

Wang et al. [43] synthesized $Gd_2O_3$ doped 4-YSZ (G-YSZ) ceramic coatings by electrophoretic deposition method followed by vacuum sintering and isothermally annealing at 1,000°C for different times. XRD was used to investigate their phase composition. SEM was employed to examine their microstructure, while EDS was used to assess the composition of the composite coatings. The results showed that YSZ coating was composed of tetragonal and monoclinic phases after vacuum sintering at 1,000°C for 2 h under vacuum ($<10^{-3}$ Pa). G-YSZ composite coatings were composed of tetragonal and monoclinic phases and a small amount of $Gd_2Zr_2O_7$ phase after vacuum sintering at 1,000°C for 2 h while the content of the monoclinic phase in G-YSZ composite coatings increased with the increase of $Gd_2O_3$ concentration. It was found that after isothermal annealing at 1,000°C in air for 100 h, G-YSZ composite coatings were composed of tetragonal $ZrO_2$ phase, monoclinic $ZrO_2$ phase, and cubic phase whereas the $Gd_2Zr_2O_7$ phase disappeared [44].

# 6. RELATIVELY NEW DEVELOPMENTS OF TBC MATERIALS

## 6.1. Ceramic Top Coat

Vassen et al. [45] investigated three zirconate materials as potential TBC materials. They deposited 150 μm Ni-Co-Cr-Al-Y bond coat on IN738 substrate before deposition of zirconate (thickness-240 μm) as top coat. They indicated that $SrZrO_3$ can not be used as a top coat in TBC systems as the coating showed a phase transition with a volume expansion at ~730°C that led to the failure of the samples. $BaZrO_3$ showed relatively poor thermal and chemical stability resulting in early failure in thermal cycling tests. On the other hand, Young's modulus of the pyrochlore $La_2Zr_2O_7$ was found to be lower than that of YSZ. Fracture toughness of this material was comparable to the toughness of plasma-sprayed YSZ coatings. Furthermore, $La_2Zr_2O_7$ has favorable thermal conductivity at elevated temperatures, which is ~20% lower than that of YSZ. Failure of $La_2Zr_2O_7$ coating was not observed after the first thermal cycling tests at temperatures >1,200°C and the coating showed thermal stability. Thus, $La_2Zr_2O_7$ is a very promising material for advanced TBCs. Moskal et al. [46] studied a double-ceramic-layered (DCL) coating consisting of monolayer coatings $Nd_2Zr_2O_7$ and 8YSZ. The coatings had ~300 μm thickness and porosities of ~5%. The chemical and phase composition analysis of the DCL layers revealed an external $Nd_2Zr_2O_7$ ceramic layer (~80 μm thick), a transitional zone (~120 μm thick), and an internal 8YSZ layer (100 μm thick). The $Nd_2Zr_2O_7$ pyrochlore phase was the only one-phase component. The surface topography of both TBC systems was typical for plasma sprayed coatings, and compressive stress state had a value in the range of ~5–10 MPa. Measurements of the thermal parameters, i.e., thermal diffusivity indicated better thermal insulation for both new types of layers as compared to the standard 8YSZ layers [46].

$Yb_2O_3$ (10 mol%) and $Gd_2O_3$ (20 mol%) doped $SrZrO_3$ was investigated by Ma et al. [47] as a material for TBC applications. Measurement of thermal expansion coefficients (TECs) of sintered bulk $Sr(Zr_{0.9}Yb_{0.1})O_{2.95}$ and $Sr(Zr_{0.8}Gd_{0.2})O_{2.9}$ displayed a positive influence on phase transformations of $SrZrO_3$ by doping $Yb_2O_3$ or $Gd_2O_3$. It was observed that both dopants can reduce the thermal conductivity of $SrZrO_3$. Dense $Sr(Zr_{0.9}Yb_{0.1})O_{2.95}$ and $Sr(Zr_{0.8}Gd_{0.2})O_{2.9}$ had lower hardness, Young's modulus, and comparable fracture toughness as compared to YSZ. At operating temperatures <1,300°C, the cycling lifetimes of plasma sprayed $Sr(Zr_{0.9}Yb_{0.1})O_{2.95}$/YSZ and $Sr(Zr_{0.8}Gd_{0.2})O_{2.9}$/YSZ double DLC were comparable to that of YSZ coating. However, at operating temperatures >1,300°C, the cycling lifetime of $Sr(Zr_{0.9}Yb_{0.1})O_{2.95}$/YSZ DLC was about 25% longer than YSZ coating, while that was shorter for $Sr(Zr_{0.8}Gd_{0.2})O_{2.9}$/YSZ DLC compared to YSZ coating [47]. The rare earth zirconates ($M_2Zr_2O_7$, $M = La \rightarrow\rightarrow Gd$) have a low intrinsic thermal conductivity and high temperature phase stability, which make them attractive candidates for TBC applications. Electron-beam evaporation, directed-vapor deposition (EB-DVD) technique was used by Zhao et al. [48] to investigate the synthesis of $Sm_2Zr_2O_7$ (SZO) coatings and to explore the relationships between the deposition conditions and the coating composition, pore morphology, structure, texture, and thermal conductivity. The coatings exhibited significant fluctuations in composition because of the vapor pressure differences of the constituent oxides. It was noticed that the coatings had a metastable fluorite structure due to kinetic limitations that hindered the formation of the equilibrium pyrochlore structure. The morphology of growth of EB-DVD SZO was identical to those of EB-DVD 7YSZ and EB-PVD $Gd_2Zr_2O_7$. The conductivity values of the as-deposited SZO coatings were nearly one-half of their DVD 7YSZ counterparts. This may be ascribed to their lower intrinsic conductivity [48].

Alumina-based ceramic coating with a composition of $La_2O_3$, $Al_2O_3$ and MgO ($MMeAl_{11}O_{19}$, M-La, Nd; Me-alkaline earth elements, magnetoplumbite structure) has been developed as TBC by the researchers [49, 50]. Lanthanum hexaaluminate (LHA) coating has long-term structural and thermo-chemical stability of up to 1673 K and significantly lower sintering rate than zirconia-based TBCs. The low thermal conductivity of LHA is ascribed to the random arrangement of LHA platelets leading to micro-porous coating. The insulating properties of the material are related to its crystallographic feature. To meet the demand of advanced turbine engines, $LaTi_2Al_9O_{19}$ (LTA) was proposed and investigated as a novel TBC material for application at 1,300°C by Xie et al. [51]. LTA showed excellent phase stability up to 1,600°C. The thermal conductivities for LTA coating were in a range of 1.0–1.3 W $m^{-1}$ $K^{-1}$ (300–1,500°C). The values of thermal expansion coefficients increased from 8.0 to $11.2 \times 10^{-6}$ $K^{-1}$ (200–1,400°C), which were comparable to those of YSZ. Both the LTA and YSZ coatings had a microhardness value of about 7 GPa, whereas the fracture toughness value was relatively lower than that of YSZ. However, the double-ceramic LTA/YSZ layer design balanced the lower fracture toughness. The LTA/YSZ TBC showed thermal cycling life of ~700 h at 1,300°C [51]. Lathanum phosphate ($LaPO_4$) is considered as a potential TBC material on Ni-based superalloys because of its high temperature stability, high thermal expansion, and low thermal conductivity [52]. Further, lanthanum phosphate is expected to have good corrosion resistance in environments containing sulfur and vanadium salts. However, plasma spraying can not be easily used to make this type of coating. Detailed research is needed to establish the suitability of $LaPO_4$ as TBC. Rare earth oxide coatings ($La_2O_3$, $CeO_2$, $Pr_2O_3$, and $Nb_2O_5$ as main phases) can be used as TBCs as they have lower thermal diffusivity and higher thermal expansion coefficient than $ZrO_2$ [53]. Most of the rare earth oxides are polymorphic at elevated temperatures [54] and their phase instability affects the thermal shock resistance of these coatings to a certain extent. When zircon is used as a TBC material, it dissociates during plasma spraying and consequently coatings are composed of a mixture of crystalline $ZrO_2$ and amorphous $SiO_2$. For diesel engines, the decomposed $SiO_2$ in the coating may cause problems due to the evaporation of SiO and $Si(OH)_2$ [55]. The thermal barrier effect is supposed to be due to the $ZrO_2$ phase in the coating [56]. However, few other silicates such as garnet almandine [$Fe_3Al_2(SiO_4)_3$], garnet pyrope [$Mg_3Al_2(SiO_4)_3$], garnet andradite-grossular [$Ca_3Al_2(SiO_4)_3$], and basalt (glass) have potential as TBC materials [57]. The composite oxide coating consisting of $2CaO.SiO_2$-10 to 30 wt% $CaO.ZrO_2$ shows excellent resistance to thermal shock and hot corrosion [58].

Researchers have conceived garnets [$Y_3Al_xFe_{5-x}O_{12}$ (x=0, 0.7, 1.4, and 5)] as TBC materials [59]. YAG ($Y_3Al_5O_{12}$) has superior high-temperature mechanical properties, low thermal conductivity, excellent phase/thermal stability up to the melting point and significantly lower oxygen diffusivity than those of zirconia. However, the major drawback of this material is its low melting point and relatively low thermal expansion coefficient [59]. Guo et al. [60] produced $BaLa_2Ti_3O_{10}$ (BLT) by solid-state reaction of $BaCO_3$, $TiO_2$, and $La_2O_3$ for 48 h at 1,500°C. BLT showed phase stability between room temperature and 1,400°C. BLT showed a linearly increasing thermal expansion coefficient with increasing temperature up to 1,200°C and the coefficients of thermal expansion (CTEs) were in the range of $1 \times 10^{-5}$–$12.5 \times 10^{-6}$ $K^{-1}$, comparable to those of 7YSZ. BLT coatings with stoichiometric composition were developed by APS technique. The coating contained segmentation cracks and had a porosity of ~13%. The microhardness for the BLT coating was in the range of 3.9–4.5 GPa. The thermal conductivity at 1,200°C was about 0.7 W/mK and thereby, revealing it as a promising material in improving the thermal insulation property of TBC. Thermal cycling results showed that the BLT TBC had a lifetime of more than 1,100 cycles of about 200 h at 1,100°C. The failure of the coating occurred by cracking at the TGO layer due to severe bond coat oxidation. Based on the experimental results BLT can be considered as a promising material for TBC applications [60]. Xu et al. [61] deposited DCL TBCs consisting of $La_2(Zr_{0.7}Ce_{0.3})_2O_7$ (LZ7C3) and YSZ by EB-PVD method. They showed that the DCL coating had a much longer lifetime than the single layer LZ7C3 coating and much longer than that of the single layer YSZ coating. Similar thermal expansion behaviors of YSZ interlayer with LZ7C3 coating and TGO layer, high sintering-resistance of LZ7C3 coating and unique columnar growth within DCL coating led to the extension of thermal cycling life of DCL coating. The failure of DCL coating occurred due to the reduction-oxidation of cerium oxide, the crack initiation, propagation and extension, the abnormal oxidation of bond coat, the degradation of $t'$-phase in YSZ coating, and the outward diffusion of Cr alloying element into LZ7C3 coating [61]. $Dy_2O_3$–$Y_2O_3$ co-doped $ZrO_2$ exhibits lower thermal conductivity and higher coefficient of thermal expansion. Thus, it is a promising ceramic thermal barrier coating material for aero-gas turbines and high temperature applications in metallurgical and chemical industry. Qu et al. [62] prepared $Dy_2O_3$–$Y_2O_3$ co-doped $ZrO_2$ ceramics using solid state reaction methods. $Dy_{0.06}Y_{0.072}Zr_{0.868}O_{1.934}$ exhibited a lower thermal conductivity and higher coefficient of thermal expansion as compared with standard 8 wt% $Y_2O_3$-stabilized $ZrO_2$ used in conventional TBCs. The compatibility between the TGO ($Al_2O_3$) and the new compositions is complicated to ensure the durability of TBCs. $Dy_{0.06}Y_{0.072}Zr_{0.868}O_{1.934}$ was found to be compatible with $Al_2O_3$ whereas $YAlO_3$ and $Dy_3Al_2(AlO_4)_3$ were formed when $Dy_{0.25}Y_{0.25}Zr_{0.5}O_{1.75}$ and $Al_2O_3$ were mixed and sintered [62].

New alternative TBC materials to YSZ for applications above 1,473 K are being explored by researchers. Zhou et al. [63] prepared $Y_4Al_2O_9$ (YAM) ceramics by solid state reaction at 1,873 K for 12 h. They investigated the phase stability, thermophysical properties, and sintering-resistance behavior of the material. XRD results revealed single monoclinic phase YAM. Even no new phase appeared after long-term annealing. The thermal conductivities of YAM ceramic decreased gradually with the increase of temperature ranges from room temperature to 1,273 K. The minimum value obtained was ~1.81 W m$^{-1}$ K$^{-1}$, which is lower than that of YSZ. YAM showed moderate thermal expansion coefficient, i.e., 8.91 $\times$ 10$^{-6}$ K$^{-1}$ in the temperature range of 300–1,473 K. In comparison to YSZ, YAM has lower density and higher sintering-resistance ability, which is very favorable for TBC applications. The results indicated that YAM is a promising ceramic material candidate for application in the TBC system [63]. YSZ is usually used as ceramic top coat for gas turbine blades and vanes. The accelerated phase transformation and the intensified sintering of the YSZ top coat at temperatures between 1,200°C and 1,300°C lead to microstructural changes resulting in higher thermal stress generation and lifetime reduction. Additionally, thermal conductivity ($\lambda$) of the top coat increases. Therefore, lanthanum zirconate (La$_2$Zr$_2$O$_7$) and gadolinium zirconate (Gd$_2$Zr$_2$O$_7$) is being suggested by researchers as a top coat because of their high phase stability up to their melting points and the lower thermal conductivity compared to YSZ. Bobzin et al. [64] deposited single-(SCL) and DCL top coats consisting of 7 wt% yttria-stabilized zirconia (7YSZ), La$_2$Zr$_2$O$_7$, or Gd$_2$Zr$_2$O$_7$ using the EB-PVD method. They wanted to investigate the temperature-dependent phase behavior and change of thermal conductivity of SCL and DCL top coats, as well as the influence of different top coat materials and architectures on the growth of the TGO. Morphology and coating thickness were determined using SEM. The SCL and DCL systems showed a columnar microstructure with a coating thickness of about 150 μm. The thermal conductivity of SCL and DCL systems was measured between 400°C and 1,300°C by laser flash technique. The XRD of SCL and DCL systems were carried out after isothermal oxidation at 1,300°C. Finally, the TGO phase was identified by XRD and EDS analysis. Correlation between morphology, architecture, coating material, and TGO behavior can give details of oxygen diffusion processes [64].

Investigation of the ZrO$_2$–YO$_{1.5}$–TaO$_{2.5}$ system reveals several promising aspects for TBC applications. Unique presence of a stable, non-transformable, tetragonal region in this ternary oxide system allows for phase stability to elevated temperatures, e.g.,1,500°C. Yttria- and tantala-containing compositions exhibited significantly high resistance to vanadate corrosion compared to 7YSZ. Further, yttria- and tantala-stabilized zirconia compositions within the non-transformable tetragonal phase field exhibited toughness values comparable or higher than those of 7YSZ and thereby, increasing their stability as TBCs. Pitek and Levi discussed about these promising attributes based on recent experimental works [65]. Liu et al. [66] prepared pyrochlore-type (La$_{0.8}$Eu$_{0.2}$)$_2$Zr$_2$O$_7$ feedstocks by spray drying and used that to produce ceramic thermal barrier coatings. DCL TBCs with a first layer of 8 wt% YSZ and a top layer of (La$_{0.8}$Eu$_{0.2}$)$_2$Zr$_2$O$_7$ were deposited by plasma spraying. Plasma-sprayed (La$_{0.8}$Eu$_{0.2}$)$_2$Zr$_2$O$_7$ coatings were composed of a defect fluorite-type phase and a t-ZrO$_2$ phase. However, after thermal shock tests at 1,250°C for 32 cycles, (La$_{0.8}$Eu$_{0.2}$)$_2$Zr$_2$O$_7$ coatings exhibited a pyrochlore-type structure. The thermal shock failure of DCL (La$_{0.8}$Eu$_{0.2}$)$_2$Zr$_2$O$_7$/YSZ coatings mainly occurred at the interface between the YSZ and $(La_{0.8}Eu_{0.2})_2Zr_2O_7$ layers. However, the TGO layer from the bond coat had no effect on the thermal shock failure [66]. Two kinds of rare earth zirconate $(Sm_{0.5}La_{0.5})_2Zr_2O_7$ and $(Sm_{0.5}La_{0.5})_2(Zr_{0.8}Ce_{0.2})_2O_7$ ceramics were prepared by Hong-song et al. [67] through solid state reaction at 1,600°C for 10 h. They investigated the phase compositions, microstructures, and thermophysical properties of these materials. XRD results confirmed the formation of single phase $(Sm_{0.5}La_{0.5})_2Zr_2O_7$ and $(Sm_{0.5}La_{0.5})_2(Zr_{0.8}Ce_{0.2})_2O_7$ with pyrochlore structure. Dense microstructures of these materials and absence of other phases among the particles were revealed by SEM studies. The TEC of the ceramic increased with the increasing temperature, while the thermal conductivity decreased. TECs of $(Sm_{0.5}La_{0.5})_2Zr_2O_7$ and $(Sm_{0.5}La_{0.5})_2(Zr_{0.8}Ce_{0.2})_2O_7$ were lower than that of $Sm_2Zr_2O_7$. The CeO$_2$ addition resulted in the higher TEC of $(Sm_{0.5}La_{0.5})_2(Zr_{0.8}Ce_{0.2})_2O_7$ than those of 8YSZ and $(Sm_{0.5}La_{0.5})_2Zr_2O_7$. Although the TEC of $(Sm_{0.5}La_{0.5})_2Zr_2O_7$ was lower than that of 8YSZ, still it can serve as a TBC. Doping with $La_2O_3$ or $CeO_2$ led to phonon scattering resulting in much lower thermal conductivities of $(Sm_{0.5}La_{0.5})_2Zr_2O_7$ and $(Sm_{0.5}La_{0.5})_2(Zr_{0.8}Ce_{0.2})_2O_7$ than that of $Sm_2Zr_2O_7$. In comparison to the thermal conductivity of $(Sm_{0.5}La_{0.5})_2Zr_2O_7$ the thermal conductivity of $(Sm_{0.5}La_{0.5})_2(Zr_{0.8}Ce_{0.2})_2O_7$ was relatively lower. The experimental results showed that $(Sm_{0.5}La_{0.5})_2Zr_2O_7$ and $(Sm_{0.5}La_{0.5})_2(Zr_{0.8}Ce_{0.2})_2O_7$ are novel candidate materials for TBCs in near future [67].

## 6.2. Composite Top Coat

A new TBC was developed by Dietrich et al. [68] from a powder mixture of metal and normal glass by using vacuum plasma spraying technique. This type of TBC material had a similar thermal expansion coefficient of a metal substrate. The thermal conductivity of this composite top coat was about two times greater than that of YSZ. Long thermal cycling life of the metal-glass TBC was attributed to high thermal expansion coefficient, good adherence to the bond coat, and absence of open porosity and thereby, preventing the bond coat oxidation from corrosive gases [68]. Majumdar and Jana [69] studied the properties of a TBC prepared from 3 wt% YSZ dispersed in a high temperature resistant alumino-borosilicate glassy matrix. The YSZ-glass composite coating was applied on stainless steel substrate by a simple and cost-effective enameling technique. The thermal gradient of 800 μm thick TBC was found to be 175–180°C after 30 min exposure at 1,000°C. Significant improvement of the gradient to 650–675°C was observed after long exposure of the coated surface at 1,000°C when compressed air cooling was utilized [69]. The spallation of ceramic coating from the bond coat is an important problem for TBC systems. Basically, the spallation is caused by the oxidation and hot corrosion at the interface of the ceramic layer and bond coat. Keyvani et al. [70] investigated the oxidation and hot corrosion behavior of plasma sprayed nanostructured $Al_2O_3$/YSZ composite TBC coatings on Ni-based (IN-738LC) superalloy substrate and compared it with the conventional YSZ. The coatings were deposited by plasma spray method. High temperature oxidation test at 1,100°C and hot corrosion test at 1,050°C using $Na_2SO_4$ and $V_2O_5$ molten salts were conducted on the coatings. The experimental data demonstrated that the nanostructured $Al_2O_3$/YSZ composite coating had higher oxidation and hot corrosion resistance than those of the conventional YSZ coating. The microstructural analysis indicated that the growth of TGO was much less for this nanostructured $Al_2O_3$/YSZ composite coating. Further, the composite top coating prevented infiltration of both oxygen and aggressive molten salt [70]. Novel YSZ (6 wt% yttria partially stabilized zirconia)–($Al_2O_3$/YAG) (alumina–yttrium aluminum garnet, $Y_3Al_5O_{12}$) DLC coatings were formed by using the composite sol-gel and pressure filtration microwave sintering (PFMS) technologies by Ren et al. [71]. The microstructural observations showed that micro-sized YAG particles were embedded in nano-sized $\alpha$-$Al_2O_3$ film. A thin $Al_2O_3$/YAG layer had good adherence with the substrate and the thick YSZ top layer. Cyclic oxidation tests at 1,000°C indicated that they can resist oxidation of alloy and improve the spallation resistance. The thermal insulation capability tests at 1,000°C and 1,100°C indicated that 250 μm coating had better thermal barrier effect than that of the 150 μm coating at different cooling gas rates. The decrease in oxidation rate for forming a TGO scale using the sealing effect of $\alpha$-$Al_2O_3$ and the reduced thermal stresses by means of nano/micro composite structure led to these beneficial effects. This double-layer coating can be considered as a promising TBC [71].

## 6.3. Glass-Ceramics as Tbc Materials

$MgO–Al_2O_3–TiO_2$ and $ZnO–Al_2O_3–SiO_2$ based glass-ceramic coatings have been developed as TBCs for gas turbine engine components by Datta and Das [72, 73]. These coatings were formed on nimonic alloy substrates using the vitreous enameling technique. $MgO–Al_2O_3–TiO_2$-based glass coating was applied on nimonic alloy substrate by spraying the glass slurry, drying, and then firing at about 1,160°C for 5–6 min. Further, the glass coating was heat treated for 1 h at 880°C followed by 1 h at 1,020°C to develop crystals such as magnesium aluminum titanate as a major phase along with magnesium silicate and aluminum titanate as minor phases in the glass matrix. The thermal shock resistance of the glass-ceramic coating was found to be more than 10 cycles when repeatedly heated to 750°C and immediately quenched in cold water. No chipping or spalling defect was observed. Slight weight gain was noted during the thermal endurance test at 1,000°C for 100 h. However, the operating temperature of this coating is limited to 750°C. Glass-ceramic coating based on $ZnO–Al_2O_3–SiO_2$ systems can operate at high working temperatures of up to 1,000°C. This type of glass coating was applied on a nimonic alloy through the spraying of a suitable glass slip, drying, and firing at 1,200°C for 5–6 min. The glass coating was subsequently heat treated at 1,000°C for 1 h to develop gahnite, willemite, and cristobalite crystalline phases. Thermal shock at 1,000°C for 10 cycles showed no chipping. During the thermal endurance test at 1,000°C for 100 h, negligible weight gain was observed. Figure 1(a) shows the oxidative weight gain of the bare substrate and $MgO–Al_2O_3–TiO_2$-based glass-ceramic coated substrate during the oxidation test at 1,000°C for 100 h. Typical SEM image of $MgO–Al_2O_3–TiO_2$-based glass-ceramic coating is shown in Figure 2(b).

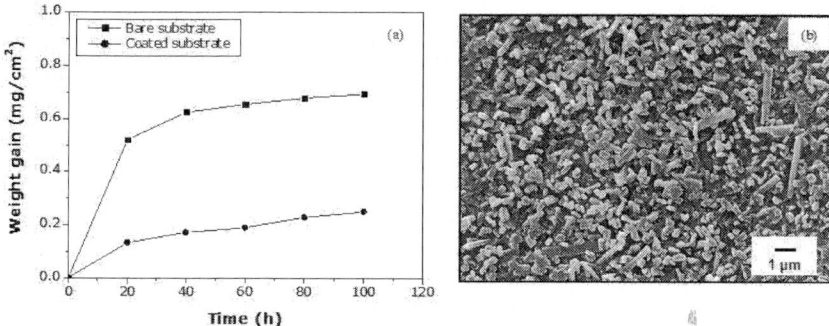

**Figure 1.**(a) Oxidative weight gain of $MgO-Al_2O_3-TiO_2$-based glass-ceramic coated substrate at $1,000°C$ for 100 h and (b) typical SEM microstructure of the corresponding coating.

# 7. PROMISING BOND COAT MATERIALS FOR TBC SYSTEMS

TBCs with ceramic top coat and MCrAlY (M=Ni, Co) bond coat are generally applied on gas turbine engine components to protect them from high temperature exposure [3]. The bond coat provides thermo-elastic relaxation to accommodate the high stresses generated in the TBC system. The chemistry and microstructure of bond coat affects the structure and morphology of the TGO [3]. The oxidation of bond coat needs to be restricted to improve the performance of the TBC system. Glass-ceramics may be used as bond coats because of several reasons. As this bond coat is basically oxide-based, failure of the TBC system from bond coat oxidation may be avoided. Further, high stress may be accommodated by the viscous flow of the glass-ceramics, which may increase the stability of the TBC system during thermal cycling at high operating temperatures. In addition, this TBC system may protect the metallic component from oxidation and creep failure more effectively because of the lower thermal conductivity of glass-ceramics compared to metals. Detailed studies on the TBC system consisting of 8 wt% YSZ (~400 μm) top coat, $BaO-MgO-SiO_2$-based glass-ceramic bond coat (~100 μm) and nimonic alloy (AE 435) substrate have been carried out by Das [74]. The glass-ceramic bond coat and YSZ top coat were applied on the nimonic alloy substrate by conventional enameling and air plasma spraying techniques, respectively. Figure 2 depicts the typical SEM cross-sectional micrograph of this kind of TBC system, which is composed of $BaO-MgO-SiO_2$-based glass-ceramic bond coat, 8-YSZ top coat, and nimonic superalloy substrate.

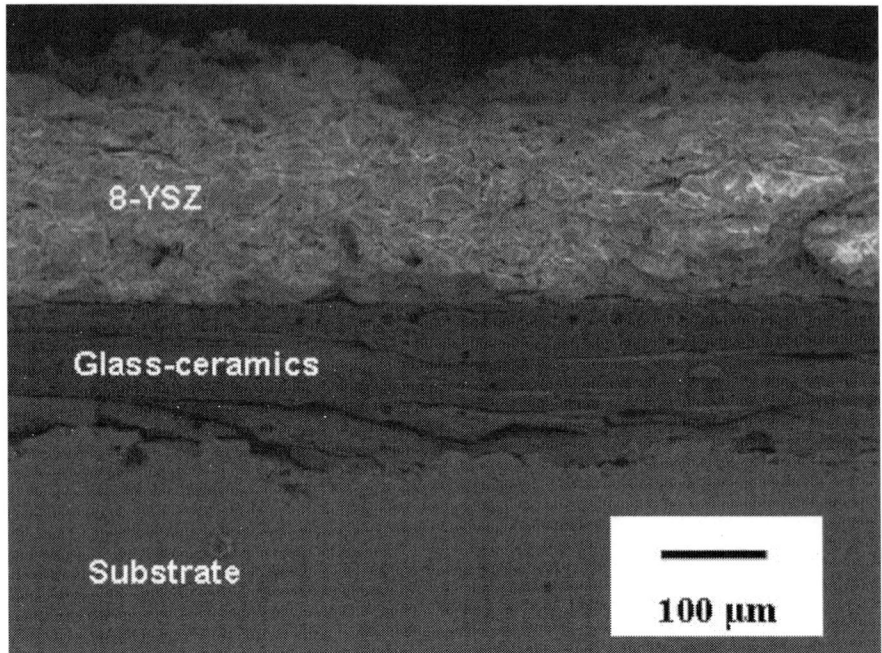

**Figure 2.**Typical TBC system consisting of glass-ceramic bond coat, 8-YSZ top coat, and nimonic superalloy substrate.

The 90° bend tests on these TBC systems showed that only a small amount of YSZ coating chipped off from the edges, indicating strong adherence of the TBC with the nimonic alloy substrate. The microhardness and Young's modulus values of YSZ coating, glass-ceramic coating, and nimonic alloy substrate of the TBC system were lower on the cross-section than those obtained on the plan-section at a load of 100 mN. The four-point bend test on the TBC system displayed low stiffness (bending elastic modulus−45−52 GPa at room temperature) that leads to low residual stresses in the TBC resulting in high thermo-mechanical stability of the TBC system [74]. Das et al. [75] studied the oxidation behavior of a TBC system consisting of 8 wt% YSZ top coat, BaO–MgO–SiO$_2$-based glass-ceramic bond coat, and nimonic alloy (AE 435) substrate wherein static oxidation test was carried out at 1,200°C for 500 h in air. Oxidation resistance of this TBC system was compared with the conventional TBC system under identical heat treatment conditions. Both TBC systems were characterized by SEM, as well as EDS analysis. The TGO layer was not found between the bond coat and the top coat in the case of glass-ceramic bonded TBC system, while the conventional TBC system showed a TGO layer of ~16 μm thickness at the bond coat-top coat interface [75].

Thermal cyclic behavior of glass-ceramic bonded TBC on nimonic alloy substrate was investigated by Das et al. [76]. In that study, a TBC system comprised of 8 wt% YSZ top coat, BaO–MgO–SiO$_2$-based glass-ceramic bond coat, and nimonic alloy (AE 435) substrate was subjected to thermal shock test from 1,000°C to room temperature for 100 cycles. Specimens held at 1,000°C for 5 min were forced air quenched, as well as water quenched from the same conditions. Microstructural changes were investigated using SEM. The phase analysis was conducted by XRD analysis and EDS analysis. Deterioration was not observed in the top coats after 100 cycles in the case of forced air quenched specimens, whereas the top coats were damaged in the water quenched specimens. After thermal cycling experiments interfacial cracks did not appear at the top coat-bond coat and bond coat-substrate interfaces both in forced air quenched and water quenched specimens. Further, the top coat retained its phase stability [76]. The mechanical properties of a glass-ceramic bonded TBC system have been reported by Ghosh [77].Glass-ceramic bonded TBC showed good thermal gradient property as both the glass-ceramic bond coat and YSZ top coat can act as a thermal barrier to the nimonic alloy substrate and reduce the substrate temperature. The thermal gradient of a TBC-coated substrate was 856°C after 45 min holding of the YSZ coating at 1,200°C. The present TBC prevents the thermal conduction to the nimonic alloy substrate as both the glass-ceramic bond coat and the YSZ top coat have low thermal conductivity. Thermal conductivity measurement showed that the ~100 μm glass-ceramic coated substrate had lower thermal conductivity (~23–27 W/m.K at 1,000°C) than that of the bare nimonic alloy substrate (~28 W/m.K at 1,000°C). Moreover, the thermal conductivity of the glass-ceramic- (~100 μm) and YSZ (~400 μm)-coated nimonic alloy substrate was much lower (17.19 W/m.K at 1,000°C) than that of the bare nimonic alloy substrate (~28 W/m.K at 1,000°C) [74, 78, 79].

Efficient gas turbines can be achieved through the use of engineered components having the capability of operating at higher metal temperatures with longer lifetimes. Gas turbine Inlet temperatures can exceed the melting temperatures of nickel-based superalloys. Advanced air cooling system in association with TBCs can decrease the underlying substrate temperature. NiCoCrAlY overlay coatings are generally used as bond coatings for industrial gas turbines. Extensive research is being carried out to find the suitable bond coat composition. Seraffon et al. [80] reported a new type of bond coat with a wide range of compositions. They focused on the oxidation behavior of the bond coatings at 950°C. A range of Ni–Co–Cr–Al coatings were deposited on sapphire substrates using the physical vapor deposition technique and magnetron sputtering method. Co-sputtering of two targets, such as Ni–10%Cr, Ni–20%Cr, Ni–50%Cr, Ni–20%Co–40%Cr, or Ni–40%Co–20%Cr target, and a pure Al target was used for the deposition of coatings. The coatings were then oxidized in air for 500 h at 950°C. All samples were characterized by measuring the change in coating thickness using pre- and post-exposure metrology only and also the change in specimen weight. Thick coatings (20–30 μm) were deposited by magnetron sputtering successfully. EDS analysis was used to determine the elemental compositions of the samples. Furthermore, XRD was used to identify the major oxides formed during thermal exposure. The selective growth of protective $Cr_2O_3$, $Al_2O_3$ or other less protective mixed oxides was observed. The oxide scale growth rate indicated the suitable coatings that produce more protective oxides and allow future optimization of the bond coating composition for service within the turbine section of industrial gas turbines [80].

In the last decade, it has been observed that Pt-rich $\gamma$–$\gamma'$ alloys and coatings have good oxidation and corrosion properties. Selezneff et al. [38] used this technique to fabricate doped Pt-rich $\gamma$–$\gamma'$ bond coatings on AM1® superalloy substrate. These TBC systems were compared with the conventional TBC system composed of a $\beta$-(Ni,Pt)Al bond coating. Most of the compositions were superior to the $\beta$-(Ni,Pt)Al bond coatings with respect to ceramic top coat adherence and better oxide scale adherence of the $\gamma$–$\gamma'$-based systems [38]. Iridium modified nickel alluminides are promising bond coats because of their ability to promote $\alpha$-$Al_2O_3$ scale growth and to form an oxygen diffusion barrier Ir layer. An innovative Al–Ni–Ir alloy was formulated by Lamastra et al. [81]. A detailed microstructural investigations of both powder and bulk samples were conducted to compare the phase composition, oxidation behavior, and thermal stability of the proposed system with those of the Ir free ones. The AlNiIr system was composed of $Al_3Ni_2$, $AlNi_3$ and $\beta$-NiAl. It was assumed that the presence of Ir promoted the alumina scale growth, which started at ~1000°C. Ni-poor and Al-rich islands were observed in both as cast and oxidized AlNiIr bulk samples. However, Ir had high concentration in Al-rich islands and thereby, suggesting higher affinity of iridium towards Al than Ni. After oxidation at 1,150°C, the $\alpha$-$Al_2O_3$ scale growth was observed increasing the TGO thickness with dwelling time. Both Ir ODB and Ir-rich islands at the interface between the alloy and the $Al_2O_3$ scale were not identified due to the low Ir amount. However, metallic Ir and the compound $Al_{2.75}Ir$ were detected in the powder after thermal treatment at 1,000°C [81].

Developing new bond coat is an effective way to extend the service life of TBCs during high temperature exposure. Yao et al. [82] prepared a novel TBC system composed of an $(Al_2O_3$–$Y_2O_3)$/ (Pt or Pt–Au) composite bond coat and a YSZ top coat and Ni-based superalloy by magnetron sputtering and EB-PVD, respectively. Cyclic oxidation tests in air at 1,100°C for 200 h showed that the YSZ top coat and alloy substrate can be bonded together effectively by the $(Al_2O_3$–$Y_2O_3)$/(Pt or Pt–Au) composite coating. So, this kind of TBC had excellent oxidation resistance and cracking/buckling resistance, which can be attributed to the sealing effect of such coating. Therefore, the interdiffusion between the bond coat and alloy substrate as well as substrate oxidation can be avoided. The toughening effect of noble metals and composite structure of bond coat resulted in inhibition of the micro-cracks propagation and relaxation of the stress in the bond coat. This ceramic/noble metal composite coating has great prospect for the TBC applications [82]. Wang et al. [83] produced NiAl and NiAlHf/Ru coatings on nickel-based single crystal superalloy in order to investigate the interdiffusion behavior and cyclic oxidation resistance at 1,100°C. Needle-like topologically close-packed phases and secondary reaction zone (~30 μm thick layer) were formed in the NiAl-coated superalloy after annealing at 1,100°C for 100 h while the precipitates of TCP and SRZ were effectively constrained in the NiAlHf/Ru-coated alloy. The NiAlHf/Ru coating exhibited superior cyclic oxidation resistance as compared to the NiAl coating. They have shown that Ru and Hf have important roles in terms of affecting interdiffusion and cyclic oxidation [83]. Zhang et al [84] developed gradient TBCs consisting of $(Gd_{0.9}Yb_{0.1})_2Zr_2O_7$–yttria-stabilized zirconia (8YSZ) and Hf-doped NiAl bond coat by EB-PVD technique. The effect of the interfacial structure between $(Gd_{0.9}Yb_{0.1})_2Zr_2O_7$ (GYbZ) and 8YSZ layers on the thermal cycling behavior was investigated by comparing the DCL coatings with gradient thermal barrier coatings (GTBCs). The thermal cycling tests showed that the GTBCs had a more extended lifetime than that of the DCL coatings. The failure of GYbZ-8YSZ DCL coating with clear interface between different ceramic layers occurred through delaminating cracking as a result of crack initiation and propagation caused by stress concentration within the ceramic layers. Further, the failure of GTBC occurred due to the thermal expansion mismatch between the Hf-doped NiAl bond coat and the TGO layer [84].

# 8. CONCLUSIONS

In the future, TBCs are required to be more suitably designed for the thermal protection of gas turbine engine components to significantly increase engine operating temperatures, fuel efficiency, and engine reliability. However, coating durability is a vital factor to increase the engine operating temperature. Therefore, the coating behavior and failure modes under high temperature, high thermal gradient cyclic conditions should be properly understood to develop next-generation advanced TBCs.

# 9. ACKNOWLEDGEMENTS

The authors are very grateful to Mr. K. Dasgupta, Director, CSIR-Central Glass and Ceramic Research Institute (CSIR-CGCRI), Kolkata-700 032, India, for his kind permission to publish this book chapter.

# REFERENCES

1. Tang F, Ajdelsztajn L, Kim GE, Provenzano V, Schoenung JM. Effects of variations in coating materials and process conditions on the thermal cycle properties of NiCrAlY/YSZ thermal barrier coatings. Materials Science and Engineering A. 2006; 425: 94-106.
2. Matsumoto M, Takayama H, Yokoe D, Mukai K, Matsubara H, Kagiya Y, Sugita Y. Thermal cycle behavior of plasma sprayed La2O3,Y2O3 stabilized ZrO2 coatings. Scripta Materialia. 2006; 54: 2035-2039.
3. Martena M, Botto D, Fino P, Sabbadini S, Gola MM, Badini C. Modelling of TBC system failure: Stress distribution as a function of TGO thickness and thermal expansion mismatch. Engineering Failure Analysis. 2006; 13: 409-426.
4. Clarke DR, Oechsner M, Padture NP. Thermal-barrier coatings for more efficient gas-turbine engines. MRS bulletin. 2012; 37: 891-898.
5. Busso EP, Lin J, Sakurai S, Nakayama M. Mechanistic study of oxidation-induced degradation in a plasma-sprayed thermal barrier coating system. Part I: Model formulation. Acta Materialia. 2001; 49: 1515-1528.

6.   Daroonparvar M, Hussain MS, Yajid MAM. The role of formation of continues thermally grown oxide layer on the nanostructured NiCrAlY bond coat during thermal exposure in air. Applied Surface Science. 2012; 261: 287-297.

7.   Xu R, Fan XL, Zhang WX, Wang TJ. Interfacial fracture mechanism associated with mixed oxides growth in thermal barrier coating system. Surface and Coatings Technology. 2014; 253: 139-147.

8.   Evans HE. Oxidation failure of TBC systems: An assessment of mechanisms. Surface and Coatings Technology. 2011; 206: 1512-1521.

9.   Haynes JA, Ferber MK, Porter WD. Thermal cycling behavior of plasma-sprayed thermal barrier coatings with various MCrAlX bond coats. Journal of Thermal Spray Technology. 2000; 9: 38-48.

10.  Sfar K, Aktaa J, Munz D. Numerical investigation of residual stress fields and crack behavior in TBC systems. Materials Science and Engineering: A. 2002; 333: 351-360.

11.  Yang L, Yang F, Long Y, Zhao Y, Xiong X, Zhao X, Xiao P. Evolution of residual stress in air plasma sprayed yttria stabilised zirconia thermal barrier coatings after isothermal treatment. Surface and Coatings Technology. 2014; 251: 98-105.

12.  Pujol G, Ansart F, Bonino J-P, Malié A, Hamadi S. Step-by-step investigation of degradation mechanisms induced by CMAS attack on YSZ materials for TBC applications. Surface and Coatings Technology. 2013; 237: 71-78.

13.  Thompson JA, Clyne TW. The effect of heat treatment on the stiffness of zirconia top coats in plasma-sprayed TBCs. Acta Materialia. 2001; 49: 1565-1575.

14.  Abubakar AA, Akhtar SS, Arif AFM. Phase field modeling of $V_2O_5$ hot corrosion kinetics in thermal barrier coatings. Computational Materials Science. 2015; 99: 105-116.

15.  Kulkarni A, Vaidya A, Goland A, Sampath S, Herman H. Processing effects on porosity-property correlations in plasma sprayed yttria-stabilized zirconia coatings. Materials and Engineering A. 2003; 359: 100-111.

16.  Zhang D, Gong S, Xu H, Wu Z. Effect of bond coat surface roughness on the thermal cyclic behavior of thermal barrier coatings. Surface & Coatings Technology. 2006; 201: 649-653.

17.  Movchan BA, Yakovchuk YK. Graded thermal barrier coatings, deposited by EB-PVD. Surface & Coatings Technology 2004;188-189: 85-92.

18.  Wu LT, Wu RT, Zhao X, Xiao P. Microstructure parameters affecting interfacial adhesion of thermal barrier coatings by the EB-PVD method. Materials Science & Engineering: A. 2014; 594: 193-202.

19. Basu SN, Ye G, Gevelber M, Wroblewski D. Microcrack formation in plasma sprayed thermal barrier coatings. International Journal of Refractory Metals & Hard Materials. 2005; 23: 335-343.

20. Bengtsson P, Johanneson TJ. Characterization of microstructural defects in plasma sprayed thermal barrier coatings. Journal of Thermal Spray Technology. 1995; 4: 245-251.

21. Chen WR, Wu X, Dudzinski D, Patnaik PC. Modification of oxide layer in plasma sprayed thermal barrier coatings. Surface & Coatings Technology. 2006; 200: 5863-5868.

22. Tang F, Schoenung JM. Evolution of Young's modulus of air plasma sprayed yttria-stabilized zirconia in thermally cycled thermal barrier coatings. Scripta Materialia. 2006; 54: 1587-1592.

23. Basu D, Funke C, Steinbrech RW. Effect of heat treatment on elastic properties of separated thermal barrier coatings. Journal of Materials Research. 1999; 14: 4643-4650.

24. Limargaa AM, Widjajab TS, Yip TH. Mechanical properties and oxidation resistance of plasma-sprayed multilayered Al2O3/ZrO2 thermal barrier coatings. Surface & Coatings Technology. 2005; 197: 93-102.

25. Guo HB, Kuroda S, Murakami H. Segmented thermal barrier coatings produced by atmospheric plasma spraying hollow powders. Thin Solid Films. 2006; 506-507: 136-139.

26. Chen X, Gu L, Zou B, Wang Y, Cao X. New functionally graded thermal barrier coating system based on LaMgAl11O19/YSZ prepared by air plasma spraying. Surface & Coatings Technology. 2012; 206: 2265-2274.

27. Préauchat B, Drawin S. Properties of PECVD-deposited thermal barrier coatings. Surface & Coatings Technology. 2001; 142-44: 835-842.

28. Vyas JD, Choy K-L. Structural characterisation of thermal barrier coatings deposited using electrostatic spray assisted vapour deposition method. Materials Science and Engineering: A. 2000; 277: 206-212.

29. Gell M, Xie L, Ma X, Jordan EH, Padture NP. Highly durable thermal barrier coatings made by the solution precursor plasma spray process. Surface and Coating Technology. 2004; 177-178: 97-102.

30. Jadhav A, Padture NP, Wu F, Jordan EH, Gell M. Thick ceramic thermal barrier coatings with high durability deposited using solution-precursor plasma spray. Materials Science and Engineering: A. 2005; 405: 313-320.

31. Viazzi C, Bonino JP, Ansart F. Synthesis by sol–gel route and characterization of yttria stabilized zirconia coatings for thermal barrier applications. Surface & Coatings Technology. 2006; 201: 3889-3893.

32. Sniezewski J, Le MY, Lours P, Pin L, Minvie BV, Monceau D, Oquab D, Fenech J, Ansart F, Bonino J-P. Sol–gel thermal barrier coatings: Optimization of the manufacturing route and durability under cyclic oxidation. Surface & Coatings Technology. 2010; 205:1256-1261.

33. Pin L, Ansart F, Bonino J-P, Maoult YL, Vidal V, Lours P. Processing,repairing and cyclic oxidation behaviour of sol–gel thermal barrier coatings. Surface & Coatings Technology. 2011; 206:1609-1614.

34. Pin L, Ansart F, Bonino J-P, Le Maoult Y, Vidal V, Lours P. Reinforcedsol–gel thermal barrier coatings and their cyclic oxidation life. Journal of the European Ceramic Society. 2013; 33: 269-276.

35. Ren C, He YD, Wang DR. Cyclic oxidation behavior and thermal barrier effect of YSZ–(Al2O3/YAG) double-layer TBCs prepared by the composite sol–gel method. Surface & Coatings Technology. 2011; 206: 1461-1468.

36. Monceau D, Oquab D, Estournes C, Boidot M, Selezneff S, Thebault Y, Cadoret Y. Pt-modified Ni aluminides, MCrAlY-base multilayer coatings and TBC systems fabricated by Spark Plasma Sintering for the protection of Ni-base superalloys. Surface & Coatings Technology. 2009; 204: 771-778.

37. Boidot M, Selezneff S, Monceau D, Oquab D, Estournès C. Proto-TGO formation in TBC systems fabricated by spark plasma sintering. Surface & Coatings Technology. 2010; 205: 1245-1249.

38. Selezneff S, Boidot M, Hugot J, Oquab D, Estournès C, Monceau D. Thermal cycling behavior of EBPVD TBC systems deposited on doped Pt-rich γ–γ′ bond coatings made by Spark Plasma Sintering (SPS). Surface & Coatings Technology. 2011; 206: 1558-1565.

39. Rousseau F, Fourmond C, Prima F, Serif MHV, Lavigne O, Morvan D, Chereau P. Deposition of thick and 50% porous YpSZ layer by spraying nitrate solution in a low pressure plasma reactor. Surface & Coatings Technology. 2011; 206: 1621-1627.

40. Ramachandran CS, Balasubramanian V, Ananthapadmanabhan PV. Synthesis, spheroidization and spray deposition of lanthanum zirconate using thermal plasma process. Surface & Coatings Technology. 2012; 206: 3017-3035.

41. Bahadori E, Javadpour S, Shariat MH, Mahzoon Fatemeh. Preparation and properties of ceramic Al2O3 coating as TBCs on MCrAly layer applied on Inconel alloy by cathodic plasma electrolytic deposition. Surface & Coatings Technology. 2013; 228: S611-S614.

42. Kim JH, Kim MC, Park CG. Evaluation of functionally graded thermal barrier coatings fabricated by detonation gun spray technique. Surface & Coatings Technology. 2003; 168: 275-280.

43. Chwa SO, Akira O. Microstructures of ZrO2-8wt.%Y2O3 coatings prepared by a plasma laser hybrid spraying technique. Surface & Coatings Technology. 2002; 153: 304-312.

44. Wang W, Li C, Li J, Fan J, Zhou X. Effect of gadolinium doping on phase transformation and microstructure of Gd2O3-Y2O3-ZrO2 composite coatings prepared by electrophoretic deposition. Journal of Rare Earths. 2013; 31: 289-295.

45. Vassen R, Cao X, Tietz F, Basu D, Sto°ver D. Zirconates as new materials for thermal barrier coatings. Journal of the American Ceramic Society. 2000; 83: 2023-2028.

46. Moskal G, Swad´zba L, Hetma´nczyk M, Witala B, Mendala B, Mendala J, Sosnowy P. Characterization of microstructure and thermal properties of Gd2Zr2O7-type thermal barrier coating. Journal of the European Ceramic Society 2012; 32: 2025-2034.

47. Ma W, Mack D, Malzbender J, Vaßen R, St°over D. Yb2O3 and Gd2O3 doped strontium zirconate for thermal barrier coatings. Journal of the European Ceramic Society. 2008; 28: 3071-3081.

48. Zhao H, Levi CG, Wadley HNG. Vapor deposited samarium zirconate thermal barrier coatings. Surface & Coatings Technology. 2009; 203: 3157-3167.

49. Vassen R, Cao X, Dietrich M, Stoever D. Improvement of new thermal barrier coating systems using layered or graded structure. In: Singh M, Jessen T (Eds.) The 25th Annual International Conference on Advanced Ceramics and Composites: An Advanced Ceramics Odyssey, Cocoa Beach of Florida: American Ceramic Society; 2001. p 435.

50. Friedrich CJ, Gadow R, Lischka MH. Lanthanum hexaaluminate thermal barrier coatings. In: Singh M, Jessen T (Eds.) The 25th Annual International Conference on Composites, Advanced Ceramics, Materials, and Structures: B, Cocoa Beach of Florida: American Ceramic Society; 2001. p 372-375.

51. Xie X, Guoa H, Gonga S, Xu H. Lanthanum–titanium–aluminum oxide: A novel thermal barrier coating material for applications at 1300∘C. Journal of the European Ceramic Society. 2011; 31: 1677-1683.

52. Sudre O, Cheung J, Marshall D, Morgan P, Levi CG. Thermal insulation coatings of LaPO4. In: Singh M, Jessen T (Eds.) The 25th Annual International Conference on Composites, Advanced Ceramics, Materials,and Structures: B, Cocoa Beach of Florida: American Ceramic Society; 2001. p 367.

53. Ding C, Xi Y, Zhang Y, Qu J, Qiao, H. Thermophysical properties of plasma sprayed rare earth oxide coatings. In: Sandmeier S, Eschnauer H, Huber P, Nicoll AR. (Eds.) The 2nd Plasma-technik-symposium (Lucerne Switzerland, June 1991), Switzerland: Plasma-Technik AG, Wohlen; 1991. p 27-32.

54. Warshaw I, Roy R. Polymorphism of the rare earth sesquioxides. The Journal of Physical Chemistry. 1965; 65: 2048-2051.

55. Kvernes I, Lugscheider E, Ladru F. Lifetime and degradation processes of TBCs for diesel engines. In: Lecomte-Beckers J, Schuber F, Ennis PJ. (Eds.) Proceedings of the 6th Lie´ge Conference on Materials for Advanced Power Engineering (Universite de Lie´ge, Belgium), Forschungszentrum Ju° lich GmbH, Ju° lich; 1998. p 997-1001.

56. Ramaswamy P, Seetharamu S, Varma KB, Rao KJ. Thermal barrier coating application of zircon sand. Journal of Thermal Spray Technology. 1999; 8: 447-453.

57. Chra´ ska P, Neufuss K, Kolman B, Dubsky J. Plasma spraying of silicates. In: Berndt CC. (Ed.) Proceedings of the 1st United Thermal Spray Conference: Thermal Spray—A United Forum for Scientific and Technological Advances, Indiana USA: ASM International, Materials Park; 1997. p 477-481.

58. Morgan PED, Marshall DB. Ceramic composites of monayite and alumina. Journal of the American Ceramic Society 1995; 78: 1553-1563.

59. Nitin PP, Klemens PG. Low thermal conductivity in garnets. Journal of the American Ceramic Society. 1977; 80: 1018-1020.

60. Guo H, Zhang H, Ma G, Gong S. Thermo-physical and thermal cycling properties of plasma-sprayed $BaLa_2Ti_3O_{10}$ coating as potential thermal barrier materials. Surface & Coatings Technology. 2009; 204: 691-696.

61. Xu Z, He, S, He L, Mu R, Huang G, Cao X. Novel thermal barrier coatings based on $La_2(Zr_{0.7}Ce_{0.3})_2O_7/8YSZ$ double-ceramic-layer systems deposited by electron beam physical vapor deposition. Journal of Alloys and Compounds. 2011; 509: 4273-4283.

62. Qu L, Choy K-L, Thermophysical and thermochemical properties of new thermal barrier materials based on $Dy_2O_3–Y_2O_3$ co-doped zirconia. Ceramics International. 2014; 40: 11593-11599.

63. Zhou X, Xu Z, Fan X, Zhao S, Cao X, He L. $Y_4Al_2O_9$ ceramics as a novel thermal barrier coating material for high-temperature applications. Materials Letters. 2014; 134:146-148.

64. Bobzin K, Bagcivan N, Brögelmann T, Yildirim B. Influence of temperature on phase stability and thermal conductivity of single- and double-ceramic-layer EB–PVD TBC top coats consisting of 7YSZ, $Gd_2Zr_2O_7$and $La_2Zr_2O_7$. Surface and Coatings Technology. 2013; 237: 56-64.

65. Pitek FM, Levi CG. Opportunities for TBCs in the $ZrO_2–YO_{1.5}–TaO_{2.5}$ system. Surface and Coatings Technology. 2007; 201: 6044-6050.

66. Liu Z-G, Zhang W-H, Ouyang J-H, Zhou Y. Novel double-ceramic-layer $(La_{0.8}Eu_{0.2})_2Zr_2O_7/YSZ$ thermal barrier coatings deposited by plasma spraying. Ceramics International. 2014; 40: 11277-11282.

67. Hong-song Z, Qiang X, Fu-chi W, Ling L, Yuan W, Xiaoge C. Preparation and thermophysical properties of (Sm0.5La0.5)2Zr2O7 and (Sm0.5La0.5)2 (Zr0.8Ce0.2)2O7 ceramics for thermal barrier coatings. Journal of Alloys and Compounds. 2009; 475: 624-628.

68. Dietrich M, Verlotski V, Vassen R, Stoever D. Metal-glass based composites for novel TBC-systems. Materials Science and Engineering Technology. 2001; 32: 669-672.

69. Majumdar A, Jana S. Yttria doped zirconia in glassy matrix useful for thermal barrier Coating. Materials Letters. 2000; 44:197-202.

70. Keyvani A. Microstructural stability oxidation and hot corrosion resistance of nanostructured Al2O3/YSZ composite compared to conventional YSZ TBC coatings. Journal of Alloys and Compounds. 2015; 623: 229-237.

71. Ren C, He YD, Wang DR. Cyclic oxidation behavior and thermal barrier effect of YSZ–(Al2O3/YAG) double-layer TBCs prepared by the composite sol–gel method. Surface and Coatings Technology. 2011; 206: 1461-1468.

72. Datta S, Das S. A new high temperature resistant glass-ceramic coating for gas turbine engine components. Bulletin of Materials Science 2005; 28: 689-696.

73. Datta S, Das S. A new high temperature resistant glass-ceramic coating developed in CGCRI, Kolkata. Transactions of the Indian Ceramic Society. 2005; 64: 25-32.

74. Das S. Study of structure and property relationship in thermal barrier coating system. Ph.D. thesis, Jadavpur University, Kolkata, 2010.

75. Das S, Datta S, Basu D, Das GC. Glass-ceramics as oxidation resistant bond coat in thermal barrier coating system. Ceramics International. 2009; 35: 1403-1406.

76. Das S, Datta S, Basu D, Das GC. Thermal cyclic behavior of glass-ceramic bonded thermal barrier coating on nimonic alloy substrate. Ceramics International. 2009; 35: 2123-2129.

77. Ghosh S. Microstructure and mechanical properties of a glass-ceramic bond coated TBC system. Procedia Materials Science. 2014; 6: 425-429.

78. Ghosh S. Thermal behavior of glass-ceramic bond coat in a TBC system. Vacuum. 2014; 101: 367-370.

79. Ghosh S. Thermal properties of glass-ceramic bonded thermal barrier coating system. Transactions of Nonferrous Metals Society of China. 2015; 25: 457-464.

80. Seraffon M, Simms NJ, Sumner J, Nicholls JR. The development of new bond coat compositions for thermal barrier coating systems operating under industrial gas turbine conditions. Surface and Coatings Technology. 2011; 206: 1529-1537.

81. Lamastra FR, Cacciotti I, Bellucci A, Nanni F. Innovative Al–Ni–Ir alloy for bond coats: Microstructure, phase analysis and oxidation behaviour. Intermetallics. 2012; 22: 241-250.

82. Yao J, He Y, Wang D, Peng H, Guo H, Gong S. Thermal barrier coatings with (Al2O3–Y2O3)/(Pt or Pt–Au) composite bond coat and 8YSZ top coat on Ni-based superalloy. Applied Surface Science. 2013; 286: 298-305.

83. Wang D, Peng H, Gong S, Guo H. NiAlHf/Ru: Promising bond coat materials in thermal barrier coatings for advanced single crystal superalloys. Corrosion Science. 2014; 78: 304-312.

84. Zhang H, Guo L, Ma Y, Peng H, Guo H, Gong S. Thermal cycling behavior of (Gd0.9Yb0.1)2Zr2O7/8YSZ gradient thermal barrier coatings deposited on Hf-doped NiAl bond coat by EB-PVD. Surface and Coatings Technology. 2014; 258: 950-955.

# CHAPTER 10

# Microwave Fast Sintering of Double Perovskite Ceramic Materials

*Penchal Reddy Matli[1], Adel Mohamed Amer Mohamed[1, 2] and Ramakrishna Reddy Rajuru[3]*

[1] Center for Advanced Materials, Qatar University, Doha, Qatar
[2] Department of Metallurgical and Materials Engineering, Faculty of Petroleum and Mining Engineering, Suez University, Suez, Egypt
[3] Department of Physics, Sri Krishnadevaraya University, Anantapur, India

**ABSTRACT**

The book chapter mainly deals with the microwave sintering of high quality crystals of La2MMnO6 (M = Ni or Co) ceramics. Double perovskite La2MMnO6 (M = Ni or Co) ceramics with average particle size of ~65 nm were manufactured using microwave sintering at 90°C for 10 min in N2 atmosphere for the first time. The morphology, structure, composition, and magnetic properties of the prepared compacts were characterized using X-ray diffraction (XRD), scanning electron microscopy (SEM), transmission electron microscopy (TEM), energy-dispersive X-ray spectroscopy (EDX), infrared spectroscopy (IR and FTIR), and physical properties measurement system (PPMS). The corresponding dielectric property was tested in the frequency range of 1 kHz–1 MHz and in the temperature range of 300–600 K, and the ceramics exhibited a relaxation-like dielectric behavior.

**Keywords:** ceramics, microwave sintering, microstructures, XPS, multiferroic properties

# 1. INTRODUCTION

Microwave sintering (MWS) is emerging and an innovative sintering technology for processing of ceramic materials and is commonly related with volumetric and uniform heating. MWS is one of the exciting new fields in material science with vast potential for preparation of novel and/or nanostructured ceramics/materials. Microwave heating has some important benefits over normal heating for ceramic processing, including reduced processing time, higher energy efficiency, selective and controlled heating, environmental friendliness, and improved product uniformity and yields. Microwave processing of materials is a relatively new technology that can be used in wide range of different materials such as ceramics, ferroelectrics, oxides, metals, and composites [1–10].

The effect of microwave radiation on the processing of several ceramic materials such as magnetic materials, superconducting materials, dielectric materials, metals, polymers, ceramics, and composite materials offers numerous benefits over conventional heating techniques. These benefits include time and energy savings, volumetric and uniform heating, considerably reduced processing time and temperature, improved product yield, fine microstructures, improved mechanical properties, lower environmental impact, reduction in manufacturing cost, and synthesis of new materials [11].

Microwave sintering has developed in recent years as a promising technology for faster, cheapest and most environmental-friendly processing of a wide variety of materials, which are regarded as significant advantages over conventional sintering procedures. Microwave radiation/heating for sintering of ceramic constituents has recently appeared as a newly motivated scientific approach [5].

Microwave sintering approach has unique advantages over conventional sintering methods in many respects. The essential difference in the conventional and microwave sintering processes is in the heating mechanism (Figure 1). In microwave heating, the materials themselves absorb microwave energy and then transform it into heat within the sample volume and sintering can be completed in shorter times. In microwave sintering, the heat is generated internally within the test sample due to the rapid oscillation of dipoles at microwave frequencies [12]. The contribution of diffusion from external sources is lesser. The internal and volumetric heating makes the sintering rapidly and uniformly. The heat generated through conventional heating is generally transferred to the sample via radiation, conduction, and convention [13]. This process takes longer duration for sintering the materials and causes some of the constituents to evaporate. This may lead to modify the desired stoichiometry and grain.

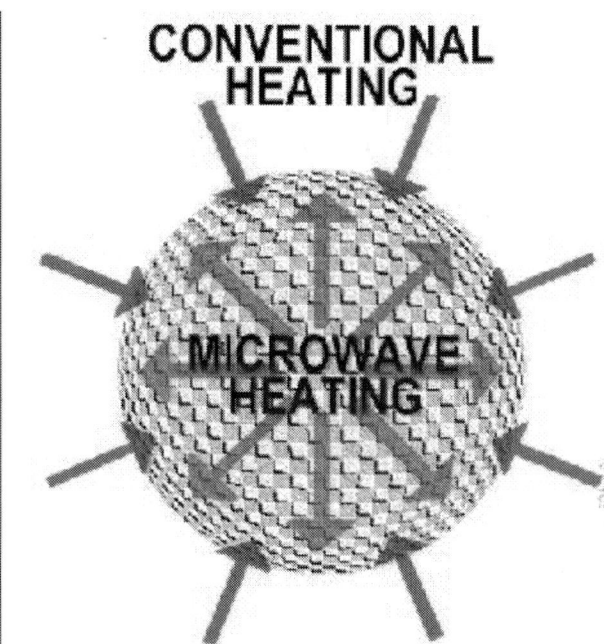

**Figure 1.** Comparison of heating mechanism in microwave and conventional sintering methods.

In contrast, microwave energy is transferred directly to the material through molecular interaction with an electromagnetic field. Microwave heating is more effective than conventional methods in terms of the usage of energy, produces higher temperature homogeneity, and is considerably faster than conventional heat sources. In the last 8 years, we have successfully sintered various ferrites/ferroelectrics, oxides, ceramics, mullite fiber, composites, and even powdered metals to full density using microwave processing [4, 5, 14–16].

Due to the energy efficient nature of microwave heating, there is a great opportunity for the application of microwaves to process metal based materials to couple the many gains of microwave heating. Recently, microwaves energy has been successfully used in different composites, metal, ceramics, melting of metals and metal ores, joining or brazing of metals, and heat treatment of metals [17].

The microwave energy is highly versatile in its application in the field of communication, and it still dominates almost all communications in space and mobile or cordless phone technology involves microwave frequencies. However, other than this communication, microwave energy has found its use for a variety of applications including rubber products industry, food processing, wood/paper/textile/ceramic drying, pharmaceuticals, polymers, printing materials, and biomedical fields over the past 50 years. These applications involve low temperature (<500°C) utilization of microwaves. The high temperature (> 800°C) applications of microwaves are a rather recent phenomenon.

Many researchers have reported that microwave heating is relatively faster than the conventional heating processes. This faster speed is manifested as a reduction in the densification time of ceramic powder compacts, often allied to lower sintering temperatures [7]. Generally, the synthesis kinetics and sintering materials are apparently upgraded by two or three orders of magnitude or even more when conventional heating is switched for microwave heating [18].

## 1.1. Interaction of Microwaves with Ceramic Materials: Theoretical Aspects

Microwaves are electromagnetic waves with the electromagnetic radiations in the frequency band of 300 MHz to 300 GHz, and their corresponding wavelength between 1 m and 1 mm can be used successfully to heat many ceramic materials. Since most of the microwave band is used for communication purpose, the Federal Communications Commission has allocated only very few specific frequencies for industrial, scientific, and medical applications. A major portion of these microwave used in the communication sector and only certain frequencies, viz., 0.915, 2.45, 5.85, and 21.2 GHz, are chosen for medical and industrial applications. Among these allowed frequencies, 2.45 GHz is the most common microwave frequency used for industrial and scientific applications. The interaction and heating generation of ceramics under microwave field depends on the dielectric, magnetic, and conductive loss of the material and temperature dependent parameters.

The ability of a material to be heated in microwave field depends on its dielectric properties, characterized by the dielectric complex constant $\varepsilon^*$:

$$\varepsilon^* = \varepsilon' - j\varepsilon''$$

Dielectric permittivity $\varepsilon'$ represents the material capacity to store electromagnetic energy and loss factor $\varepsilon''$ to dissipate it.

The dielectric constant of a material varies with its temperature, frequency, and composition. The power P absorbed in the material is proportional to the loss factor $\varepsilon''$, the frequency $f$ [Hz], and the electric field intensity $E$ [V/m]:

$$P = 2\Pi f \varepsilon^0 \varepsilon'' E^2$$

(or)

$$P = 55.63 \times 10^{-2} f \varepsilon' \tan \delta E^2$$

where the dielectric loss angle (loss tangent) is

$$\tan \delta = \frac{\varepsilon''}{\varepsilon'},$$

Eq. (3) shows that for a fixed value of electric field (E), the power in microwave absorbed in the material mass is proportional to the frequency ($f$) (which is practically 2.45 GHz), the dielectric permittivity ($\varepsilon'$), and loss factor ($\varepsilon''$) (through the loss tangent tan $\delta$), which vary with the materials temperature and humidity in their turn.

The diffusion of electromagnetic power into the absorber is characterized by skin depth ($D$) and expressed as

$$D = \left( \pi f \mu \sigma \right)^{-1/2}$$

where $\mu$ and $\sigma$ are magnetic permeability and electrical conductivity and $f$ is the frequency, respectively.

The effective penetration depth decreases with increase in frequency which in turn causes less heating. Hence, a suitable combination of parameters in Eqs. (2) and (4) is required for achieving optimum coupling. It can be inferred from this discussion that low dielectric loss materials take longer time and high dielectric loss materials take shorter duration in the microwave sintering.

On a microscopic scale, the phenomenon of dielectric heating is the effect of impurity dipolar relaxation in the microwave frequency region. When the vacancy jumps around the impurity ion to align its dipole moment with the electric field the internal friction of the rapidly oscillating dipole cause a homogeneous (volumetric) heating. Where the maximum absorption of microwave energy at the frequency or temperature at which the loss factor (tan $\delta$) attains its maximum. This is equivalent to an elastic relaxation resulting in damping of mechanical vibrations in solids.

The efficiency of the microwave dielectric heating is dependent on the ability of a specific material (powder, solvent, or reagent or anything else) to absorb microwave energy and convert it into heat. The heat is generated by the electric component of the electromagnetic field through two main mechanisms, i.e., dipolar polarization and ionic contribution [19]. According to the electromagnetism, the effect of a material upon heat transfer rates is often expressed as

$$\frac{\Delta T}{t} = \frac{0.56 \times 10^{-10} \varepsilon_{eff}'' f E^2}{\rho C_p}$$

Where $\varepsilon''_{eff}$ is the effective relative dielectric loss factor, $f$ is the frequency of microwave, $E$ is the magnetic fields of microwave within the material, $\rho$ is the mass density of the sample, and $C_P$ is the isotonic specific heat capacity [19]. In this case, the energy efficiency can easily reach 80–90% utilization and higher than the conventional heating methods [20, 21]. However, the essential nature of the interaction between microwaves and reactant molecules during the preparation of materials is fairly uncertain and speculative.

## 1.2. Benefits of Microwave Sintering Comparison To Conventional Sintering Method

In recent years, microwave sintering has shown significant advantages against conventional sintering for the synthesis of ceramic materials. Microwave sintering has attained worldwide attention due to its major advantages against conventional sintering methods, especially in ceramic materials.

Microwave sintering can significantly shorten the sintering time leading to consume much lower energy than conventional sintering.

There are major potential and real advantages using microwave energy for material processing over conventional heating. These include the following:

- Time and energy savings
- Reduced processing time and temperature
- Rapid, volumetric, and selective heating
- Fine microstructures
- Improved physical and functional properties
- Lower environmental impact
- Controllable electric field distribution

## 1.3. Applications of Microwave Sintering of Ceramic Materials

Now microwave processing has been found that this technique can also be applied as efficiently and effectively to powdered metals as to many ceramics. Finally, The MWS operational expenses are less than 50–80% to the conventional sintering techniques. The MWS technique works 20 times faster than the conventional sintering method and takes only few minutes for processing than the conventional ones (takes hours).

## 1.4. Brief Introduction of Multiferroics

Multiferroic materials exhibits both ferroelectric and magnetic in nature and have much attracted research interest due to their potential application in multistate data storage and electric field controlled spintronics. Among all the studies related to the materials, transition metal oxides with perovskite structure are noteworthy [22, 26].

Multiferroic materials with double-perovskite structure $\left(AA'BB'O_6\right)$ are solid solutions of two perovskites: $\left(ABO_3\right)$ and $\left(A'B'O_3\right)$. In $\left(AA'BB'O_6\right)$, A and A' represent alkaline rare earth cations (La, Y, and Ce), while B and B' are transition metal cations (Ni and Co). If A and A' represent the same chemical element, the double perovskite has the general formula $\left(A_2BB'O_6\right)$ and the crystal structure of $A_2BB'O_6$ -type perovskite, as shown in Figure 2. Alkali-earth and lanthanide (smaller ion) ions are alone usually occupied in the A site [27, 28]. If the A ion is too small, the common expected distortions are cation displacement with $BO_6$ and octahedral ones [29].

The most representative $\left(A_2BB'O_6\right)$ ferromagnetic double perovskites are $La_2NiMnO_6$ [30–33], $La_2CoMnO_6$ [5, 34, 35], $La_2BMnO_6$ [36–48], and $La_2FeMnO_6$ [41, 42].

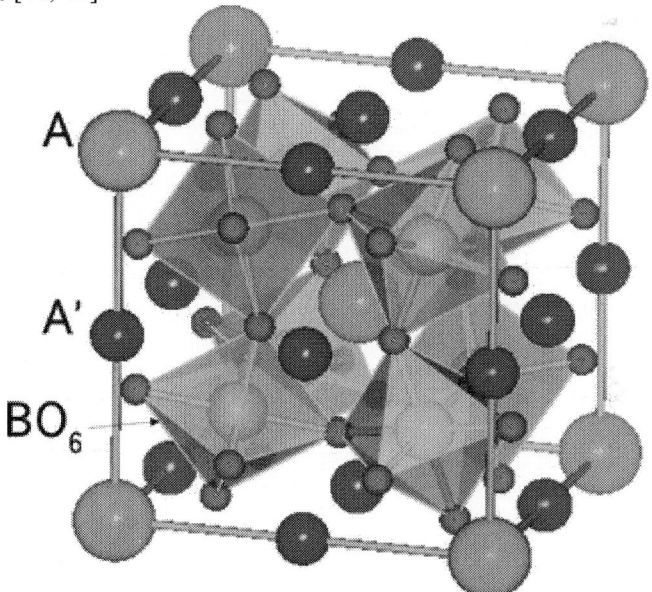

**Figure 2.** Crystal structure of $A_2BB'O_6$ type perovskite. The spheres at A and A'-site are for La and at B'-site are for Ni, Co. The network of corner-sharing $BO_6$ octahedra isare shown where oxygen atoms are in the corner of octahedra.

$La_2NiMnO_6$ (LNMO) has gained more attention as a rare example of a single-material platform with multiple functions, such as ferromagnetic (FM) semiconducting properties up to room temperature, magnetocapacitance, and magnetoresistance effects. The spin lattice coupling characteristics of LNMO exhibits a larger magnetodielectric (MD) effect close to room temperature. It has been well documented that the spins, electric charge, and dielectric functions in LNMO are turned by magnetic or electric fields. LNMO is considered as an FM semiconductor and shows Curie transition temperature ($T_c$) very close to room temperature. This property in LNMO makes the $Ni^{2+}$ and $Mn^4$ ions ordered and occupied the centers of $BO_6$ (corner shared) and $B'O_6$ structures respectively. This arrangement leads to the distribution of ideal double perovskite.

LNMO's structural system, $La_2CoMnO_6$ (LCMO), possesses an FM $T_c \sim 225$ K with an insulating behavior. The magnetic properties of the LCMO are strongly depending on the cation ordering, valences, defects, and synthetic conditions.

Among them, double perovskite $La_2NiMnO_6$ (LNMO) and $La_2CoMnO_6$ (LCMO) ceramics are attractive due to their impressive properties and potential on industrial applications [30–33, 42, 43]. LNMO is a ferromagnetic semiconductor with high critical temperature of $T_c \sim 280$ K, which may be used in commercial solid-state thermoelectric (Peltier) coolers [42]. LCMO is also a ferromagnetic semiconductor with critical temperature of $T_c \sim 230$ K [35–37]. Several crystal structures have been identified, and it is confirmed that the ferromagnetic semiconductors LNMO and LCMO with high $T_c$ are $P^2_{1/n}$ monoclinic structure, in which octahedra with Ni (or Co) and Mn centers alternately stacking along (111). Recent reports indicate LNMO and LCMO have considerable magnetodielectric effects at room temperature, which is very useful for future electronic device [29, 35, 44, 45].

The double perovskites $La_2MMnO_6$ (M = Co and Ni) are one of the most commonly occurring and important in all of materials science because they can exhibit novel magnetic, electric, and optical properties [28–44]. $La_2MMnO_6$ crystallizes in a double perovskite structure with rock salt configuration of $MO_6$ and $MnO_6$ octahedra. The ordering of $M^{2+}$ and $Mn^{4+}$ gives rise to 180° super exchange interactions based on Goodenough–Kanamori rules and consequently high ferromagnetic Curie transition temperature [43].

It is familiar that the properties of double perovskite compounds are strongly influenced by the materials composition and microstructure, which are sensitive to the preparation technique employed for their synthesis [46]. Various synthesis techniques such as sol–gel [30, 32, 35], coprecipitation [31], solid-state reaction method [33, 34], microwave sintering process [5], molten-salt synthetic process [26,27] sol-gel autocombustion [41], and chemical solution deposition method [47] have been reported in the preparation of double perovskite compounds. Each of the techniques has its own merits and limitations. For instance, conventional sintering is a simple and relatively cheap method with a long holding time (several hours), formation of lots of undesirable intermetallic compounds, and nonhomogeneous pore-size distribution. In the recent years, microwave sintering has emerged as a new sintering method for ceramics, semiconductors, metals, and composites.

Microwave sintering (MWS) technique has gained a lot of significance in recent times for materials (metals, composites, ceramics/nanoparticles) synthesis and sintering mainly because of its intrinsic advantages [5] such as rapid heating rates, reduced processing times, substantial energy savings novel and improved properties, finer microstructures, and being environmentally more clean. Therefore, it is viewed as one of the most advanced sintering techniques in material processing [5, 48] and improved physical and mechanical properties [7]. It has been shown that microwave sintering technique may provide enhanced densification in sintering of metal, oxides and non-oxide ceramics [5, 48, 49, 50].

However, to the best of our knowledge in the open literature, there have been only a few reports so far on the fabrication of double perovskite nanoparticles by microwave sintering approach [5, 51]. The purpose of the current chapter will focus on fabrication of the double perovskite $La_2MMnO_6$ (M = Ni, Co) ceramics and in order to further improve their magnetic and dielectric properties for practical spintronic applications through microwave sintering approach.

# 2. EXPERIMENTAL PROCEDURE

## 2.1. Materials

All the chemical reagents were of analytical pure grade (99.99%) and used without further purification. The versatile chemical coprecipitation–microwave sintering process [15] employed in present investigation is two-step process which consists of coprecipitation method is the first step of synthesis followed by microwave sintering in second half of experiment. High-purity $La(NO_3)_3.5H_2O$ (Merck), $Ni(NO_3)_2.6H_2O$ (Sigma-Aldrich), $Co(NO_3)_2.6H_2O$ (Sigma-Aldrich), and $Mn(NO_3)_3.4H_2O$ (Sigma-Aldrich) were used as starting materials. In a typical experimental process, the high purity stoichiometric amounts of $La(NO_3)_3.5H_2O$, $Ni/CO(NO_3)_2.6H_2O$, and $Mn(NO_3)_3.4H_2O$ were dissolved in appropriate amounts of deionized water and magnetically stared vigorously for 2 h at 80°C. The ammonia solution was used until to get 8.5 PH value. The stirring will continue for about 30 min, and the suspension was ball milled for about 24 h with ethanol as a milling media. The reactants were to be mixed well and then dried at 80°C in a cabinet drier for 24 h to obtain precursor powder sample. Then the powder was subjected to microwave sintering under uniform heating to get dense ceramics.

## 2.2. Microwave Sintering Setup

Microwave processing systems usually consist of a microwave source, for generation of microwaves, a circulator, an applicator to deliver the power to the load, and systems to control the heating and the experimental diagram of the microwave sintering set up used is shown in Figure 3. Most applicators are multimode, where different field patterns are excited simultaneously.

Further, In order to achieve pure double perovskite phases, the precursor samples were put into 2.45 GHz, 6 kW continuously adjustable microwave equipment (HAMiLab-HV3, Syno-Therm), the maximum operating temperature up to 1400°C, and 0.5–3 kW. The multimode microwave furnace consists of a cubical stainless steel chamber with a side of 30 cm. Two magnetrons (microwave source), each with a maximum rated power of 1100 watts, are situated opposite to each other. A box made of alumina, zirconia, and silica mixed cardboard is used as a thermal insulator. The material is positioned in the center of box and is surrounded by silicon carbide (susceptor) plates.

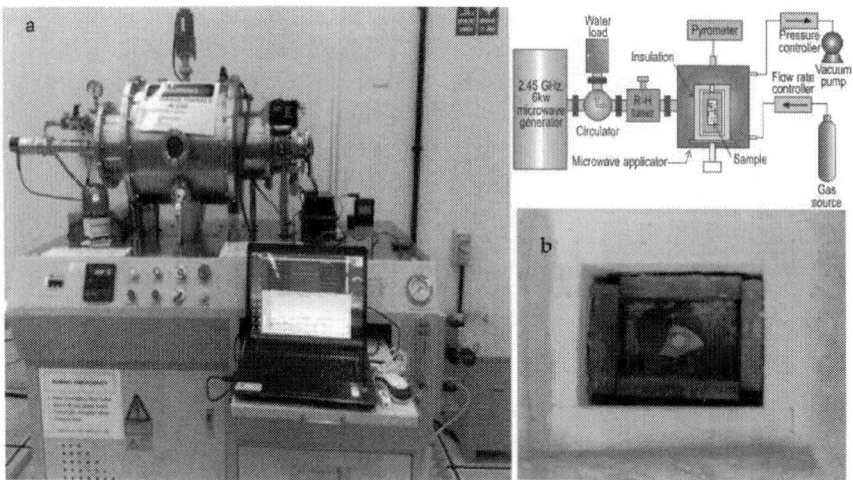

**Figure 3.**Experimental setup of multimode microwave furnace [3] (a) and Ssusceptor (b).

During the sintering process, the microwave sintering chamber was filled with high purity nitrogen gas flow (99.999%). An adjustable programmed electrical control system was used to deliver the required energy to the sample. The employed heating chamber was made up with stainless steel double walled tubular cavity with water-cooled facility, and the maintained processing temperature is about 1400°C. A high purity quartz crystal cylinder arrangement is available inside the chamber, where the samples were loaded for processing; the temperature of the sample was measured using infrared pyrometer during the MW sintering. The SiC plates surrounded in the crucible were served as susceptors and provide initial heating to be compact disc samples. Once the materials received absorb sufficient MW heat including the core and will get uniform heating. The secondary purpose of SiC is to maintain the surface temperature. The crucible was positioned at the center of the furnace so it provides strong MW radiation. The green compacted disks for heated at 900°C for 10 min in atmospheric $N_2$ ambient temperature and heating rate of 20°C/min is maintained by varying magnetron power between 1000 and 2500 W followed by normal frequency cooling.

## 2.3. Characterization and Property Measurements of La$_2$mmno6 (M = Ni, Co) Ceramics

The crystal structure of the microwave sintered products was characterized by X-ray diffraction (XRD) using a Shimadzu X-ray diffractometer with Cu-K$_\alpha$ radiation $20^\theta$ range of 20 to 80°. Raman spectra were carried out on an RM-1000 micro-Raman spectrometer with the 514.53 nm line of an argon laser under ambient conditions. The composition, morphology, and microstructures of the products were characterized by transmission electron microscope (TEM FEI Tecnai F20 microscope, Japan) and field emission scanning electron microscope (FESEM, Hitachi S-4800, Japan) equipped with an energy-dispersive X-ray spectrometer (EDS). Fourier transform infrared spectroscopy (FTIR) was performed on a Nicolet 5700 spectrometer in the wave number range of 400–4000 $cm^{-1}$. The spectroscopic grade KBr pellets were used for collecting the spectra with a resolution of 4 $cm^{-1}$ performing 32 scans. X-ray photoelectron spectroscopy (XPS) was performed on an ESCA-UK XPS system with an Mg $K_\alpha$ excitation source (hv = 1486.6 eV), where the binding energies were referenced to the C1s peak at 284.6 eV of the surface adventitious carbon. The magnetic properties were measured using a physical property measurement system (PPMS-9, Quantum Design, Inc., San Diego, CA, USA) at room temperature under a maximum field of 20 kOe. Silver paste was applied on both sides of the pellet for the electrical measurements. The variation of dielectric constant and dielectric loss as a function of frequency at room temperature and as a function of temperature at different frequencies was measured using computer interfaced HIOKI 3532-50 LCR-HITESTER.

# 3. MICROWAVE-SINTERING OF ENGINEERED DOUBLE PEROVSKITE CERAMIC MATERIALS

## 3.1. Crystal Structure of La₂mmno6 (M = Ni, Co) Ceramics

The phase structure of the microwave sintered LNMO and LCMO nanoparticles was characterized by X-ray diffraction (XRD). As shown in Figure 4, no extra reflection peaks other than those of pure perovskite phase are detected, indicating the high purity of nanoparticles can be obtained in 10 min by this microwave sintering approach, which confirms the formation of single phase compositions of LNMO and LCMO double perovskites [30].

The crystallite size was calculated from XRD patterns using the Debye–Scherrer formula [7], described by the Eq. (7):

$$D = \frac{0.94 \times \lambda}{\beta_{1/2} \times \cos\theta}$$

where $D$ = crystallite size, $\lambda$ = radiation wavelength (1.5405 Å), $\beta_{1/2}$ = half-width of diffraction profile, and $\theta$ = diffraction angle.

The average crystal size was found to be 23 nm for LNMO and 28 nm for LCMO, which are 2–3 times smaller than the particle/grain sizes measured by TEM as shown in below section.

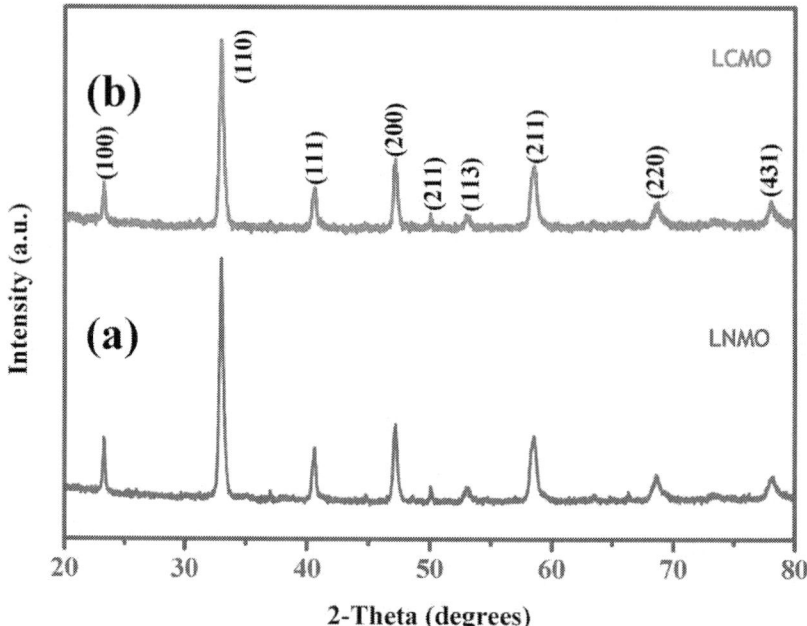

**Figure 4.**XRD patterns of the microwave sintered (a) LNMO and (b) LCMO ceramics.

Raman spectroscopy is one of the most important tools to attain the information about the structure of the samples. The Raman spectra of microwave sintered LNMO and LCMO ceramics are shown inFigure 5. The Raman spectra display two characteristics peaks at around 514, 653 cm$^{-1}$ for LNMO and 488, 670 cm$^{-1}$ for LCMO ceramics, corresponding to the well-documented A band and B band, respectively. Martín–Carron et al. have assign the two peaks to the $A_g$antisymmetric stretching (or Jahn–Teller stretching mode) and $B_g$symmetric stretching vibrations of the MnO$_6$ octahedra, respectively [34–36, 52–54]. It is well known that the $A_g$band is usually assigned to antisymmetric stretching (or Jahn–Teller stretching mode), while the $B_g$ band distributed to symmetric stretching vibrations. A noticeable difference is seen between our LNMO/LCMO ceramics and the bulk sample: the $A_g$and $B_g$ peaks for the nanoparticles shift to higher binding energy, 13 and 25 cm$^{-1}$, respectively, when compared to the bulk crystal. The shifting may occur due to surface strain of the crystal structure [55].

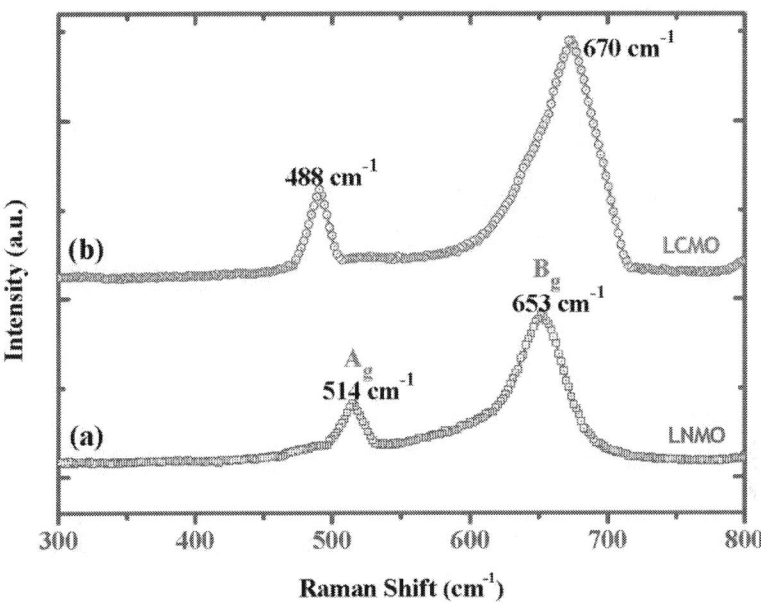

**Figure 5.**Raman of the microwave sintered (a) LNMO and (b) LCMO ceramics.

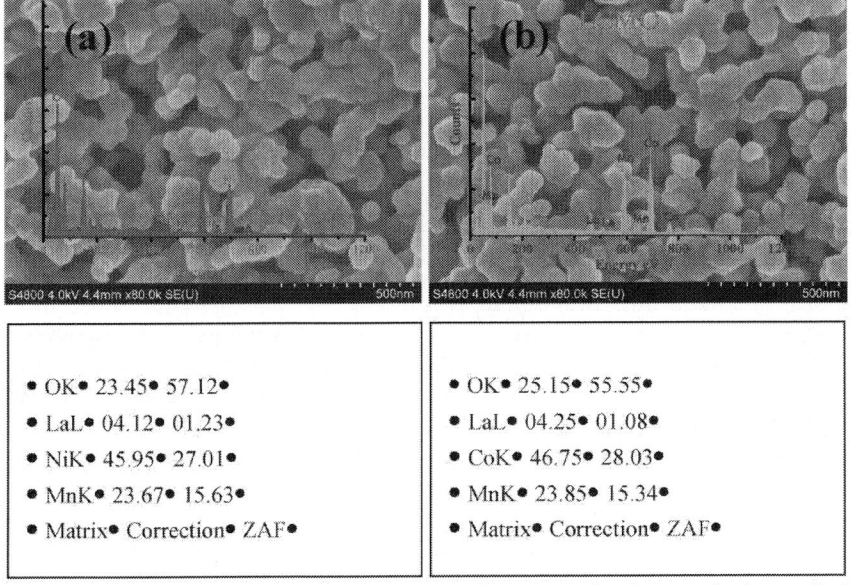

| | | | | | |
|---|---|---|---|---|---|
| • OK• 23.45• 57.12• | | | • OK• 25.15• 55.55• | | |
| • LaL• 04.12• 01.23• | | | • LaL• 04.25• 01.08• | | |
| • NiK• 45.95• 27.01• | | | • CoK• 46.75• 28.03• | | |
| • MnK• 23.67• 15.63• | | | • MnK• 23.85• 15.34• | | |
| • Matrix• Correction• ZAF• | | | • Matrix• Correction• ZAF• | | |

**Figure 6.**FESEM images, EDX spectra (inset) and elemental data of the microwave sintered (a) LNMO and (b) LCMO ceramics.

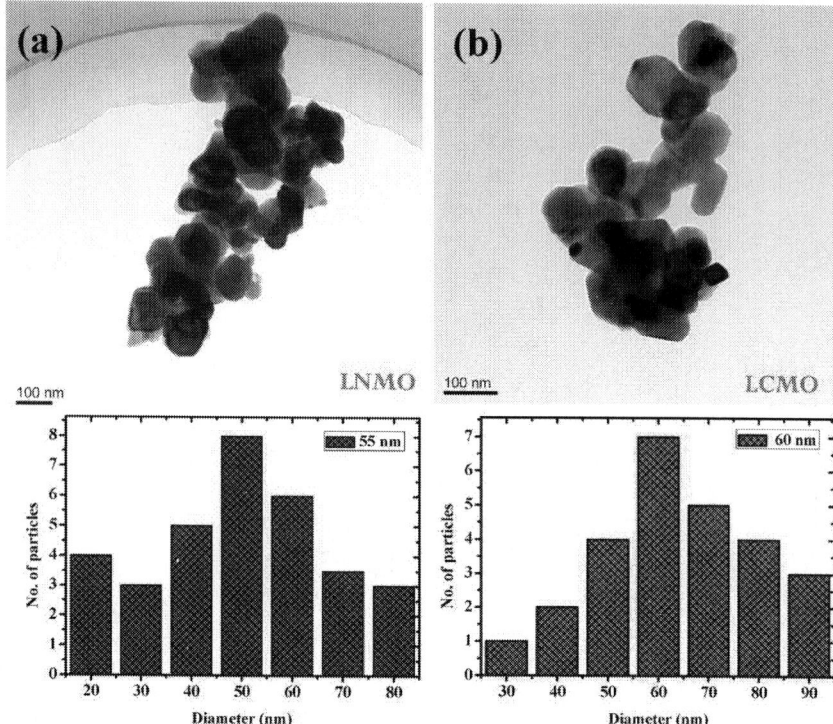

**Figure 7.**TEM images and particle size distributions of the microwave sintered (a) LNMO and (b) LCMO ceramics.

The microstructure and morphology of microwave sintered LNMO and LCMO ceramics were investigated by FESEM and TEM techniques. Typical SEM images of $La_2MMnO_6$ (M = Ni, Co) nanoparticles are shown in Figure 6, the average grain size is about 52 nm and 58 nm for $La_2NiMnO_6$and $La_2CoMnO_6$, respectively. The grain size of $La_2CoMnO_6$ is bigger than that of $La_2NiMnO_6$, which obeys the rule that relatively large ionic radius id benefit to the diffusion in the microwave sintering process.

From the morphologies of both samples, the grains seem to be homogeneous and form a group of cluster phenomenon. The perovskite material has better microwave absorption capability [5, 51] and leads to fine grain growth during the sintering process.

The EDX spectra (inset of 6a and 6b) and their corresponding tables confirm the presence of the constituent elements (La, Ni, Co, Mn, and O), the composition being nearly the same as that of stoichiometric $La_2NiMnO_6$ and $La_2CoMnO_6$, respectively.

As shown in Figures 7a and 7b, transmission electron microscopy (TEM) was applied for all samples to determine particle size and confirmed that the particle sizes are about 53 ± 12 and 60 ± 15 nm for LNMO and LCMO, respectively, which agrees good agreement with the SEM results. From the TEM micrograph, nanosized grains with quasi spherical shape can be observed.

The formation mechanism of the perovskite type structure in the microwave sintered LNMO and LCMO ceramics is further supported by FT-IR spectrum shown in Figure 8. The FTIR spectrum is used to characterize the phase composition and purity of the prepared samples. The intense peak around 3423 cm$^{-1}$ is referring to the stretching vibration of hydroxyl group. In addition, the bands at about 1629 cm$^{-1}$ can be ascribed to the asymmetric COO$^-$ stretching vibrations. The bands at 1450 and 1357 cm$^{-1}$ attributed to the asymmetric stretching of $CO_3^{2-}$. The intensive absorption band observed at 597cm$^{-1}$ can be assigned to Fe–O stretching vibrations formed by the octahedral MnO$_6$ group.

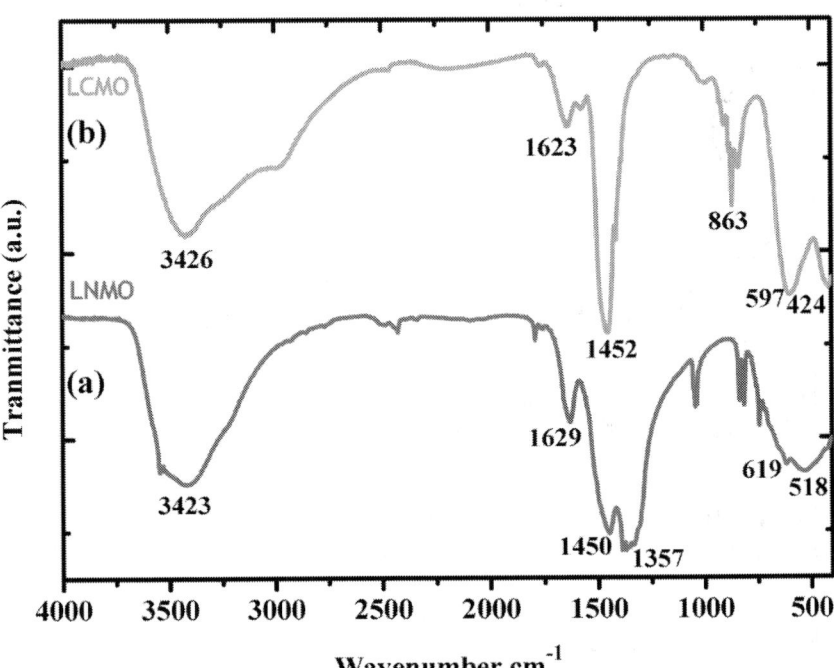

**Figure 8.** FTIR of the microwave sintered (a) LNMO and (b) LCMO ceramics.

The chemical states of elements of Ni, Mn in LNMO, and Co, Mn in LCMO ceramics was further investigated by X-ray photoelectron spectroscopy. The XPS core level spectra of Ni 2p, Co 2p, and Mn 2p of La$_2$MMnO$_6$ (M = Ni, Co) are presented in Figure 9. A Ni 2P$_{3/2}$ signal was observed at 851.3 eV along with a satellite peak at 858.5 eV. Another peak was noticed at 869.5 eV and ascribed to the Ni 2P$_{1/2}$ level. Auger electron peak of Ni of [Figure 9a] explains the presence of +2 oxidation state of the nickel in LNMO ceramics. The Mn 2p$_{3/2}$ peak of LNMO is at 638.4 eV, while the same Mn 2p$_{3/2}$ peak is at 641.5 eV for Mn$_2$O$_3$ [56]. In the spectrum of Co2p (Figure 6c), the peaks of Co 2p$_{3/2}$ and Co 2p$_{1/2}$states were located at 777.3 and 782.7 eV, respectively [57]. The Mn 2p$_{3/2}$ peak of LCMO shown inFigures 6 b and 6d is found at 637.6 eV, close to that in Mn$_2$O$_3$ [56].

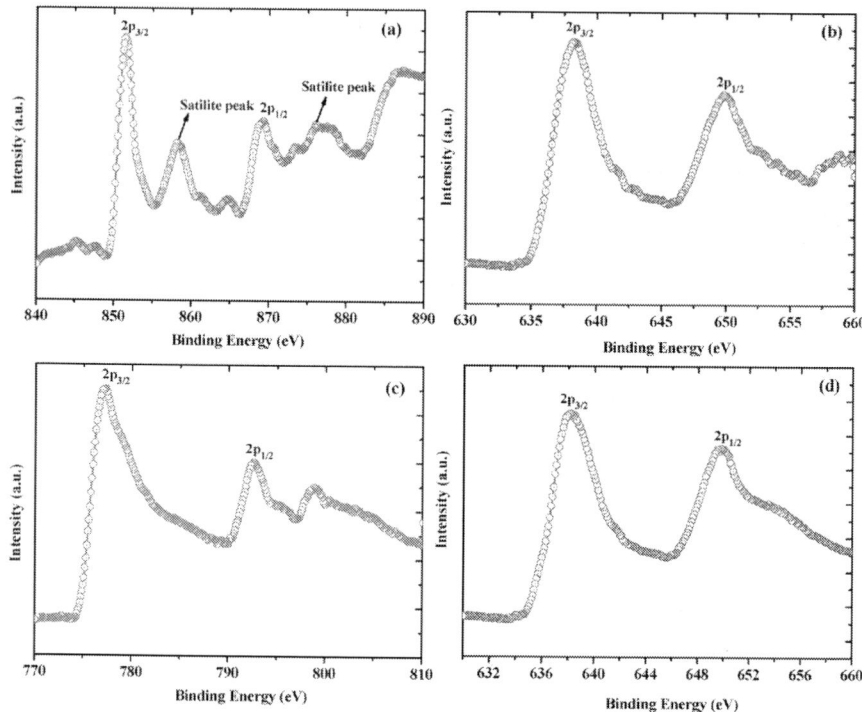

**Figure 9.** High-resolution XPS spectra's of (a) Ni 2p peaks, (b) Mn 2p peaks for LNMO and (c) Co 2p peaks, (d) Mn 2p peaks for LCMO ceramics.

### 3.1.1. Magnetic Analysis Of La2mmno6 (M = Ni, Co) Ceramics

The magnetization characteristics were measure both as a function of applied magnetic field at fixed temperatures and as a function of temperature at fixed fields. The room temperature hysteresis loops of the microwave sintered $La_2MMnO_6$ (M = Ni, Co) ceramics were measured using physical property measurement system (PPMS). The magnetization curves, as shown in Figures 10a and 10c, display relatively high saturation magnetization. The magnetic saturation ($M_s$) values of LNMO and LCMO are 42.9 and 65.4 emu/g, respectively, which is lower their theoretical values of 47.5 and 71.21 emu/g reported in the literature [40]. One can note that MWS products saturation magnetization was higher than for the conventionally sintering products [36], indicating that MWS method is efficient to fabricated high quality double perovskite material.

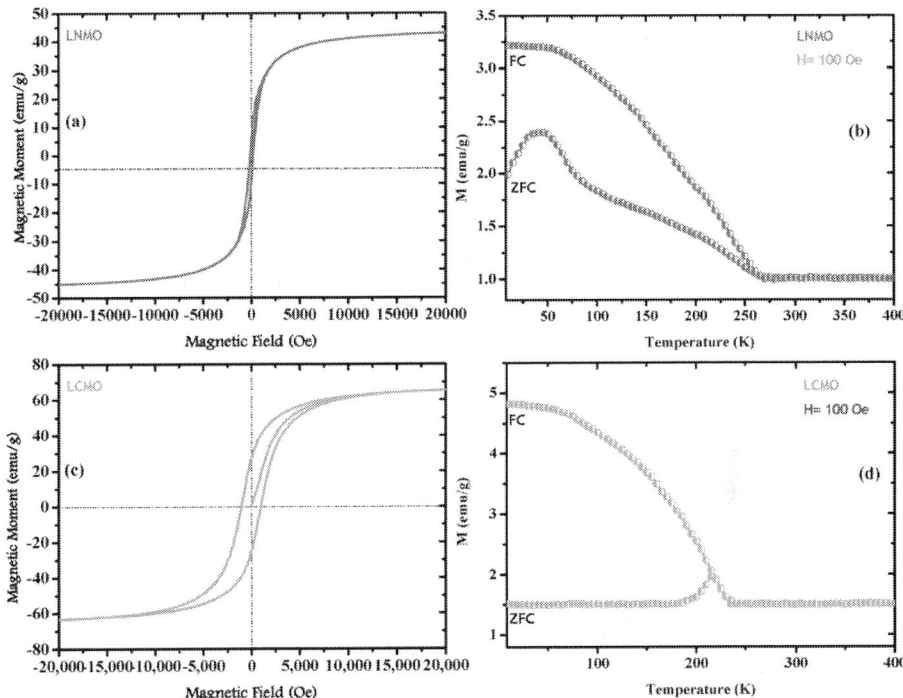

**Figure 10.**Magnetic field dependent magnetization data at 5 K (a and c), zZero-Ffield-Ccooled (ZFC) and Ffield-Ccooled (FC) magnetization as a function of temperature (b and d) the microwave sintered LNMO and LCMO ceramics.

The frequency dependence of saturation magnetization hysteresis curves was recorded at room temperature for the LNMO ceramics as shown in Figure 10a. A hysteresis loop has been observed at 5 K with a coercive field of ~282 Oe and remnant magnetization of ~7.7 emu/g, which show that the LNMO sample exhibit typical ferromagnetic behavior. Figure 10c shows the variation of magnetization as a function of magnetic field for LCMO ceramics. A hysteresis loop has been observed at 5 K with a coercive field of ~972 Oe and remnant magnetization of ~8.14 emu/g, which show that the LCMO sample exhibit typical super paramagnetic behavior. Apart from the magnetic characteristics presented here, detailed examination is in progress and the extensive and expected results will be reported elsewhere shortly.

Figures 10b and 10d show the temperature dependence of magnetization measurements for LNMO and LCMO under an applied field was carried out by field-cooled (FC) and zero-field-cooled (ZFC) processes at an applied magnetic field of 100 Oe in the temperature range of 5–400 K. For the LNMO, It could be observed from ZFC as well as FC magnetization that the material shows two ferromagnetic transitions around 270 K and 240 K, which is reliable with the presence of two phases as showed by the X-ray diffraction studies. As the ferromagnetic transition temperature in the pure monoclinic phase is found to be near 255 K, we attribute the ferromagnetic transition at 240 K to the rhombohedral phase. FC magnetization reaches a maximum value of ~3.2 emu/g at 5 K. The magnetic transition at ~255 K indicates the onset of FE long-range ordering, very close to the magnetic transition temperature ($T_c =$ ~280 K) reported earlier in the literature [41]. It is pertinent to maintain that there is a divergence between ZFC and FC magnetization curves below 220 K for the LCMO. Noticeable difference has also been observed in the case of low field ZFC and FC magnetization for LNMO particles. These LCMO nanoparticles possess a single magnetic transition at about 225 K under 100 Oe field. This observation is very close to the behavior of bulk LCMO ceramics [58]. The maximum FC magnetization is noticed about 4.8 emu/g at 5 K.

## 3.2. Dielectric Properties of La$_2$mmno$_6$ (M = Ni, Co) Ceramics

The temperature variation of dielectric constant ($\varepsilon'$) and loss tangent (tan $\delta$) at different frequencies ranging from 1 kHz to 1 MHz for the microwave sintered La$_2$MMnO$_6$(M = Ni, Co) ceramics is shown in Figures 11a–11d. Noticeably, the dielectric constant ($\varepsilon'$) decreases significantly with increasing frequency. An interesting Maxwell–Wager relaxation [59]-type dielectric behavior (at high dielectric constant) has been noticed around 450 K in these materials and also strong dispersion in the relaxation spectra. The dielectric constant is gradually increased first along with the increase in temperature and attains significant growth at a critical temperature. The critical temperature value shifts toward higher side as and when the measuring frequency increases. These features indicate the thermally activated process [59]. This phenomenon has been most widely described in various earlier reports [24, 59–61].

Such dielectric performance could be attributed to the cationic disorder prompted by the exchange of B sites [62]. In the present systems, Ni$^{3+}$/Co$^{3+}$ and Mn$^{4+}$ ions instantly exist in B sites, which results in two kinds of BO$_6$ octahedra in the structure of La$_2$MMNO$_6$. Therefore, the ion disorder in the unit cell should be one of the causes for this behavior.

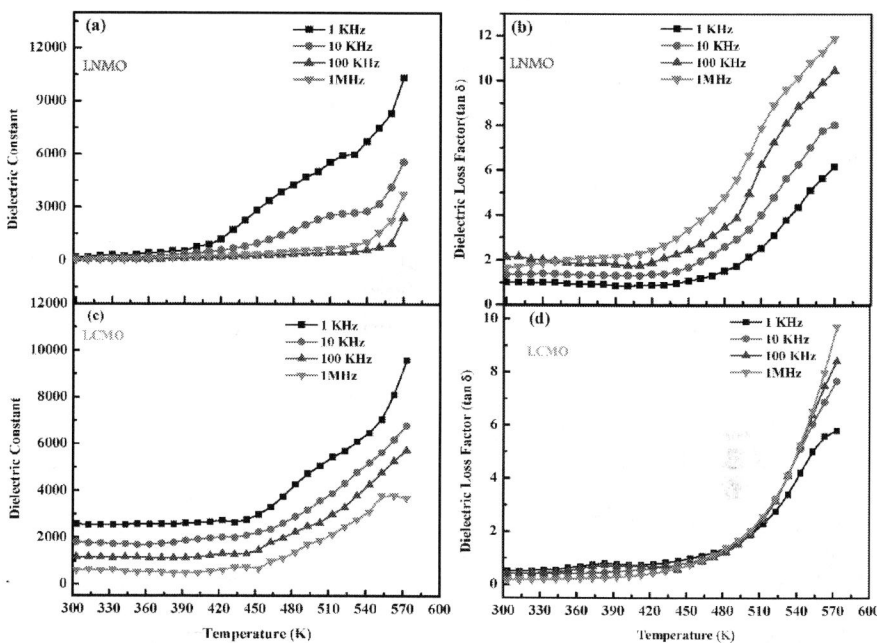

**Figure 11.**Temperature dependence of dielectric constant and dielectric loss of the microwave sintered (a & b) LNMO and (c & d) LCMO ceramics at various frequencies.

Figures 12a and 12b show the dielectric constant ($\varepsilon'$) as a function of frequency of LNMO and LCMO ceramics at different temperatures. It can be observed that the dielectric constant of both ceramics decreases as frequency increases. The decrease in the dielectric constant with increase in frequency can be explained by the behavior on the basis of electron happing from $Fe^{2+}$ to $Fe^{3+}$ ions or on basis of decrease in polarization with the increase in frequency. Polarization of a dielectric material is the quantity of the contributions of ionic, electronic, dipolar, and interfacial polarizations [63]. At low frequencies, polarization mechanism is keenly observed at low frequencies to the time varying electric fields. As the frequency of the electric field increases, different polarization contributions are filter out under leads to the decrement in net polarization under dielectric constant. Similar behavior has also been reported by different investigators earlier in the literature [60, 64]. The physical, magnetic, and dielectric properties of LMNO and LCMO are summarized in Table 1.

**Table 1.**Physical and multiferroic characteristics of the microwave sintered LNMO and LCMO nanoparticles

| Parameters | LNMO | LCMO |
|---|---|---|
| $d_{XRD}$ (nm) | 23 | 28 |
| $d_{TEM}$ (nm) | 53 | 60 |
| $A_g$ (cm$^{-1}$) | 514 | 653 |
| $B_g$ (cm$^{-1}$) | 488 | 670 |
| $M_s$ (emu/g) | 42.9 | 65.4 |
| $M_r$ (emu/g) | 7.7 | 8.14 |
| $H_c$ (Oe) | 272 | 972 |
| $T_N$ (K) | 225 | 240 |
| DC (1 MHz) | 161 | 19.7 |
| DL (1 MHz) | 1.235 | 0.172 |

# 4. CONCLUSIONS AND SCOPE OF THE FUTURE WORK

In this chapter, double perovskite La$_2$MMnO$_6$ (M = Ni, Co) ceramics were successfully prepared using microwave sintering technique. The microwave sintering approach shows obviously better promise over the conventional heating method, in terms of higher efficiency of heating, significantly shorter reaction time, smaller and more regular size, and stronger magnetization of the products. IR, XRD, and SEM-EDX analysis confirmed the formation of single phase for La$_2$MMnO$_6$ (M = Ni, Co) double perovskites. XPS gives information about the oxidation state and chemical stoichiometric composition of the perovskite samples. The oxidation states of transition metals are Ni$^{2+}$ and Mn$^{4+}$ in the samples. The variation with temperature of a sample's magnetization can give Neel temperatures of about 240 K for LNMO and 225 K for LCMO. Multiferroic properties of microwave sintered ceramics were found to be higher than that of the values reported for conventionally sintered samples. Microwave processing greatly reduced the processing time and improved the magnetic and dielectric properties, which hinted its superiority over conventional processing. Hence, the microwave sintering technique is more facile approach for the preparation of the industrially important perovskite-type ceramics.

Furthermore, one can study the correlation between different magnetic and electric order parameter for this material, to use as a potential candidate for multiferroics. The magnetization dynamics can be studied to know the spin–spin interactions. Low temperature dielectric spectroscopy can also be studied to know whether this material undergoes any transition or not and if so what is the inside story of this transition and much more work can be done related to this material.

# REFERENCES

1. Vasylkiv O, Demirskyi D, Sakka Y, Ragulya A, Borodianska H. Densification kinetics of nanocrystalline zirconia powder using microwave and spark plasma sintering—a comparative study. Journal of Nanoscience and Nanotechnology 2012;12: 4577–82.

2. Demirskyi D, Borodianska H, Grasso S, Sakka Y, Vasylkiv O. Microstructure evolution during field-assisted sintering of zirconia spheres. Scripta Materialia 2011;65: 683–6.

3. Shannigrahi SR, Pramoda KP, Nugroho FAA. Synthesis and characterizations of microwave sintered ferrite powders and their composite films for practical applications. Journal of Magnetism and Magnetic Materials 2012;324: 140–145.

4. Penchal Reddy M, Madhuri W, Balakrishnaiah G, Ramamanohar Reddy N, Siva Kumar KV, Murthy VRK, Ramakrishna Reddy, R. Microwave sintering of iron deficient Ni–Cu–Zn ferrites for multilayer chip inductors. Current Applied Physics 2011;11: 191–8.

5. Penchal Reddy M, Zhou XB, Jing L, Huang Q. Microwave sintering, characterization and magnetic properties of doubleperovskiteLa2CoMnO6 nanoparticles. Material Letters 2014;132: 55–8.

6. Demirskyi D, Agrawal D, Ragulya A. Neck growth kinetics during microwave sintering of copper. Scripta Materialia 2010;62: 552–5.

7. Roy R, Agrawal D, Cheng J, Gedevanishvili S. Full Sintering of powdered-metal bodies in a microwave field. Nature 1999;399: 668–70.

8. Oghbaei M, Mirzaee O. Microwave versus conventional sintering: a review of fundamentals, advantages and applications. Journal of Alloys and Compounds 2010;494:175–89.

9. Demirskyi D, Ragulya A, Agrawal D. Initial stage sintering of binderless tungsten carbide powder under microwave radiation. Ceramics International 2011;37: 505–12.

10. Demirskyi D, Borodianska H, Agrawal D, Ragulya A, Sakka Y, Vasylkiv O. Peculiarities of the neck growth process during initial stage of spark-plasma, microwave and conventional sintering of WC spheres. Journal of Alloys and Compounds 2012;523: 1–10.

11. Plapcianu C, Agostino A, Badica P, Aldica GV, Bonometti E, Ieluzzi G, Popa S, Truccato M, Cagliero S, Sakka Y, Vasylkiv O, Vidu R. Microwave synthesis of fullerene-doped MgB2. Industrial and Engineering Chemistry Research 2012;51: 11005–10.

12. Borrell A, Salvador MD, Rayon E, Penaranda-Foix FL. Improvement of microstructural properties of 3Y-TZP materials by conventional and nonconventional sintering techniques. Ceramics International 2012;38: 39–43.

13. Costa ACFM, Tortella E, Morelli MR, Kiminami RHGA. Microstructure and magnetic properties of Ni–Zn ferrites. Journal of Magnetism and Magnetic Materials 2003;256: 174–82.

14. Phani AR, Santucci S. Evaluation of structural and mechanical properties of aluminum oxide thin films deposited by a sol–gel process: comparison of microwave to conventional anneal. Journal of Non-Crystalline Solids 2006;352: 4093–100.

15. Zhou XB, Qiu F, Penchal Reddy M, Han YH, Lee J, Huang Q. Comparative study of conventional and microwave sintered mullite fibers: a structural study. Advances in Applied Ceramics: Structural, Functional and Bioceramics 2015;114: 139–43.

16. Penchal Reddy M, Madhuri W, Venkata Ramana, M, Ramamanohar Reddy N, Siva Kumar KV, Murthy VRK, Ramakrishna Reddy R. Effect of sintering temperature on structural and magnetic properties of NiCuZn and MgCuZn ferrites. Journal of Magnetism and Magnetic Materials 2010;322: 281923.

17. Praveena K, Bououdina M, Penchal Reddy M, Srinath S, Sandhya R, Sadhana K. Structural, magnetic, and electrical properties of microwave sintered Cr3+ doped Sr hexa ferrites. Journal of Electronic Materials 2012;44: 1–7.

18. Gupta M, Wong WLE. Microwaves and Metals. Singapore: John Wiley & Sons; 2007.

19. Upadhya DD, Ram Prasad. DAE-BRNS Symposium, IT-03, 9. Shivaji University, Kolhapur.

20. Jin Q. Microwave Chemistry. Beijing: China Science Press; 1999.

21. Berteaud AJ, Badet JC. High temperature microwave heating in refractory materials. Journal of Microwave Power Electromagnetic Energy 1976;11: 315–20.

22. Liu P, Wang H, Cheng X, Shui A, Zeng L. China Ceramics 2005;41(4): 12–15.

23. Kimura T, Goto T, Shintani H, Ishizaka K, Arima T, Tokura Y, Magnetic control of ferroelectric polarization. Nature 2004;26: 55–58.

24. Ishiwata S, Tokunaga Y, Taguchi Y, Tokura Y. High-pressure hydrothermal crystal growth and multiferroic properties of a perovskite YMnO3. Journal of American Chemical Society 2011;133: 13818–13820.

25. Chai YS, Oh YS, Wang LJ, Manivannan N, Feng SM, Yang YS, Yan YQ, Jin CQ, Kim KH. Intrinsic ferroelectric polarizations of orthorhombic manganites with E-type spin order. Physical Review B 2012;85: 184406.

26. Lee JH, Jeong YK, Park JH, Oak MA, Jang HM, Son JY, Scott JF. Spin-canting-induced improper ferroelectricity and spontaneous magnetization reversal in SmFeO3. Physical Review Letters 2011;107: 117201.

27. Weber MC, Kreisel J, Thomas PA, Newton M, Sardar K, Walton RI. Phonon Raman scattering of RCrO3 perovskites (R=Y, La, Pr, Sm, Gd, Dy, Ho, Yb, Lu). Physical Review B 2012;85: 054303.

28. Retuerto M, Lope MJM, Hernandez GM, Munoz A, Diaz MTF, Alonso JA. Crystal and magnetic study of the disordered perovskites Ca(Mn0.5Sb0.5)O3 and Ca(Fe0.5Sb0.5)O3. Materials Research Bulletin 2010; 45: 1449–1454.

29. Ivanov SA, Nordblad P, Tellgren R, Hewat A, Temperature evolution of structure and magnetic properties in the perovskite Sr2MnSbO6. Materials Research Bulletin 2009;44: 822–830.

30. Lufaso MW, Barnes PW, Woodward PW. Structure prediction of ordered and disordered multiple octahedral cation perovskites using SPuDS. Acta Crystallographica Section B: Structural Science 2006;62: 397–410.

31. Zhao S, Shi L, Zhou S, Zhao J, Yang H, Guo Y. Journal of Applied Physics 2009;106:123901.

32. Wu ZY, Ma CB, Tang XG, Li R, Liu QX, Chen BT, Double-perovskite magnetic La2NiMnO6 nanoparticles for adsorption of bovine serum albumin applications. Nanoscale Research Letters 2013;8: 207–211.

33. Li C, Liu B, He Y, Lv C, He H and Y. Xu. Preparation, characterization and dielectric tunability of La2NiMnO6 ceramics. Journal of Alloys and Compounds 2014;590: 541–545.

34. Kumar P, Ghara S, Rajeswaran B, Muthu DVS, Sundaresan A, Sood AK. Temperature dependent magnetic, dielectric and Raman studies of partially disordered La2NiMnO6. Solid State Communications 2014;184: 47–51.

35. Lin YQ, Chen XM. Local structure evolution in Ba-substituted Pb(Fe1/2Nb1/2)O3 ceramics. Journal of American Society 2011;94(3): 782–787.

36. Murthy JK, Chandrasekhar KD, Murugavel S, Venimadhav A. Investigation of intrinsic magnetodielectric effect in La2CoMnO6: role of magnetic disorder. Journal of Material Chemistry C 2015; 3, 836–843.

37. Mao Y, Facile molten-salt synthesis of double perovskite La2BMnO6 nanoparticles. RSC Advances 2012;2: 12675–12678.

38. Mao Y, Parsons J, McCloy JS. Magnetic properties of double perovskite La2BMnO6 (B=Ni or Co) nanoparticles. Nanoscale 2013;5: 4720–4728.

39. Zhu M, Lin Y, Lo EWC, Wang Q, Zhao Z, Xie W. Electronic and magnetic properties of La2NiMnO6 and La2CoMnO6 with cationic ordering. Applied Physics Letters 2012;100: 062406.

40. Barrozo P, Moreno NO, Aguiar JA. Ferromagnetic cluster on La2FeMnO6. Advanced Materials Research 2014;975: 122–127.

41. Qian Y, Wu H, Kan E, Lu J, Lu R, Liu Y, Tan W, Xiao C, Deng K. Biaxial strain effect on the electronic and magnetic phase transitions in double

perovskite La2FeMnO6: a first-principles study. Journal of Applied Physics 2013;114: 063713.

42. Rogado NS, Li J, Sleight AW, Subramanian MA. Magnetocapacitance and magnetoresistance near room temperature in a ferromagnetic semiconductor:La2NiMnO6. Advanced Materials 2005;17: 2225–2227.

43. Lufaso MW, Woodward PM. Jahn–Teller distortions, cation ordering and octahedral tilting in perovskites. Acta Crystallographica Section B: Structural Science 2004;60: 10–20.

44. Dass RI, Goodenough JB. Multiple magnetic phases of La2CoMnO6. Physical Review B 2003;67: 014401.

45. Wolf SA, Awschalom DD, Buhrman RA, Daughton JM, Molnar S, Roukes ML, Chtchelkanova AY, Treger DM. Spintronics: a spin-based electronics vision for the future. Science 2001;294:1484–88.

46. Awschalom DD, Flatte ME, Samarth N. Scientific American. 2002, 286, 66–73.

47. Prasatkhetragarn A, Kaowphong S, Yimnirun R. Synthesis, structural and electrical properties of double perovskite Sr2NiMoO6 ceramics. Applied Physics A: Materials Science and Processing 2012;107(1): 117–121.

48. Gu Y, Wang Y, Wang T, Shi W. Structure and current-induced effect on the resistivity of La2CoMnO6 thin films. Materials Chemistry and Physics 2012;132: 466–70.

49. Agrawal D. Microwave sintering of ceramics, composite, metals, and transparent materials. Journal of Material Education 1999;19: 49–58.

50. Agrawal DK, Microwave processing of ceramics: a review. Current Opinion in Solid State Material Science 1998;3: 480–486.

51. Penchal Reddy M, Madhuri W, Ramamanohar Reddy N, Siva Kumar KV, Murthy VRK, Ramakrishna Reddy R. Influence of copper substitution on magnetic and electrical properties of MgCuZn ferrite prepared by microwave sintering method. Materials Science and Engineering: C 2010;30: 1094–1099.

52. Uvarov U, Popov I. Metrological characterization of X-ray diffraction methods for nanocrystallite size determination. Material Characterization 2007;58: 883–91.

53. Carron LM, Andres A, Lopez MJM, Casais MT, Alonso JA. Raman phonons as a probe of disorder, fluctuations, and local structure in doped and undoped orthorhombic and rhombohedral manganites. Physical Review B 2002;66: 174303.

54. Carron LM, Andres A. Excitations of the orbital order in RnMnO3 manganites: light scattering experiments. Physical Review Letters 2004;92: 175501.

55. Guo H, Burgess J, Street S, Gupta A, Calvarese TG, Subramanian MA. Structural and magnetic properties of epitaxial thin films of the ordered double perovskite La2CoMnO6. Applied Physics Letters 2006;89: 022509.

56. Burgess J, Guo H, Gupta A, Street S. Raman spectroscopy of La2NiMnO6 films on SrTiO3 (100) and LaAlO3 (100) substrates: observation of epitaxial strain. Vibrational Spectroscopy 2008;48: 113–117.

57. Kim KJ, Lee HJ, Park. Cationic behavior and the related magnetic and magnetotransport properties of manganese ferrite thin films. Journal of Magnetism and Magnetic Materials 2009;321: 3706–3711.

58. Zhou G, Lee DK, Kim YH, Kim CW, Kang YS. Preparation and spectroscopic characterization of Ilmenite-Type CoTiO3 nanoparticles. Bulletin of Korean Chemical Society 2006;27: 368–372.

59. Schnorr JM, Swager TM. Emerging applications of carbon nanotubes. Chemical Materials 2011;23: 646–657.

60. Zhang Z, Jian H, Tang X, Yang J, Zhu X, Sun Y. Synthesis and characterization of ordered and disordered polycrystalline La2NiMnO6 thin films by sol–gel. Dalton Transaction 2012;41: 11836–11840.

61. Qing TY, Meng Y, Mei HY. Structure and colossal dielectric permittivity of Ca2TiCrO6 ceramics. Journal of Physics D: Applied Physics 2013;46: 015303.

62. Liu F, Li J, Li Q, Wang Y, Zhao X, Hu Y, Wang C, Liu X. High pressure synthesis, structure, and multiferroic properties of two perovskite compounds Y2FeMnO6 and Y2CrMnO6. Dalton Transaction 2014;43: 1691–1698.

63. Subramanian MA, Li D, Duan N, Reisner BA, Sleight AW. High dielectric constant in ACu3Ti4O12 and ACu3Ti3FeO12 phases. Journal of Solid State Chemistry 2000;151: 323–325.

64. Singh AK, Goel TC, Mendiratta RG., Thakur OP, Prakash C. Dielectric properties of Mn-substituted Ni–Zn ferrites. Journal of Applied Physics 2002;91: 6626–6630.

65. Singh AK, Choudhary RNP. Structural, dielectric and electrical properties of Pb5−xLa1+xTi3+xNb7−xO30 (x = 0, 1 and 2) ceramics. Journal of Physics and Chemistry of Solids, 2003;64: 1185–1193.

# CHAPTER 11

# Surface Coating of Oxide Powders: A New Synthesis Method to Process Biomedical Grade Nano-Composites

*Paola Palmero [1,*], Laura Montanaro[1], Helen Reveron[2] and Jérôme Chevalier[2,3]*

[1]Department of Applied Science and Technology, Politecnico di Torino, INSTM R.U. PoliTO, Laboratorio di Tecnologia ed Ingegnerizzazione dei Materiali Ceramici (LINCE), Corso Duca degli Abruzzi, 24, Torino 10129, Italy
[2]Université de Lyon, INSA–Lyon, MATEIS UMR CNRS 5510, Bât. Blaise Pascal 7, Av. Jean Capelle, Villeurbanne 69621, France
[3]Institut Universitaire de France, 103 bd Saint-Michel, Paris 75005, France

## ABSTRACT

Composite and nanocomposite ceramics have achieved special interest in recent years when used for biomedical applications. They have demonstrated, in some cases, increased performance, reliability, and stability *in vivo*, with respect to pure monolithic ceramics. Current research aims at developing new compositions and architectures to further increase their properties. However, the ability to tailor the microstructure requires the careful control of all steps of manufacturing, from the synthesis of composite nanopowders, to their processing and sintering. This review aims at deepening understanding of the critical issues associated with the manufacturing of nanocomposite ceramics, focusing on the key role of the synthesis methods to develop homogeneous and tailored microstructures. In this frame, the authors have developed an innovative method, named "surface-coating process", in which matrix oxide powders are coated with inorganic precursors of the second phase. The method is illustrated into two case studies; the former, on Zirconia Toughened Alumina (ZTA) materials for orthopedic applications, and the latter, on Zirconia-based composites for dental implants, discussing the advances and the potential of the method, which can become a valuable alternative to the current synthesis process already used at a clinical and industrial scale.

**Keywords:**Ceramic composites; biomedical applications; elaboration method; dental and orthopedic applications; alumina-zirconia composites

# 1. INTRODUCTION: A BRIEF HISTORY FROM PURE MONOLITHIC TO COMPOSITE BIOCERAMICS

The use of ceramics has been growing in the past 40 years in the biomedical field. Their main applications include orthopedic and dental devices, being mainly applied as replacements for hip, knee, teeth, and as bone gaps filler. Depending on their *in vivo* behavior, they are classified into bioinert, bioactive, and/or bioresorbable [1]. In the first case, as a rule, the material has a high chemical stability *in vivo*, as well as a high mechanical strength, and, when incorporated in a living bone, it shows a minimum interaction with the surrounding tissue. On the other hand, bioactive ceramics present the capability of inducing a favorable response from the host tissues (e.g., inducing strong chemical bonds with bone and favoring bone ingrowth), but their mechanical properties are generally lower than those of bioinert materials. Finally, bioresorbable ceramics have the unique ability to degrade gradually, being replaced by the natural tissue.

This wide spectrum of biological interactions leads to a broad range of engineering design and strategies. In fact, bioactive materials (such as hydroxyapatite and tricalcium phosphate) have poor mechanical properties, so their applicability is confined to implants that do not have to bear significant loadings, and the main requirement is to provide favorable surfaces for biological bonding and bone ingrowth. Otherwise, the critical conditions in joint replacement restrict the choice of materials to the harder and stronger ones, such as alumina ($Al_2O_3$) and zirconia ($ZrO_2$), even though they are not able to create (at least at the moment) a bone-implant interface and cannot be used as bone filler [1].

According to the previous classification, this review focuses on bioinert oxide ceramics, and, most of all, on alumina and zirconia-based materials for load-bearing applications, such as hip and knee bearings and dental devices. These ceramics present drastically reduced wear rates and excellent long-term biocompatibility, which can increase the longevity of the prosthetic joints. This benefit is clinically important, as hip and knee replacement has become a very common surgical procedure, being performed in increasingly young and active patients. The main characteristics for materials to be used in total hip arthroplasty (THA) are [2]:

    a.   High mechanical properties (in terms of flexural strength, elastic modulus, fracture toughness, and fatigue resistance). It should be kept in mind that the loads in the body can vary from three times the body weight (~3 kN) for normal walking to eight times the body weight (~8 kN) for jogging or stumbling;

    b.   High corrosion resistance and biocompatibility *in vivo*;

c.  High hardness and good surface finish, for ensuring long term wear resistance;

d.  Good wetting, for providing good lubrication between the implant surface and the synovial fluids.

According to such properties, $Al_2O_3$ was introduced as a candidate material for orthopedic bearings in the 1970s, whereas partially stabilized-$ZrO_2$ appeared in the mid-1980s.

Alumina is, by far, the most widely used ceramic in THA. However, the early clinical applications showed a high fracture rate (~13%) of alumina devices, later imputed to an incomplete densification of the material [3]. In order to improve, as much as possible, the mechanical properties of alumina, efforts addressed the use of high purity raw powders and the application of unconventional sintering techniques, such as hot isostatic pressing (HIP), for reaching full densification. Such advances gradually increased the sintered density and decreased the alumina average grain size, giving rise to the last generation of alumina ceramic components, characterized by excellent mechanical properties (flexural strength higher than 550 MPa, Vickers hardness of 1800–2000 HV and a low fracture rate of about 0.004%–0.015%) [3,4]. Technical data of first, second, and last-generation polycrystalline alumina are reported in Table 1 [2], showing that the third-generation materials completely fulfill the International Standard Organization (ISO) requirements for medical-grade alumina implants [1,4], reported in the same Table.

**Table 1.** Properties of various medical-grade Alumina ceramics and International Standard Organization (ISO) requirements [1,2,4].

| Property | Alumina: 1970s | Alumina: 1980s | Alumina: 1990s | ISO Alumina Standard 6474 |
|---|---|---|---|---|
| Density (g/cm³) | 3.94 | 3.96 | 3.98 | >3.90 |
| Mean grain size (µm) | 4.5 | 3.2 | 1.8 | <7 |
| Bending strength (MPa) | 400 | 500 | 580 | >400 (in Ringer's solution) |
| Vickers Hardness (HV) | 1800 | 1900 | 2000 | >2000 |
| Hot Isostatic pressed | No | No | Yes | --- |

Despite the mentioned good properties, there are two major concerns for the reliability of alumina bearings: the low fracture toughness (~4 MPa·m$^{1/2}$) and the susceptibility of alumina to failure through slow crack growth at stresses well below to the ultimate fracture strength. The explanation of this last behavior will be given later. Alumina is, therefore, restricted to standard designs (for example ball heads with a diameter larger than 28 mm), for which the risk of failure is now exceptionally low. The development of new concepts or products will require new ceramics with better intrinsic crack resistance. Alumina has been also used for dental devices, starting from Dr. Sami Sandhaus, a Swiss dentist, who developed a medical device entirely made of polycrystalline alumina, the CBS® dental implant [5]. This dental device enjoyed good clinical success, thus leading to the development of other highly pure alumina implants. In addition to the advantages of aesthetic and biocompatibility, alumina dental implants have shown several shortcomings in terms of design, stiffness, and modest fracture toughness, similarly to what already has been observed for THA. The low flexural strength and the flaws introduced during the surface grinding have caused failures on these devices leading to their demise in favor of titanium dental fixtures [5].

Zirconia femoral heads were introduced in the 1980s in response to the brittleness of alumina and the consequent potential failure of implants [6,7]. Biomedical grade zirconia exhibits the best mechanical properties of oxide ceramics, well superior of those of alumina. In fact, the zirconia fracture toughness is almost the double of that of alumina (6–8 MPa·m$^{1/2}$ *versus* 3–4 MPa·m$^{1/2}$) and the strength can reach values around 2 GPa for a fine-grained material [2]. This exceptional mechanical strength is due to the phase transformation toughening mechanism, able to increase the crack propagation resistance in zirconia. To explain such phenomenon, we should introduce the three polymorphic zirconia forms, being progressively the monoclinic (*m*), tetragonal (*t*), and cubic (*c*) phases as temperature increases. By adding suitable stabilizers, such as yttrium oxide (Y$_2$O$_3$), the *t* phase can be retained at room temperature in a metastable phase, being able to re-transform into the stable *m* phase under an applied stress. This martensitic transformation (*t* to *m* phase) involves a volume expansion of ~3%–5% and large shear strains (~7%) [8] able to induce compressive stresses at the crack tip that avoid the crack propagation, leading to the well-known toughening effect. A scheme of the transformation toughening mechanism is given in Figure 1a [9].

The mostly used biomedical grade zirconia typically contains 3 mol% yttria (3Y-TZP). The sintered material commonly consists of a single phase (the metastable *t* phase), fine-grained (0.1–1 μm), fully densified microstructure.

Despite the excellent properties of zirconia ceramics, some compositions, such as 3Y-TZP, present a major drawback in moist atmosphere, since they undergo Low-Temperature Degradation [9,10,11], as shown in Figure 1b. This is an ageing phenomenon that causes loss of strength and generation of micro-cracking in presence of water. It consists of a slow transformation of metastable *t* zirconia to the *m* phase (without any applied stress) in an important temperature range, typically from room temperature up to around 400 °C, thus including the temperature used for steam sterilization (~140 °C) and the human body temperature (37 °C) [9,10]. Although Low Temperature Degradation has been studied for more than 20 years, the precise mechanism by which moisture catalyzes the phase transformation is still not fully clarified [9,10]. Several experimental results show that moisture, in the form of OH$^-$ ions, diffuses into the zirconia lattice during exposure to humid atmosphere. Most probably, the oxygen of environmental water is located on vacancy sites, whereas the hydrogen is placed on adjacent interstitial sites. This highlights the role of oxygen vacancies initially present in zirconia on water diffusion rate. This also emphasizes the effect of stabilizers inside the zirconia lattice: in Y-TZP, many oxygen vacancies are generated by the trivalent cation (Y$^{3+}$), making the water diffusion rate higher than in other zirconia-based ceramics stabilized with tetravalent cations, such as CeO$_2$-doped ZrO$_2$ (Ce-TZP), in which Ce$^{4+}$ does not induce vacancies [12].

Schubert and Frey [13] showed that the penetration of OH⁻ leads to a lattice contraction, inducing tensile stresses in the surface of the grains. This destabilizes the *t* phase and favors its martensitic transformation to the *m* phase. The mechanism proceeds then in an autocatalytic way, since the transformation of one grain leads to a volume increase, stressing up the neighboring grains, and leading to micro-cracking. This offers a path to water to penetrate inside the sample [12].

Low Temperature Degradation has been known for more than 20 years [14], but its effects were underestimated at ambient temperature until the end of the 1990s, when Chevalier *et al.* showed that it could proceed under *in vivo* conditions with time scales in the order of only some years [15]. In 1997, the US Food and Drug Administration (FDA) reported on the critical effects of standard steam sterilization procedure (134 °C, 2 bar pressure) on the surface roughness of zirconia implants [16,17]. In 2001, the Therapeutic Goods Administration in Australia announced a large series of failures (more than 800) in batches of implants processed with a new furnace technology by Saint-Gobain Desmarquest Prozyr® [16,17]. This event had a catastrophic impact for the use of zirconia, and the market sale decreased of more than 90% between 2001 and 2002, with no evidence of clear renew.

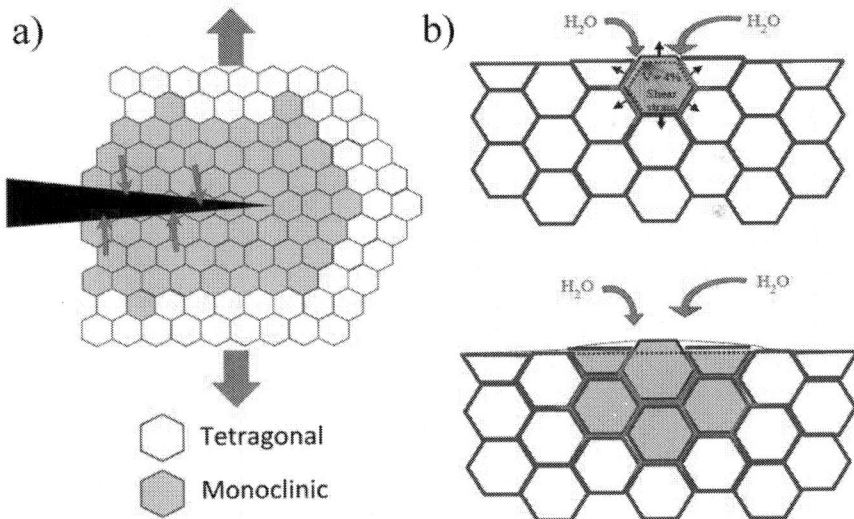

**Figure 1.** Schematic illustration of (**a**) phase transformation toughening and (**b**) ageing [16]. In (**b**): Nucleation on a particular grain at the surface, leading to micro-cracking and stresses to the neighbors (**top**); growth of the transformed zone, leading to extensive micro-cracking and surface roughening (**bottom**). Red path represents the penetration of water due to micro-cracking around the transformed grains.

Quite surprisingly, in the same years the dental community discovered the use of zirconia [18,19,20] and, apparently, without concerns about aging [11,21,22]. Inlays, onlays, single crowns, and fixed partial dentures have been realized by using zirconia core. Moreover, also implant abutment and implants are today available in zirconia [7]. Such a wide application explains the continuous increase (of more than 12% per year) of the zirconia market in the dental field [16], the success being imputable to the zirconia esthetic properties, in addition to its mechanical specifications. Once again, Y-TZP ceramics exhibits the best combination of mechanical and esthetical properties among polycrystalline oxide ceramics. 3Y-TZP materials with a translucently of 12%–15% are available, and the color can be adjusted by doping, for example, with iron or rare earth elements, meeting the demand of natural-like restoration [17]. If ageing has been well documented in the orthopedic field, the lack of studies for dental applications is striking. Even though a few general papers devoted to dental zirconia underline the need to "keep in mind that some forms of zirconia are susceptible to ageing and that processing conditions can play a critical role on the Low Temperature Degradation of zirconia" [23] the problem of ageing in dental zirconia is still underestimated. In part, this is due to the availability of new ageing resistant 3Y-TZP ceramics (such as TZ3Y-E developed by Tosoh), but also to a lack of exchanges from one community to another. The ageing consequences may seem less important in dental applications, especially when restorations are concerned, but if large-scale failure events such as those of Prozyr® heads occur, it would be a critical issue for zirconia and for ceramics in general [17].

In order to overcome the low toughness of alumina and the sensitivity to ageing of zirconia, the trend today is to develop alumina-zirconia composites. This can be the way to get benefits from the zirconia transformation toughening mechanism without the major drawback associated with its transformation under steam or body fluid conditions. A recent literature overview on alumina-zirconia composites developed for biomedical applications show that different compositions, from the zirconia rich side, to the alumina rich one have been tested.

Concerning the alumina-rich compositions (alumina in the range 60–95 vol%), Zirconia Toughened Alumina (ZTA) composites have been widely investigated. Starting from June 2000, ZTA became available as femoral head material, commercialized by CeramTec AG (Plochingen, Germany) under the trade name of BIOLOX® delta. After approval by FDA in 2003, ZTA started to be widely used in THA: in the last 10 years, approximately one million CeramTec ZTA femoral heads and over 700,000 inserts have been implanted worldwide. A second ZTA product, AZ209, was introduced in the Japan market by KYOCERA Medical (Osaka, Japan) during 2011 [24].

In ZTA, the zirconia particles can exhibit a $m$-phase or a $t$-phase, but the higher mechanical properties (strength and fracture toughness) have been observed in the latter case. A critical factor is, however, the $ZrO_2$ grains size, which should be optimized by controlling the composition and processing conditions. In fact, the zirconia grain size must range between two critical values, the $D_c'$ below which the transformation is completely hindered (even under stresses at the crack tip), and the tetragonal phase is stabilized by its very small size, and the $D_c$ above which spontaneous transformation occurs during cooling the composite system from the sintering temperature [25].

These critical sizes depend on the zirconia amount into the ZTA composites and on the oxide stabilizer nature and content: for instance, values in the range 0.6–2.0 μm for the upper limit, and 0.1–0.4 μm for the lower one have reported in literature for ZTA containing 8–15 vol% of un-stabilized $ZrO_2$ [26].

Because of the $t$-particles are present within the alumina matrix, the strength degradation by hydrothermal ageing in ZTA is limited, as compared to monolithic Y-TZP. However, remembering that ageing occurs by nucleation and growth mechanisms, starting at the surface, the Y-TZP or zirconia particles should not form a continuous network in the alumina matrix, meaning that their content should be below the percolation limit (~16 vol%) [2,27].

When un-stabilized zirconia particles are added to the alumina matrix, the metastabilization of the $t$ zirconia grains is achieved thanks to their very fine grain size and to the presence of the stiffer alumina phase. Therefore, if the addition of yttria is avoided, the formation of oxygen vacancies inside the zirconia lattice is reduced, thus, limiting the diffusion of water radicals in the zirconia ceramics [15].

Hip simulator tests indicate that the wear of ZTA-on-ZTA is lower than that of Alumina-on-Alumina. The ZTA/ZTA couple displayed the highest resistance to wear and the lowest amount of surface roughness after five million cycles [24]. However, ageing of ZTA containing 14 vol% Y-TZP particles was observed when tested for long time (19 months) in Ringer's solution, promoting the formation of a $m$-$ZrO_2$ surface layer which leads to a significant reduction of the flexural strength [2].

These results demonstrate that, in spite of the advances shown by ZTA composites with respect to monolithic alumina and zirconia materials, there is still room for further optimization, in terms of both composition and architecture of the composites.

An example is given by Biolox delta® manufactured by CeramTec AG (Plochingen, Germany). This company produces a ZTA material, which also contains small amounts of SrO and $Cr_2O_3$ [28]. During sintering, these additives react with alumina, leading to the *in situ* formation of plate-like aluminates. This approach, which is becoming increasingly popular [29,30,31,32], brings to a fracture toughness increment, since the elongated grains (with high aspect ratios) can induce additional toughening mechanisms by crack deflection and crack bridging [33]. By optimizing the microstructure and the sintering conditions, materials with remarkable mechanical properties are produced, being the flexural strength higher than 1200 MPa, the fracture toughness of 6.5 MPa·m$^{1/2}$ and the Vickers hardness of 1975 HV [2].

In the zirconia-rich side, alumina-toughened zirconia (ATZ) composites are currently produced. The addition of $Al_2O_3$ to Y-TZP up to 0.25%, allows increasing the bending strength from 1100 to 1200 MPa and improving the resistance to ageing [20]. However, when the alumina content was increased to 25% and the composite was sintered by HIP, the highest strength—over 2000 MPa, was yielded. ZIRALDENT®, manufactured by Metoxit AG and having the above composition, is at the present the strongest biomedical ceramic known.

In the case of Ce-TZP/Al$_2$O$_3$ composites, the first investigation belong to Sato in 1989 [34], proving an increase of the hardness and of the Young's modulus of this material with respect to pure Ce-TZP ceramics. However, the best results were reached by Nawa [35,36], in 1997, when he developed an intra-granular microstructure, in which several nano-sized (10–100 nm) Al$_2$O$_3$ particles were trapped into zirconia grains. Nawa observed that the hardness, the elastic modulus and the fracture strength increased, while increasing the alumina content. At the same time, he observed a decrease of the fracture toughness, in agreement with the decrease of transformability. Thanks to the precise control of the microstructure and composition, the mechanical properties were optimized (bending strength and fracture toughness of 1290 MPa and 8.62 MPa·m$^{1/2}$, respectively), giving rise to the commercial product named NANOZR® (Panasonic Electric Works, Tokyo, Japan [5,37]). More recently, a new type of zirconia matrix composite based on Ce-TZP/magnesium spinel (MgAl$_2$O$_4$) system was developed [38] (Figure 2), showing a good combination of high strength (about 900 MPa) and toughness (15 MPa·m$^{1/2}$).

**Figure 2.** Sintered microstructure of 10 Ce-TZP/16 vol% MgAl$_2$O$_4$ composite [38].

In addition, for this class of composites, elongated grains inside the zirconia matrix have been already used: Ce-TZP composites reinforced by LaNbO$_4$ [39], by strontium [40], barium [41], and lanthanum [42,43] hexaaluminate were previously reported in literature. More complex quasi-ternary aluminates, such as BaMnAl$_{11}$O$_{18}$ and CeMnAl$_{11}$O$_{19}$, were investigated as well [44,45]. In spite of such publications, the rule of the elongated grains inside a zirconia matrix is still unclear: they have been supposed to provide crack bridging and deflection effects, but these mechanisms are not yet clearly demonstrated.

What here described highlights that the major drawbacks of pioneering pure monolithic bioceramics can be overcome by the use of composites. However, the ability to tailor the microstructure of ceramic composites is essential in order to fulfill the requirements of biomedical materials: for this reason, the following paragraph reports and briefly discusses the microstructure-properties relationships, already observed in ceramics used in structural biomedical applications.

## 2. MICROSTRUCTURE-PROPERTIES RELATIONSHIP IN COMPOSITE BIOCERAMICS

In order to better understand the role of the microstructure, we should recall that failure of ceramics *in vivo* commonly results from slow crack growth under static or repetitive loading experienced in the body, until fracture. As extensively described in literature (see for example [2,46]), this phenomenon can be understood as a corrosion-assisted crack propagation process, leading to a progressive loss of strength. In fact, under an applied tensile stress $\sigma$, the stress at the crack tip can be described as the stress intensity $K_I$ given by:

$$K_I = \sigma\sqrt{\pi a}$$

where $2a$ is the length of the crack. It is generally assumed that fast failure occurs in brittle materials if $K_I$ becomes equal to (or greater than) the critical stress intensity factor, $K_{IC}$, commonly known as fracture toughness. The fracture strength of a brittle material can be written as:

$$\sigma_f = \frac{K_{IC}}{\sqrt{\pi a}}$$

According to Equation (2), it is evident that the fracture strength is determined by the fracture toughness and the flaw size. This equation represents a critical value for *fast crack growth*. Additionally, it is now well recognized that brittle materials can be also susceptible to *slow crack growth* (SCG, also referred to as *subcritical crack growth*), meaning that they are sensitive to the applied stress, as well as to environmental factors, such as water, water vapor, and temperature. The combined effect of high stress at the crack tip and reaction with water (or body fluids) allows the crack to grow, until reaching the critical length for failure at the stress level (Equation (2)), where the material fails. Thus, many efforts have been devoted to determine a threshold in stress intensity factor ($K_{I0}$), corresponding to the condition of zero crack velocity, at which crack propagation does not occur. Being $K_{I0}$ the safe range for using ceramics in THA, this parameter has been determined for both monolithic and composite bioceramics [46].

Chevalier *et al.* [46] clearly demonstrated a higher threshold for subcritical crack propagation ($K_{I0}$) for Alumina-10 vol% $ZrO_2$ composite, as compared to both pure alumina and 3Y-TZP materials, indicating a higher reliability and expected lifetime for the composite. More interestingly, the same authors [47] compared the SCG propagation behavior of the previous ZTA micro-composite with that of a nano-composite containing only 1.7 vol% of nano-sized $ZrO_2$ grains (average size of about 150 nm) embedded in a micron-sized alumina matrix (average size of about 5 µm) [47], where they predominantly occupied transgranular positions. The microstructure of the ZTA micro- and nano-composites are shown inFigure 3, whereas the SCG behavior of the monolithic and composite materials is depicted in Figure 4, determined by the crack velocity ($V$) *versus* the stress intensity factor ($K_I$).

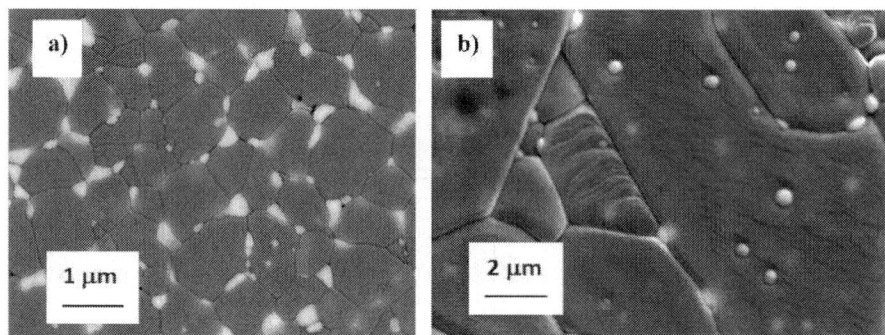

**Figure 3.** Sintered microstructures of (**a**) micro-; and (**b**) nano-ZTA composites [46,47].

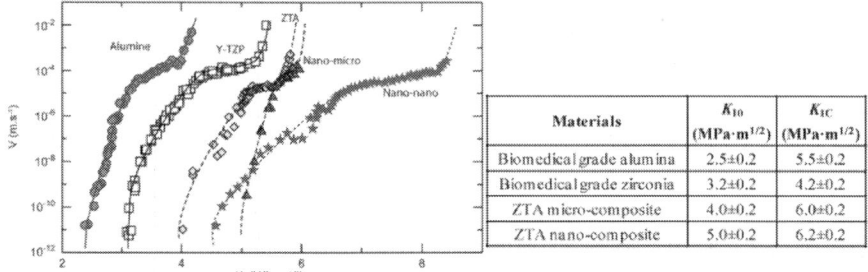

| Materials | $K_{I0}$ (MPa·m$^{1/2}$) | $K_{IC}$ (MPa·m$^{1/2}$) |
|---|---|---|
| Biomedical grade alumina | 2.5±0.2 | 5.5±0.2 |
| Biomedical grade zirconia | 3.2±0.2 | 4.2±0.2 |
| ZTA micro-composite | 4.0±0.2 | 6.0±0.2 |
| ZTA nano-composite | 5.0±0.2 | 6.2±0.2 |

**Figure 4.** Crack velocity ($V$) versus stress intensity factor ($K_I$) for biomedical grade alumina, yttria stabilized zirconia, ZTA micro-composite, and ZTA nano-composite. Enclosed to the Figure, the values of $K_{I0}$ and $K_{IC}$ for the same materials [46,47].

From this figure, it is possible to estimate the threshold ($K_{I0}$) for subcritical crack propagation, determined from the points on the $V$-$K_I$ diagram below which there is an abrupt drop of the crack velocity (V < $10^{-12}$ m· s$^{-1}$). On the other hand, the toughness ($K_{IC}$) was determined by extrapolation of the $V$-$K_I$ curve to high crack velocities ($10^{-2}$ m· s$^{-1}$). These values, for the different ceramics, are given in the Table within the Figure 4.

From the above figure, we can see that zirconia exhibits a higher $K_{IC}$ than alumina, but that their thresholds are quite close, meaning that the necessary crack tip stress to initiate slow crack growth is quite similar in both materials. The higher toughness of zirconia can be attributed to its transformation toughening effect; on the other side, zirconia bonds are prone to chemisorption of the polar water molecules, more than the alumina ones, explaining the lower slope of the $V$-$K_I$ curve of zirconia as compared to that of alumina. The ZTA micro-composite has both larger $K_{IC}$ and $K_{I0}$ values than the two monolithic ceramics. In fact, thanks to the alumina matrix, this composite shows a low susceptibility to stress assisted corrosion by water (or body fluid). However, as it is reinforced by transformable zirconia particles, its $V$-$K_I$ curve slop is similar of that of alumina, but shifted towards higher $K_I$ values. Finally, we can see that the ZTA nano-composite is characterized by similar $K_{IC}$ as the micro-composite, but a significantly higher $K_{I0}$ value. In the nanocomposite, only a small fraction of nanosized $ZrO_2$ grains into the alumina matrix is needed to dramatically increase its crack resistance, even if the transformation toughening mechanism has to be ruled out since the $t$ grains transformation is hindered by their small size. The behavior of the ZTA nanocomposites was instead attributed to the residual compressive stress field inside the alumina grains, generated by the transgranular zirconia particles, and having a strong impact on the SCG resistance.

Hydrothermal ageing also critically depends on the microstructure, and, hence, on the processing. A key point is producing a fully dense microstructure and, more important, free from percolative porosity [17]. We recall that the failure of Prozyr® femoral heads, i.e., the spontaneous disintegration because of accelerated ageing, occurred in batches processed in a new tunnel furnace causing a lack of densification at the center of the head [17]. For the same reason, the current industrial production of ceramic femoral heads implies two sintering steps to assure full densification: pressureless sintering followed by post hot isostatic pressing, being the first step unable to guarantee to the bodies the end product specifications (see Table 1).

When zirconia grains are stabilized by yttria, the control of the size and distribution inside the ZTA composite is of primary importance. It has been shown that, if the zirconia amount is above the percolation limit (~16 vol%), the composite might be sensitive to aging. Below this critical amount, only the largest zirconia grains (or the zirconia clusters eventually present in non-homogeneous materials) can undergo ageing, the smaller ones being over-stabilized by the stiffer alumina matrix. In order to deepen the role of zirconia aggregates on the Low Temperature Degradation behavior, Gutknecht et al.[48] investigated the ageing sensitivity of two batches of ZTA composites (containing 13 vol% of 3Y-TZP) both processed by the powder mixing method, but using different polyelectrolyte dispersants. Under optimal electrostatic dispersion [48], the composite was completely free from aggregates and did not exhibit any ageing phenomena. On the opposite, when a not appropriate dispersant was used [49], the final microstructure presented a severe aggregation of zirconia grains and ageing was observed especially during the first two hours of autoclave treatment. In agreement with this result, a colloidal method was used to produce a ZTA composite [47], made by zirconia nano-particles uniformly distributed in the alumina matrix, which was almost ageing-free.

Many other specifications of biomedical grade ceramics depend on their microstructure. For instance, to gain benefit from the transformation toughening mechanism, the zirconia particle size should range between two critical values as previously mentioned: the highest relates to the size for spontaneous transformation of the *t*-phase to the *m*-phase during cooling, and the lowest to the size for which no transformation to the *m*-phase is possible. Both critical sizes depend on the stiffness of the matrix and on the amount and composition of the zirconia particles. Generally, zirconia particle sizes of a few hundred are observed [17]. More in general, the toughness-strength relationship depends on the microstructure, including the amount and type of zirconia, grain size of matrix and second phase, location of the zirconia grains inside the matrix [46,50,51,52,53].

Finally, also the optical properties of the ceramics depend on the microstructure: translucency of technical ceramics may be achieved with a very fine grain size (submicron) and low porosity content (lower than 1% or even 0.1%). An example is given by the fully dense translucent Y-TZP ceramics when processed with a grain size lower than 0.5 μm, meeting the demand for both natural-teeth-looking restoration and high mechanical strength [12].

## 3. SYNTHESIS METHODS TO PRODUCE BIOMEDICAL-GRADE CERAMIC COMPOSITES: A FOCUS ON THE SURFACE-COATING STRATEGY

Zirconia- and alumina-based composite powders have been produced by several elaboration techniques, from the traditional milling–mixing [48,49,54] to wet chemical routes (such as alkoxide-based route [55], sol-gel [56,57,58,59,60], co-precipitation [61,62], and hydrothermal methods [63]).

As stated in literature, milling-mixing procedures are the most applied method, in which the powders are generally mixed in form of oxides to produce bi- and tri-phasic composite materials [35,38,40,54,64,65,66]. When powder mixing is performed, it is mandatory to check the characteristic of the starting powders and to carefully control the dispersion and stabilization degree of the mixed suspension.

Concerning the characteristics of the powder, size and size distribution, shape and morphology, agglomeration degree, phase composition and surface properties have to be considered. In fact, a wide size distribution on one hand leads to a higher packing density in the green bodies. On the other hand, the precise control of the microstructural development during sintering could be difficult to achieve because the larger grains can grow at the expense of the smaller ones. At the same time, the particle size influences the final grain size and the densification rate. Due to the higher specific surface area, the densification rate increases as the particle size decreases [67]. In addition, if the powder is agglomerated, the particle packing in the green body will be heterogeneous, giving rise to differential sintering rates and to heterogeneous microstructures [68].

When wet-forming methods are used, the characteristics of the mixed suspension have to be carefully controlled. In fact, it is well recognized that microstructural defects, aggregates or irregular phase distribution of the second phase in the ceramic matrix can arise from a non-optimized or unstable mixed suspension [49]. Thus, the dispersion degree (and eventually the nature of dispersant selected), the homogeneity and the rheological properties of the suspension have to be investigated.

Furthermore, when the powder mixture contains a relevant fraction of nanocrystalline particles, additional complications can arise. The use of high shear forces during dispersion is sometimes necessary to homogeneously distribute the nanoscale fraction into the liquid. Separation by selective agglomeration or sedimentation during the shaping process can also occur, thus, affecting the final microstructure. Finally, when prolonged milling is carried out, contamination of the powder from jars and milling spheres could occur.

In order to overcome such drawbacks, several wet chemical routes were exploited in the last few years. Among them, sol-gel and co-precipitation are the mostly employed.

In the sol-gel method, inorganic salts or metal-organic compounds are used for the sol preparation. Then, by hydrolysis and condensation reactions, the sol is converted into a gel that has to be dried, to eliminate the exceeding liquid. Advantages of this technique include the precise control of the particles morphology and size, the relatively low processing and crystallization temperatures of the final phases, the product homogeneity, the chemical purity and the full control of stoichiometry. However, the complexity of the method increases when multi-cation materials (such as in the alumina-zirconia system) have to be prepared. In fact, the hydrolysis and condensation reactions have to be even more carefully controlled (solution pH, temperature, and reactant concentration), in order to avoid selective segregation of one, or more, metal ions. Concerning the other disadvantages of wet-chemical methods, it is known that the chemical precursors are expensive and often sensitive to moisture; the dried product is often hard agglomerated and sometimes it is difficult to preserve the characteristics of a lab-scale production when large batches of powders are concerned [69].

In the precipitation technique, the solubility of the metal ions dissolved in aqueous solution is exceeded by evaporation of the liquid or by adding a chemical reactant. Thus, the precipitation of metal hydroxides is promoted. The modification of the pH and temperature of the solution allows controlling nucleation and growth mechanisms and consequently the particles morphology [70,71]. In addition, in the case of co-precipitation, it is necessary to achieve the suitable conditions for the simultaneous and quantitative precipitation of all the species involved [72].

An example of the role of various wet-chemical synthesis methods on the final microstructure of $Al_2O_3$-$ZrO_2$ composites is given in the work of Rana [62]. Here, three different processing routes were employed to prepare composite powders, namely gel precipitation (GPT), precipitation (PPT), and washed precipitation (WPT). In all three cases, starting raw materials were zirconium and aluminum chloride and aqueous ammonia was added to induce precipitation. During the synthesis, the pH was maintained in the range 6–6.5 (gelation point) for GPT, whereas for the other two routes precipitation was carried out at a higher pH, in the range 8.7–9.1. The difference between the PPT and WPT routes was the washing process: in the former case, the precipitate was separated from the liquid and then dried; in the latter, it was washed with hot water and alcohol before drying. These different processing routes affected the crystallization temperature of the amorphous powder, as well as the phase evolution of $Al_2O_3$ and $ZrO_2$ phases during calcination. In fact, while GPT and PPT powders crystallized at 350 °C, no crystallization was observed in WPT powder until 650 °C, this latter being the only product able to produce pure $t$-phase. The powder agglomerates size was largest for GPT than for WPT. Moreover, GTP route produced hard agglomerates, while low agglomeration strength was observed in WPT powders. As expected, soft agglomerated powders gave rise to compacts with good green density and able to sinter to a higher density at lower sintering temperatures as compared to GPT and PPT products.

Hydrothermal oxidation is an alternative wet-chemical method to prepare alumina-zirconia composites powders. An example is provided by the work of Chen and Chiao [73], where $ZrAl_3$ and $Zr_5Al_3$ alloys have been used as starting materials. It was found that the sizes of the aluminum and zirconium oxides were lower than those of the respective powders obtained by the hydrothermal oxidation of aluminum and zirconium metals separately, this reduction of size being attributed to the retarding effect of the alumina particles on the growth of the zirconia particles.

In the last years, innovative procedures consisting in the surface modification of a commercial powder with the second phase precursors have been exploited. They can be considered as a compromise between the powder mixing technique and the colloidal ones, allowing a deeper control of the final microstructure. In fact, the close mixing between the matrix ceramic nanoparticles and the metal ions, precursors of the second phases, is realized at the atomic/nano level, assuring an excellent distribution of the second phases in the composite powders. On the opposite, in the case of milling methods, the second phases are in the form of a solid precipitates (such hydroxides) or solid particles, leading to a less effective mixing with the matrix powder.

Schehl et al. [55] developed a modified colloidal method for the production of alumina composites. Briefly, this method consists in grafting alumina commercial powders by using organic precursors (metal-alkoxides) of the second phases. Thanks to the use of organic media (typically ethanol), the addition of metal alkoxides initiates a substitution reaction between the metal alkoxide and the OH groups located on the particles surface. After drying and calcination, second-phase nanoparticles are formed in situ on the alumina particles surface. This method was applied to various alumina-based composite materials, with Zirconia, Yttrium Aluminum Garnet (YAG) or mullite second phases, showing in all cases a fine and homogeneous microstructure. The mechanical properties of the ZTA composites obtained by this route were compared to those of the same materials processed by classical mechanical mixing. The colloidal-route processed composites showed finer microstructures and better mechanical properties, particularly a higher $K_{I0}$ value [46] and a slower degradation rate [49].

Beside the method investigated by Schehl *et al.* [55] based on the use of organic precursors, also inorganic salts were used to coat the surface of commercial alumina particles and to induce the crystallization of second phases upon calcination. Yuan *et al.* [74], for instance, added cerium and aluminum nitrates to the isopropanol suspension of zirconia powder, in order to obtain 12Ce-TZP/2 wt% $Al_2O_3$ composites. The as-obtained suspension was mixed for 48-h on a multidirectional mixer and then water and alcohol liquid media were removed by means of a rotating evaporator at 95 °C, thus promoting the formation of the desired final phases. The dried powder was subsequently calcined in air at 800 °C for 1 h in order to obtain the $Al_2O_3$-doped $CeO_2$-coated $ZrO_2$ nanopowder.

More recently, we developed an alternative method to produce alumina- and zirconia-based composites, inspired on the concept of "surface modification" of commercial oxide powders [75,76,77,78], which shows some progress with respect to the previous described techniques. In fact, in our process, only inorganic precursors and aqueous media are used, making this strategy much more simple and potentially transferable to a pre-industrial scale production. A second difference lays in the mixture drying method, which we perform by means of a "flash" drying technique, such as atomization, in which the liquid medium is converted into fine droplets and instantaneously evaporated. This step has a key role in the process, since the homogeneity of the mixture is "frozen" in the dried products, completely avoiding the segregation of the metallic dopants, as can occur by slow drying in an oven. The method was successfully applied to alumina-based bi- and tri-phasic composites and, more recently, successfully exploited for the elaboration of zirconia-based composites with complex compositions and microstructures, containing both equiaxial and elongated second phases.

In order to show the potential of this new technique, in the following we present two case studies, the former devoted to alumina-zirconia composites to be potentially used for orthopedic applications, the latter focusing on zirconia-based tri-phase composites for dental industry. The process is being developed in the frame of the European Project *"Longlife"* (FP7, grant agreement No. 280741) and recently registered as an Italian Patent (TO2014A000145, registered on 21 February 2014).

# 4. CASE STUDIES

## 4.1. ZTA for Orthopedic Applications

The surface modification route was used to prepare ZTA composites, in which the zirconia content ranged between 5 and 20 vol%. The starting raw material was a commercial α-alumina powder (TM-DAR, supplied by TAIMEI Chemicals Co., Tokyo, Japan) characterized by nanosized primary particles (average size of about 100 nm [79]) but presenting a certain agglomeration degree, since by laser granulometry we determined agglomerates of about 30 μm in size [76]. Zirconium (IV) chloride (ZrCl$_4$, >99.5% purity, supplied by Sigma-Aldrich) was selected as zirconia precursor. Further details on the elaboration method can be found elsewhere [75,80,81].

In this process, a key point to obtain a homogeneous distribution of the zirconia second phase in the alumina matrix is, first of all, to achieve a close and homogeneous mixing between the matrix powder and the zirconium ions contained in the starting aqueous suspension. This can be obtained on one side by reducing the size, or eliminating, the alumina agglomerates from the starting powder, and on the other side by avoiding the formation of second-phase precipitates, such as Zr(OH)$_4$.

For this reason, the starting alumina powder was first dispersed alone. Aqueous suspensions at 50 wt% solid loading were ball-milled using α-Al$_2$O$_3$ milling spheres, the powder/sphere ratio being 1/5. Few drops of dilute hydrochloric acid were added to the suspension, lowering the slurry pH from the starting value (about 6.5) to 4.5, far from the isoelectric point of alumina [82]. In this way, a good de-agglomeration degree was achieved after 3 h of ball milling, reaching an average particle size of about 0.17 μm [83]. A zirconium chloride aqueous solution was then drop-wise added to the well-dispersed alumina suspension. Since Zr$^{4+}$ ions show a strong hydrolysis, when ZrCl$_4$ was dissolved in distilled water and added to the alumina slurry, the suspension pH decreased to <1. To increase the pH to acceptable values, tribasic ammonium citrate acting as a chelating agent for Zr$^{4+}$ was also added. In this way, it was possible to maintain the suspension pH to about 4.5, without inducing the precipitation of Zr(OH)$_4$ [84].

The mixed suspension was spray-dried and the product was submitted to various thermal pre-treatments, in the range 400–1000 °C, in order to decompose the synthesis by-products (mainly, ammonium chloride) and to induce the crystallization of zirconia grains onto the surface of the alumina particles.

The precise follow up of the zirconia crystallization at the surface of alpha-alumina grains during the thermal pre-treatments was deeply described in a previous paper [85]. We report here some results, with the aim of showing the potential of the surface coating route in tailoring the size and distribution of the ZrO$_2$ nanocrystals on the surface of alumina particles. In fact, as the process involves the *in situ* crystallization of the ZrO$_2$ grains, by carefully controlling the thermal treatments it is possible to control the crystalline fraction, the crystallites size, the kinetics, and mechanisms of nucleation and growth.

By TEM observations, it was found that powders treated up to 500 °C are composed by alumina particles surrounded by an amorphous layer, which progressively disappeared by increasing the calcination temperature. At the same time, the zirconia crystallites started to nucleate at 500 °C (Figure 5a) into the amorphous layer, following a mechanism of homogeneous nucleation. HRTEM observations showed the presence of small zirconia crystallites, almost completely detached from the alumina surface thanks to the remaining amorphous phase. However, when a low-temperature pre-treatment (at 500 °C or 600 °C) was prolonged (up to 10 h), the

ZrO$_2$ crystallites were dragged by the amorphous phase flow into discrete pockets between the alumina grains, where the crystallites segregated (Figure 5b, see the arrow), leading to the aggregation and growth of the zirconia grains during sintering. On the other hand, at higher temperatures, such as 800 °C or 1000 °C, the ZrO$_2$ crystallization was faster, growth predominated over nucleation, and the amorphous phase disappeared without the draining phenomenon. The result is a very different ZrO$_2$ grain distribution on the alumina particle surfaces: in fact, in this case, the zirconia grains were larger, but not aggregated (see Figure 5c). This behavior was maintained during the high-temperature isothermal pre-treatments: as we can see in Figure 5d, the ZrO$_2$ crystallites further grew, but were still very homogeneously distributed on the parent material. Such different sizes and distributions of the ZrO$_2$ grains on the alumina particle surface gave rise to different sintered microstructures [83,85]. In fact, SEM observation carried out on ZTA composites obtained by powders calcined at 600 °C for 1 h or for 20 h showed different microstructural features. The former composite presented an even distribution of fine zirconia grains, with a narrow size distribution, inside a sub-micronic, homogeneous alumina matrix; in the latter, a less homogeneous microstructure was obtained, in which zirconia aggregates were observed.

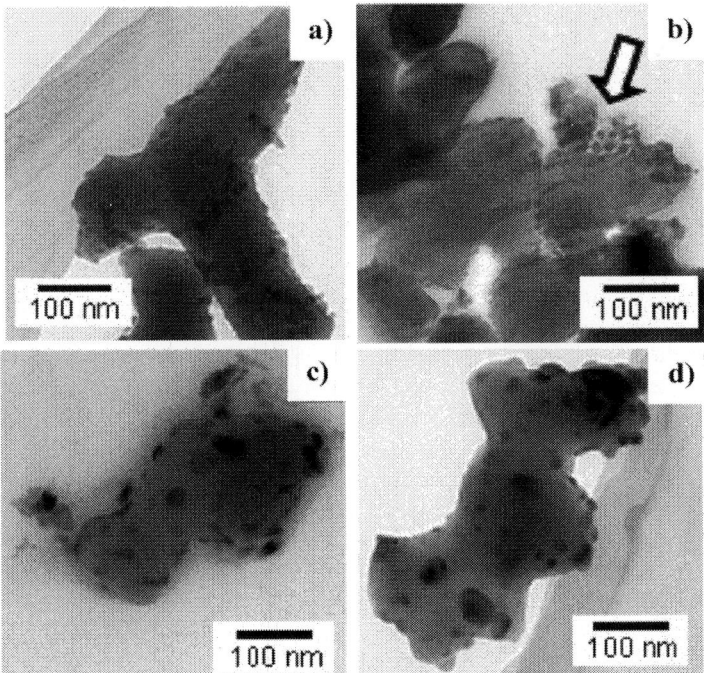

**Figure 5.** TEM images of ZTA powder after treatments at different temperatures and times: (**a**) 600 °C–1 h; (**b**) 600 °C–10 h; (**c**) 1000 °C–1 h; (**d**) 1000 °C–10 h. The arrow indicates the zirconia nuclei dragged into discrete pockets [83].

These results show that the selection of the pre-treatment temperature is a key feature to control the phase distribution and size in the composite powders and, subsequently, in the sintered material. As an example, Figure 6 shows a typical microstructure of a ZTA composite (alumina-5 vol% $ZrO_2$), obtained by pre-treating the as-dried powder at 600 °C for 1 h, after sintering at 1500 °C for 3 h. We can appreciate the very good distribution of the second phase inside the alumina matrix, and observe the lack of any alumina or zirconia aggregates. XRD analysis showed that almost all the zirconia grains crystallized under the tetragonal form, even if no phase stabilizer was added, due to the ultrafine size of the $ZrO_2$ grains.

In a recent paper [86], we showed that the surface coating method is compatible with the industrial protocol commonly used for manufacturing ZTA orthopedic devices. Such a protocol implies the granulation of the composite powders, the cold isostatic pressing of the granules, their natural sintering followed by a post-HIP step, carried out to assure full densification of the components.

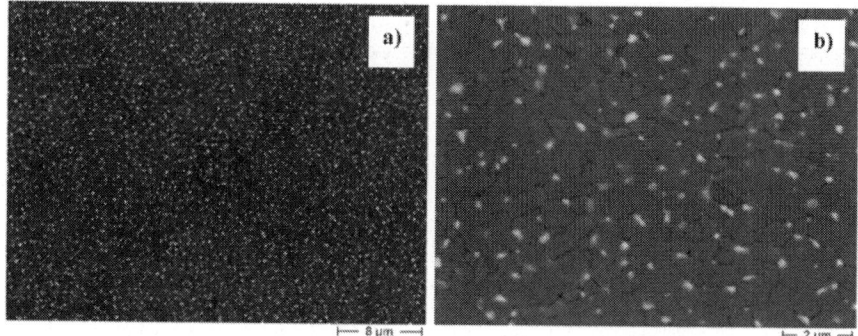

**Figure 6. (a)** Lower and **(b)** higher magnification FESEM images of a ZTA sintered composite.

According to this protocol, the granulation of the composite powder was set-up, by investigating the relationship between the rheological properties of the suspension and the morphological features of the spray-dried granules. In particular, optimal granules for pressing were obtained by controlling the viscosity of the suspension, through a flocculation step, and properly selecting type and amount of binder [86]. The prototypes were then obtained by cold isostatic pressing at 350 MPa, producing blank green bodies of spherical shape, whose mean density was about 2.43 $g/cm^3$, corresponding to 58.2% of the theoretical density (TD) [83]. After a debinding thermal treatment (at 600 °C for 1 h), samples were pressureless sintered at 1500 °C for 3 h, reaching a mean fired density of 99.3% TD. The post-HIP step, carried out at 1520 °C for 2 h under a pressure of 190 MPa, allowed the bodies to achieve full densification. In Figure 7a, the photograph of one of the femoral head prototypes produced (diameter of 28 mm) is shown [83].

One sample was cut along its cross section: SEM observations, as well as Vickers hardness and fracture toughness measurements, were performed at increasing distances from the external surface of the ball. As an example, in Figure 7b the microstructure of the material at 10 mm far from the external surface is depicted, showing that it was fully dense, highly homogeneous, almost defect and agglomerates-free.

In agreement with the microstructural homogeneity observed, also the mechanical properties were almost constant across the material section: for instance, the Vickers hardness at 3 mm from the external surface was 1956.5 ± 80.6 HV and 1946.0 ± 44.8 HV at the two diametrically opposite sides. In the center of the specimens the measured hardness was 1969.5 ± 82.9 HV, showing almost constant hardness inside the ceramic sphere, independently from the position. Accordingly, an almost constant $K_{I0}$ was determined all along the material section, whose mean value was about 3.8 ± 0.3 MPa·m$^{1/2}$ [83]. This value is comparable to that obtained by slip-cast pellets (3.8 ± 0.3 MPa·m$^{1/2}$), containing the same amount of zirconia (5 vol%) [78], showing that the results obtained by slip casting for small samples (~15 mm diameter and 3–5 mm height) can be reproduced by CIP of spray dried granules, in larger spherical components.

As a conclusion, these results highlight the potential of the surface modification route, appearing as an effective tool for tailoring the microstructural features of ZTA composite powders and sintered materials, which are maintained when the process is scaled-up to a higher level (semi-industrial), in order to prepare prototypes, and when integrated into industrial manufacturing protocols of femoral heads devices.

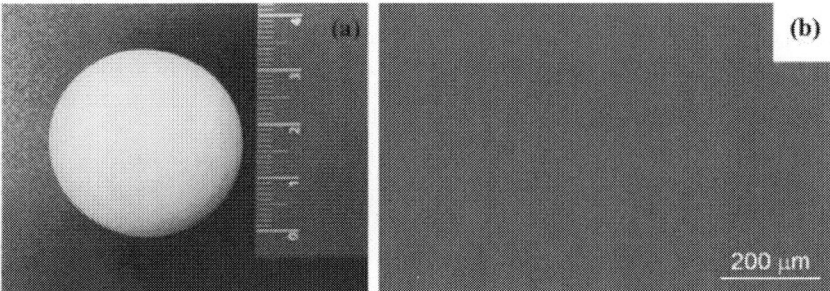

**Figure 7. (a)** Photograph of a ZTA femoral head prototype (diameter of 28 mm) [81]; **(b)** SEM micrograph of the cross section of the prototype, at 10 mm from the external surface of the ball [81].

## 4.2. Zirconia-Based Composites for Dental Applications

The results here described refer to the activities carried out in the frame of the European Project named *Longlife*("Advanced multifunctional zirconia ceramics for long-lasting implants", 7$^{th}$ Framework Program), of which the aim is the production of dental and spine implants characterized by increased strength and toughness and improved ageing stability with respect to the state-of-the-art materials.

In order to reach this goal, the materials were first properly designed, in terms of composition and microstructure.

In particular, we decided to use Ce-TZP as the matrix for the composites, since it has a lower sensitivity to Low Temperature Degradation as compared to Y-TZP. However, being aware of the key role of the stabilizing oxide content on the mechanical and physical properties of the materials, composites containing different amounts of ceria were tested, ranging from 10.0 to 11.5 mol%. Finally, on the ground of the scientific literature previously discussed, two kinds of second phases were selected, characterized by different morphologies (rounded and elongated grains) each of them playing a specific role on the materials physical and mechanical properties. The $\alpha$-$Al_2O_3$ rounded grains were used to retain the Ce-TZP grain growth during sintering, increasing the strength, the hardness, the wear resistance [35,36]. On the other hand, elongated strontium hexa-aluminate ($SrAl_{12}O_{19}$) grains were chosen, as they could provide additional toughening effects by crack deflection and bridging mechanisms [40].

Considering the complexity of the phase composition and architecture of the designed composites, the surface coating route was selected as preferred method for the elaboration of the composite powders. The starting raw zirconia powder was supplied by Daiichi Kigenso, Kagaku Kogio Co. Ltd. (Japan). It is stabilized by 10 mol% of ceria, has an average particle size of 0.5–1 $\mu$m (by laser diffraction method), and a specific surface area of 14.3 $m^2/g$, as declared by the supplier.

De-agglomeration of the starting powder was carried out by ball-milling: aqueous suspensions, at 33 wt% solid loading, were dispersed for about 15 h by using $ZrO_2$ milling spheres, lowering the starting agglomerate size from about 1 $\mu$m to about 0.4 $\mu$m. Dilute hydrochloric acid was added, to decrease the slurry pH from the starting value of about 6.5 (close to the powder isoelectric point [87]) to about 3.

In order to obtain $\alpha$-$Al_2O_3$ and $SrAl_{12}O_{19}$ second phases, $Al(NO_3)_3 \cdot 9H_2O$ ($\geq$98% purity, Sigma Aldrich) and $Sr(NO_3)_2$($\geq$99.0% purity, Sigma Aldrich) were selected as precursors, and added in suitable amount to produce 8 vol% of both second phases in the composite. In order to obtain different ceria contents in the zirconia lattice, ammonium cerium (IV) nitrate ($(NH_4)_2[Ce(NO_3)_6]$, $\geq$98.5% purity, Sigma Aldrich), was selected as the ceria precursor. The composites will be hereafter referred to as ZA$_8$Sr$_8$-CeX, where X refers to the ceria mol% inside the zirconia phase. Precisely, four composites were produced, only differing on the ceria content, being X = 10.0 (i.e., no extra-ceria added during synthesis), 10.5, 11.0, and 11.5 mol%.

The nitrates were dissolved in distilled water and then drop-wise added to the dispersed zirconia suspension. After mixing for 2 h, the suspension was spray-dried.

The powder was then pre-treated at 600 °C, for 1 h, in order to decompose the synthesis by products, and then thermally treated at 1150 °C for 30 min, in order to approach the crystallization temperature of the second phases [88,89]. The calcined powders were than dispersed by ball-milling, and green bodies were produced by slip casting. Samples were then sintered at 1450 °C, for 1 h, all achieving full densification.

Figure 8 shows the micrograph of the sintered ZA₈Sr₈-Ce11 composite. In the lower magnification image (a) we can see, once again, that the surface coating route is successful in producing fully dense composites with a completely homogeneous microstructure, free from aggregates.

**Figure 8.** FESEM micrographs at (**a**) lower; (**b**) intermediate; and (**c**) higher magnification of ZA₈Sr₈-Ce11 composite sintered at 1450 °C for 1 h.

The higher magnification image (b) shows that all the desired microstructural features were obtained in the composite, since both rounded grains and elongated ones can be observed. Image analysis revealed an average size of 0.6 ± 0.3 μm and 0.3 ± 0.1 μm for Ce-TZP and alumina phases, respectively. The elongated (strontium aluminate) grains were characterized by a mean length of 0.6 ± 0.2 μm, and aspect ratio of 4 ± 2. The highest magnification image (c) shows that, in spite the predominant location of the second grains in intergranular position, some intra-type alumina grains can be also observed. The ceria content had no effects on the microstructure, since very similar microstructural and morphological features were observed in all composites.

Contrarily, the ceria amount played a major role on other physical and mechanical properties of the materials. In fact, by XRD analyses carried out on the surface of all sintered materials, we can observe a progressive decrease of the m-ZrO₂ phase by increasing the ceria amount (Figure 9).

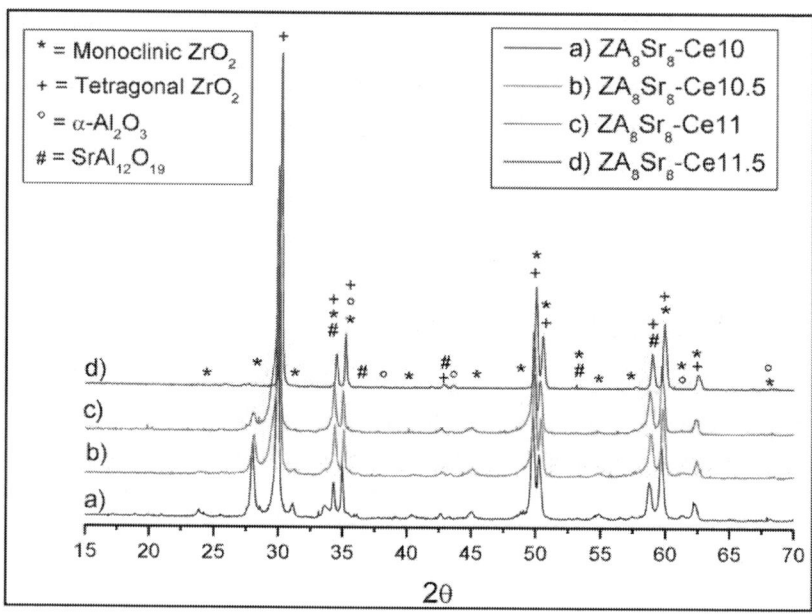

**Figure 9.** XRD patterns of $ZA_8Sr_8$-CeX composites sintered at 1450 °C for 1 h.

The monoclinic volume fraction ($V_m$) was calculated using the method proposed by Garvie and Nicholson [90], by which we estimated a progressive decreasing of $V_m$ from 5 to 1, by increasing the ceria content.

Accordingly, the stress-induced $t$-$m$ transformation of the sintered composites appeared to be clearly dependent on the ceria amount. This behavior was evaluated by performing Vickers indentations on the polished, sintered surfaces, at different loads (5, 10 and 30 kg$_f$). Then, the size of the transformed area around each indentation was measured by means of optical microscopy, as reported in Figure 10. As expected, for all the samples, the transformed area increased by increasing the applied load. In addition, moving from the sample with the lowest ceria content ($ZA_8Sr_8$-Ce10) to the highest amount ($ZA_8Sr_8$-Ce11.5), the transformed area decreased. In fact, the higher zirconia stabilization degree in this last sample accounts for its lower transformability.

In the same figure, the optical micrographs of $ZA_8Sr_8$-Ce10 and $ZA_8Sr_8$-Ce11.5 indented samples are shown. The brighter zone around the indentation is associated to the zone in which the tetragonal to monoclinic transformation took place, providing a volume increase. In addition, such images show a different morphology of the transformed zone, depending on the ceria content. In fact, smaller round-shaped transformed areas were observed in $ZA_8Sr_8$-Ce11.5 sample, while large number of deformation branches, which radially propagated from the indentation area were observed in $ZA_8Sr_8$-Ce10 material. This last morphology evidences the autocatalytic nature of the $t$-$m$ phase transformation of zirconia, meaning that once initiated, it can stimulate further transformation to propagate rapidly over an extended region.

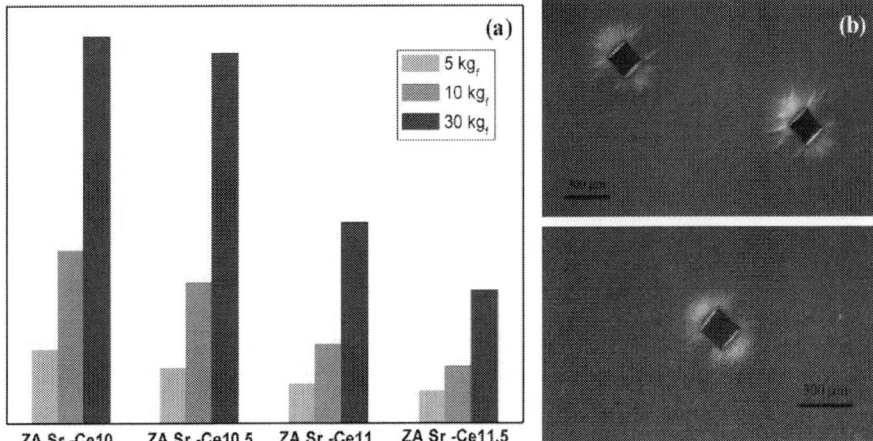

**Figure 10. (a)** Transformed area, for the four sintered composites, after indentation at different loads; **(b)** optical micrographs (with Nomarski contrast) of the transformed area of $ZA_8Sr_8$-Ce10 **(top)** and $ZA_8Sr_8$-Ce11.5 **(bottom)** samples.

These results have a key importance because they clearly demonstrate that the surface coating route allows an effective tuning of the ceria amount inside the zirconia lattice, where very small increases in the ceria content (of 0.5 mol%) can give rise to completely different behaviors.

A paper reporting the overall mechanical properties (flexural strength, fracture toughness, Vickers hardness) and ageing behavior of these composites is currently in progress. The here reported aims to highlight the effectiveness of the surface coating route in the elaboration of composite materials with features corresponding to the designed ones. The process here presented allows in fact the production of complex systems, leading to the fine and simultaneous control of many parameters, such as the cerium oxide stabilizer content in the zirconia lattice, the chemical compositions of the phases, their morphology, the microstructure, the final properties. For this reason, this process represents a significant advance over the existing technologies, being hopefully the next steps in its further development and scaling-up even at the industrial and clinical scale.

## 5. CONCLUSIONS

Since the early hip replacements in the 1960s, there is a continuous search for biomedical ceramic implants characterized by increasingly higher mechanical properties, reliability and *in vivo* stability.

This has allowed a constant spread of new materials, particularly focusing on composite and nanocomposite structural ceramics, where the design of new compositions and architectures seems the key to further enhance the material characteristics. The challenge is tailoring the micro- and nanofeatures in the composite structures, through a careful control of any step of manufacturing, being the synthesis of the composite powders the first and fundamental step of this processing chain.

In this frame, we have presented here an innovative elaboration method, named "surface coating route", to process ceramic nanocomposite powders, developed in the recent years, but already demonstrating its feasibility to develop complex structures suitable for biomedical applications.

The method is illustrated into two case studies, the former concerning ZTA materials for orthopedic applications, the latter zirconia-based composites for dental implants. The two examples well prove the capability of the process to simultaneously and perfectly tailor all the compositional and microstructural features in the composite structures, leading to the fine control of the oxide stabilizer content in the zirconia lattice, the chemical composition of the phases, their size and morphology, and the overall microstructure.

Such careful tailoring of the microstructural features allows yielding good and constant mechanical properties inside the biomedical components, showing that the surface coating method can well integrate in the industrial manufacturing protocols of biomedical devices, potentially becoming an evaluable alternative to the current synthesis process already used at a clinical and industrial scale.

## ACKNOWLEDGMENTS

Part of the research leading to these results has been performed within the LONGLIFE project (http://longlife-project.eu) and received funding from the European Community's Seventh Framework Programme (FP7/2007-2013) under grant agreement No. 280741.

## AUTHOR CONTRIBUTIONS

The manuscript has been edited by Paola Palmero thanks to the contribution of the other authors. The experimental results here reported are the output of the scientific collaboration between the authors, particularly in the frame of two European projects (IP Nanoker and Longlife) in which all the authors have been involved. Laura Montanaro and Jérôme Chevalier have been mainly involved in the design of the materials, Paola Palmero in the development of the composite powders and sintered materials and Helen Reveron in their physical and mechanical characterization. All authors have equally contributed to analyse and discuss the results.

## REFERENCES

1. Hench, L.L. Bioceramics. *J. Am. Ceram. Soc.* **1998**, *81*, 1705–1728.
2. Rahaman, M.N.; Yao, A.; Bal, B.S.; Garino, J.P.; Ries, M.D. Ceramics for prosthetic hip and knee joint replacement. *J. Am. Ceram. Soc.* **2007**, *90*, 1965–1988.
3. Willmann, G. Ceramic femoral head retrieval data. *Clin. Orthop. Relat. R.* **2000**, *379*, 22–28.

4. Willmann, G. Development in medical-grade alumina during the past two decades. *J. Mater. Process. Technol.* **1996**, *56*, 168–176

5. Piconi, C.; Condo, S.C.; Kosmač, T. Chapter 11. Alumina and zirconia-based ceramics for load-bearing applications. In*Advanced Ceramic for Dentistry*; Shen, J., Kosmač, T., Eds.; Butterworth-Heinemann (Elsevier): Wathmann, MA, USA, 2014; pp. 219–253.

6. Rieger, W. Ceramics in orthopedics. 30 years of evolution and experience. In *World Tribology Forum in Arthoplasty*; Rieker, C., Oberjolzer, S., Wyss, U., Eds.; Verlag Hans Huber: Bern, Germany, 2001; pp. 1–13.

7. Piconi, C.; Maccauro, G. Zirconia as a ceramic biomaterial. *Biomaterials* **1999**, *20*, 1–25

8. Marshall, D.B.; Ritter, J.E. Reliability of advanced structural ceramics and ceramic matrix composites. A review.*Ceram. Bull.* **1987**, *66*, 309–317.

9. Chevalier, J.; Gremillard, L.; Virkar, A.V.; Clarke, D.R. The Tetragonal-monoclinic transformation in zirconia: Lesson learned and future trends. *J. Am. Ceram. Soc.* **2009**, *92*, 1901–1920.

10. Lawson, S. Environmental degradation of zirconia ceramics. *J. Eur. Ceram. Soc.* **1995**, *15*, 485–502.

11. Lughi, V.; Sergo, V. Low temperature degradation—aging—of zirconia. A critical review of the relevant aspects in dentistry. *Dent. Mater.* **2010**, *26*, 807–820.

12. Chevalier, J.; Gremillard, L. Ceramics for medical applications: A picture for the next 20 years. *J. Eur. Ceram. Soc.* **2009**,*29*, 1245–1255

13. Schubert, H.; Frey, F. Stability of Y-TZP during hydrothermal treatment: Neutron experiments and stability considerations. *J. Eur. Ceram. Soc.* **2005**, *25*, 1597–1602.

14. Lange, F.F.; Dunlop, G.L.; Davis, B.L. Degradation during aging of transformation-toughened $ZrO_2$-$Y_2O_3$ materials at 250 °C. *J. Am. Ceram. Soc.* **1986**, *69*, 237–240

15. Chevalier, J.; Cales, B.; Drouin, J.M. Low-temperature ageing of Y-TZP ceramics. *J. Am. Ceram. Soc.* **1999**, *82*, 2150–2154

16. Chevalier, J. What future for zirconia as a biomaterial? *Biomaterials* **2006**, *27*, 535–543.

17. Chevalier, J.; Gremillard, L. Zirconia as a biomaterial. In *Comprehensive Biomaterials*; Ducheyne, P., Ed.; Elsevier: Amsterdam, The Netherlands, 2011; volume 1, pp. 95–108.

18. Guazzato, M.; Albakry, M.; Ringer, S.P.; Swain, M.V. Strength, fracture toughness and microstructure of a selection of all-ceramic materials. Part II. Zirconia-based dental ceramics. *Dent. Mater.* **2004**, *20*, 449–456.

19. Denry, I.; Holloway, J.A. Ceramics for dental applications: A review. *Materials* **2010**, *3*, 351–368.

20. Rieger, W.; Kobel, S.; Weber, W. Processing and Properties of Zirconia Ceramics for Dental Applications. Spectrum Dialogue, Reprint, March 2008. Available online: http://www.lastruttura.it/Contents/Documents/ 3%20Premium%20HT/Letteratura%20zirconia%20dentale.pdf (accessed on 26 June 2014).

21. Cattani-Lorente, M.; Scherrer, S.S.; Ammann, P.; Jobin, M.; Wiskott, A. Low temperature degradation of a Y-TZP dental ceramic. *Acta Biomater.* **2011**, *7*, 858–865.

22. Kohorst, P.; Borchers, L.; Strempel, J.; Stiesch, M.; Hassel, T. Low-temperature degradation of different zirconia ceramics for dental applications. *Acta Biomater.* **2012**, *8*, 1213–1220.

23. Denry, I.; Kelly, J.R. State of art of zirconia for dental applications. *Dent. Mater.* **2008**, *24*, 299–307.

24. Kurtz, S.M.; Kocagöz, S.; Arnholt, C.; Huet, R.; Ueno, M.; Walter, W.L. Advances in zirconia toughened alumina biomaterials for total joint replacement. *J. Mech. Behav. Biomed.* **2014**, *31*, 107–116.

25. Evans, A.G.; Burlingame, N.; Drory, M.; Kriven, W.M. Martensitic transformations in zirconia—Particle size effects and toughening. *Acta Metall.* **1981**, *29*, 447–456.

26. Heuer, A.H.; Claussen, N.; Kriven, W.M.; Rühle, M. Stability of tetragonal $ZrO_2$ particles in ceramic matrices. *J. Am. Ceram. Soc.* **1982**, *65*, 642–650.

27. Pecharroman, C.; Bartolomé, J.F.; Requena, J.; Moya, J.S.; Deville, S.; Chevalier, J.; Fantozzi, G.; Torrecillas, R. Percolative mechanism of ageing in zirconia containing ceramics for medical applications. *Adv. Mater.* **2003**, *15*, 507–511.

28. Stewart, T.D.; Tipper, J.L.; Insley, G.; Streicher, R.M.; Ingham, E.; Fisher, J. Long-term wear of ceramic matrix composite materials for hip prostheses under severe swing phase microseparation. *J. Biomed. Mater. Res. B.* **2003**, *66B*, 567–573.

29. Guo, R.; Guo, D.; Chen, Y.; Yang, Z.; Yuan, Q. *In situ* formation of $LaAl_{11}O_{18}$ rodlike particles in ZTA ceramics and effect on the mechanical properties. *Ceram. Int.* **2001**, *28*, 699–704.

30. Liu, X.Q.; Chen, X.M. Effects of $Sr_2Nb_2O_7$ additive on microstructure and mechanical properties of $3Y$-$TZP/Al_2O_3$ceramics. *Ceram. Int.* **2002**, *28*, 209–215.

31. Sktania, Z.D.I.; Azharb, A.Z.A.; Ratnamc, M.M.; Ahmada, Z.A. The influence of *in-situ* formation of hibonite on the properties of zirconia toughened alumina (ZTA) composites. *Ceram. Int.* **2014**, *40*, 6211–6217.

32. Jin, X.; Gao, L. Effects of powder preparation method on the microstructure and mechanical performance of ZTA/LaAl$_{11}$O$_{18}$ composites. *J. Eur. Ceram. Soc.* **2004**, *24*, 653–659.

33. Chen, Z.; Chawla, K.K.; Koopman, M. Microstructure and mechanical properties of *in situ* synthesized alumina/Ba-β-alumina/zirconia composites. *Mater. Sci. Eng. A* **2004**, *367*, 24–32.

34. Sato, T.; Endo, T.; Shimada, M. Postsintering hot isostatic pressing of ceria-doped tetragonal zirconia/alumina composites in an argon oxygen gas atmosphere. *J. Am. Ceram. Soc.* **1989**, *72*, 761–764.

35. Nawa, M. Tough and strong Ce-TZP/alumina nanocomposites doped with titania. *Ceram. Int.* **1998**, *24*, 497–506.

36. Nawa, M. The effect of TiO$_2$ addition on strength and toughening in intragranular type 12Ce-TZP/Al$_2$O$_3$nanocomposites. *J. Eur. Ceram. Soc.* **1998**, *18*, 209–219.

37. Panasonic Biomedical. Available online: http://www.die-modellmacher.de/wp-content/uploads/Nano-ZR-brochure.pdf (accessed on 26 June 2014).

38. Apel, E.; Ritzberger, C.; Courtois, N.; Reveron, H.; Chevalier, J.; Schweiger, M.; Rothbrust, F.; Rheinberger, V.M.; Höland, W. Introduction to a tough, strong and stable Ce-TZP/MgAl$_2$O$_4$ composite for biomedical applications. *J. Eur. Ceram. Soc.* **2012**, *32*, 2697–2703.

39. Maschio, S.; Pezzotti, G.; Sbaizero, O. Effect of LaNbO$_4$ addition on the mechanical properties of ceria-tetragonal zirconia polycrystal matrices. *J. Eur. Ceram. Soc.* **1998**, *18*, 1779–1785.

40. Cutler, R.A.; Lindemann, J.M.; Ulvensoen, J.H.; Lange, H.I. Damage-resistant SrO-doped Ce-TZP/Al$_2$O$_3$ composites.*Mater. Des.* **1994**, *15*, 123–133.

41. Ori, S.; Kojima, T.; Hara, T.; Uekawa, N.; Kakegawa, K. Fabrication of Ce-TZP/β-hexaaluminate composites using amorphous precursor of the second phase. *J. Ceram. Soc. Jpn.* **2012**, *120*, 111–115.

42. Miura, M.; Hongoh, H.; Yogo, T.; Hirano, S.; Fujll, T. Formation of plate-like lanthanum-β-aluminate crystal in Ce-TZP matrix. *J. Mater. Sci.* **1994**, *29*, 262–268.

43. Kern, F. A Comparison of microstructure and mechanical properties of 12Ce-TZP reinforced with alumina and *in situ*formed strontium- or lanthanum-hexaaluminate precipitates. *J. Eur. Ceram. Soc.* **2014**, *34*, 413–423.

44. Yamaguchi, T.; Sakamoto, W.; Yogo, T.; Fujii, T.; Hirano, S. *In situ* formation of Ce-TZP/Ba-hexaaluminate composites.*J. Ceram. Soc. Jpn.* **1999**, *107*, 814–916.

45. Tsai, O.F.; Chon, U.; Ramachandran, N.; Shetty, D.K. Transformation plasticity and toughening in $CeO_2$-partially-stabilized zirconia–alumina (Ce-TZP/$Al_2O_3$) composites doped with MnO. *J. Am. Ceram. Soc.* **1992**, *75*, 1229–1238.

46. De Aza, A.H.; Chevalier, J.; Fantozzi, G.; Schehl, M.; Torrecillas, R. Crack growth resistance of alumina, zirconia and zirconia toughened alumina ceramics for joint prostheses. *Biomaterials* **2002**, *23*, 937–945.

47. Chevalier, J.; Taddei, P.; Gremillard, L.; Deville, S.; Fantozzi, G.; Bartolomé, J.F.; Pecharroman, C.; Moya, J.S.; Diaz, L.A.; Torrecillas, R.; *et al.* Reliability assessment in advanced nanocomposite materials for orthopaedic applications. *J. Mech. Behav. Biomed.* **2011**, *4*, 303–314.

48. Gutknecht, D.; Chevalier, J.; Garnier, V.; Fantozzi, G. Key role of processing to avoid low temperature ageing in alumina zirconia composites for orthopaedic application. *J. Eur. Ceram. Soc.* **2007**, *27*, 1547–1552.

49. Deville, S.; Chevalier, J.; Fantozzi, G.; Bartolomé, J.F.; Requena, J.; Moya, J.S.; Torrecillas, R.; Díaz, L.C. Low-temperature ageing of zirconia-toughened alumina ceramics and its implication in biomedical Implants. *J. Eur. Ceram. Soc.* **2003**, *23*, 2975–2982.

50. Claussen, N. Fracture toughness of $Al_2O_3$ with an unstabilized $ZrO_2$ dispersed phase. *J. Am. Ceram. Soc.* **1976**, *59*, 49–51.

51. Lange, F.F. Transformation toughening, Part 4: Fabrication, fracture toughness and strength of $Al_2O_3$-$ZrO_2$ composites. *J. Mater. Sci.* **1982**, *17*, 247–254.

52. Szutkowska, M. Fracture resistance behaviour of alumina-zirconia composites. *J. Mater. Process. Technol.* **2004**, *153–154*, 868–874.

53. Shin, Y.S.; Rhee, Y.W.; Kang, S.J. Experimental evaluation of toughening mechanism in alumina-zirconia composites. *J. Am. Ceram. Soc.* **1999**, *82*, 1229–1232.

54. Thuan, W.H.; Chen, Z.; Wang, T.C.; Cheng, C.H.; Kuo, P.S. Mechanical properties of $Al_2O_3$/$ZrO_2$ composites. *J. Eur. Ceram. Soc.* **2002**, *22*, 2827–2833.

55. Schehl, M.; Diaz, J.A.; Torrecillas, R. Alumina nanocomposites form powder-alkoxide mixtures. *Acta Mater.* **2002**, *50*, 1125–1139.

56. Sarkar, D.; Mohapatra, D.; Ray, S.; Bhattacharyya, S.; Adak, S.; Mitra, N. Nanostructured $Al_2O_3$/$ZrO_2$ composite synthesized by sol-gel technique: Powder processing and microstructure. *J. Mater. Sci.* **2007**, *42*, 1847–1855.

57. Jayaseelan, D.D.; Rani, D.A.; Nishikawa, T.; Awaji, H.; Gnanam, F.D. Powder characteristics, sintering behavior and microstructure of sol-gel derived ZTA composites. *J. Eur. Ceram. Soc.* **2000**, *20*, 267–275.

58. Lee, B.T.; Han, J.K.; Saito, F. Microstructure of sol-gel synthesized $Al_2O_3$-$ZrO_2(Y_2O_3)$ nano-composites studied by transmission electron microscopy. *Mater. Lett.* **2005**, *59*, 355–360.

59. Naga, S.M.; Abdelbary, E.M.; Awaad, M.; El-Shaer, Y.I.; Abd-Elwaha, H.S. Effect of the preparation route on the mechanical properties of yttria–ceria doped tetragonal zirconia/alumina composites. *Ceram. Int.* **2013**, *39*, 1835–1840.

60. Boulle, A.; Oudjedi, Z.; Guinebretière, R.; Soulestine, B.; Dauger, A. Ceramic nanocomposites obtained by sol-gel coating of submicron powders. *Acta Mater.* **2001**, *49*, 811–816.

61. Matsumoto, Y.; Hirota, K.; Yamaguchi, O.; Inamura, S.; Miyamoto, H.; Shiokawa, N.; Tsuji, K. Mechanical properties of isostatically pressed zirconia-toughened alumina ceramics prepared from coprecipitated powders. *J. Am. Ceram. Soc.* **1993**, *76*, 2677–2680.

62. Rana, R.P.; Pratihar, S.K.; Bhattacharyya, S. Powder processing and densification behaviour of alumina–high zirconia nanocomposites using chloride precursors. *J. Mater. Sci. Process.* **2007**, *190*, 350–357.

63. Prete, F.; Rizzuti, A.; Esposito, L.; Tucci, A.; Leonelli, C. Highly homogeneous $Al_2O_3$–$ZrO_2$ nanopowder via microwave-assisted hydro- and solvothermal synthesis. *J. Am. Ceram. Soc.* **2011**, *94*, 3587–3590.

64. Freim, J.; McKittrick, J. Modeling and fabrication of fine-grain alumina–zirconia composites produced from nanocrystalline precursors. *J. Am. Ceram. Soc.* **1998**, *81*, 1773–1780.

65. Claussen, N.; Ruhle, M. Design of transformation-toughened ceramics. In *Advances in Ceramics, Vol. 3, Science and Technology of Zirconia*; Heuer, A.H., Hobbs, W.L., Eds.; The American Ceramics Society: Columbus, OH, USA, 1981; pp. 137–163.

66. Tsukuma, K.; Shimada, M. Strength, fracture toughness and vickers hardness of $CeO_2$-stabilized tetragonal $ZrO_2$polycrystals (Ce-TZP). *J. Mater. Sci.* **1985**, *20*, 1178–1184.

67. Groza, J.R. Nanosintering. *Nanostruct. Mater.* **1999**, *12*, 987–992.

68. Groza, J.R.; Jhwding, R.J. Nanoparticulate materials densification. *Nanostruct. Mater.* **1996**, *7*, 749–768.

69. Wang, J.; Stevens, R. Review. Zirconia-toughened alumina (ZTA) ceramics. *J. Mater. Sci.* **1989**, *24*, 3421–3440.

70. Lee, M.-H.; Tai, C.Y.; Lu, C.-H. Synthesis of spherical zirconia by precipitation between two water/oil emulsions. *J. Eur. Ceram. Soc.* **1999**, *19*, 2593–2603.

71. Hu, M.Z.-C.; Hunt, R.D.; Payzant, E.A.; Hubbard, C.R. Nanocrystallization and phase transformation in monodispersed ultrafine zirconia particles from various homogeneous precipitation methods. *J. Am. Ceram. Soc.* **1999**,*82*, 2313–2320.

72. Upadhyaya, D.D.; Ghosh, A.; Gurumurthy, K.R.; Prasad, R. Microwave sintering of cubic zirconia. *Ceram. Int.* **2001**,*27*, 415–418

73. Chen, I.W.; Chiao, Y.H. Martensitic transformation in $ZrO_2$ and $HfO_2$— An assessment of small-particles experiments with metals and ceramic matrices. In *Advances in Ceramics*; Heuer, A.H., Hobbs, W.L., Eds.; The American Ceramics Society: Columbus, OH, USA, 1984; volume 12, pp. 23–40.

74. Yuan, Z.; Vleugels, J.; van Der Biest, O. Synthesis and characterization of $CeO_2$-coated $ZrO_2$ powder-based TZP.*Mater. Lett.* **2000**, *46*, 249–254.

75. Palmero, P.; Naglieri, V.; Chevalier, J.; Fantozzi, G.; Montanaro, L. Alumina-based nanocomposites obtained by doping with inorganic salt solutions: Application to immiscible and reactive systems. *J. Eur. Ceram. Soc.* **2009**, *29*, 59–66.

76. Naglieri, V.; Palmero, P.; Montanaro, L. Preparation and characterization of alumina-doped powders for the design of multi-phasic nano-microcomposites. *J. Therm. Anal. Calorim.* **2009**, *97*, 231–237.

77. Kern, F.; Palmero, P. Microstructure and mechanical properties of alumina 5 vol% zirconia nanocomposites prepared by powder coating and powder mixing routes. *Ceram. Int.* **2013**, *39*, 673–682.

78. Naglieri, V.; Palmero, P.; Montanaro, L.; Chevalier, J. Elaboration of alumina-zirconia composites: Role of the zirconia content on the microstructure and mechanical properties. *Materials* **2013**, *6*, 2090–2102.

79. Palmero, P.; Esnouf, C. Phase and microstructural evolution of yttrium-doped nanocrystalline alumina: A contribution of advanced microscopy techniques. *J. Eur. Ceram. Soc.* **2011**, *31*, 507–516.

80. Palmero, P.; Naglieri, V.; Spina, G.; Lombardi, M. Microstructural design and elaboration of multiphase ultra-fine ceramics. *Ceram. Int.* **2011**, *37*, 139–144.

81. Palmero, P.; Sola, A.; Naglieri, V.; Bellucci, D.; Lombardi, M.; Cannillo, V. Elaboration and mechanical characterization of multi-phase alumina-based ultra-fine composites. *J. Mater. Sci.* **2012**, *47*, 1077–1084.

82. Gizowska, M.; Konopka, K.; Szafran, M. Properties of water-based slurries for fabrication of ceramic-metal composites by slip casting method. *Arch. Metall. Mater.* **2011**, *56*, 1105–1110.

83. Naglieri, V. Alumina-Zirconia Composites: Elaboration and Characterization in View of the Orthopaedic Applications. Ph.D. Thesis, Politecnico di Torino, Italy, INSA of Lyon, France, 11 February 2010.

84. Chuah, G.K.; Jaenicke, S.; Cheong, S.A.; Chan, K.S. The influence of preparation conditions on the surface area of zirconia. *Appl. Catal. A Gen.* **1996**, *145*, 267–284.

85. Naglieri, V.; Joly-Pottuz, L.; Chevalier, J.; Lombardi, M.; Montanaro, L. Follow-up of zirconia crystallization on a surface modified alumina powder. *J. Eur. Ceram. Soc.* **2010**, *30*, 3377–3387.

86. Naglieri, V.; Gutknecht, D.; Garnier, V.; Palmero, P.; Chevalier, J.; Montanaro, L. Optimized slurries for spray drying: Different approaches to obtain homogeneous and deformable alumina-zirconia granules. *Materials* **2013**, *3*, 5382–5397.

87. Greenwood, R.; Bergstriim, L. Electroacoustic and rheological properties of aqueous Ce-ZrO$_2$ (Ce-TZP) suspensions. *J. Eur. Ceram. Soc.* **1997**, *17*, 537–548.

88. Douy, A.; Capron, M. Crystallisation of spray-dried amorphous precursors in the SrO–Al$_2$O$_3$ system: A DSC study. *J. Eur. Ceram. Soc.* **2003**, *23*, 2075–2081.

89. Vishista, K.; Gnanam, F.D. Microstructural development of SrAl$_{12}$O$_{19}$ in alumina–strontia composites. *J. Eur. Ceram. Soc.* **2009**, *29*, 77–83.

90. Garvie, R.C.; Nicholson, P.S. Phase analysis in zirconia systems. *J. Am. Ceram. Soc.* **1972**, *55*, 303–305.

# CHAPTER 12

# New Coating Technique of Ceramic Implants with Different Glass Solder Matrices for Improved Osseointegration-Mechanical Investigations

*Enrico Mick [1,\*], Jana Markhoff [1], Aurica Mitrovic [2], Anika Jonitz [1] and Rainer Bader [1]*

[1] Department of Orthopaedics, Research Lab for Biomechanics and Implant Technology, University Medicine Rostock, Doberaner Strasse 142, Rostock 18057, Germany
[2] ZM Praezisionsdentaltechnik GmbH, Breite Strasse 16, Rostock 18055, Germany

## ABSTRACT

Ceramics are a very popular material in dental implant technology due to their tribological properties, their biocompatibility and their esthetic appearance. However, their natural surface structure lacks the ability of proper osseointegration, which constitutes a crucial process for the stability and, thus, the functionality of a bone implant. We investigated the application of a glass solder matrix in three configurations—consisting mainly of $SiO_2$, $Al_2O_3$, $K_2O$ and $Na_2O$ to TZP-A ceramic specimens. The corresponding adhesive strength and surface roughness of the coatings on ceramic specimens have been analyzed. Thereby, high adhesive strength ($70.3 \pm 7.9$ MPa) was found for the three different coatings. The obtained roughness ($R_z$) amounted to $18.24 \pm 2.48$ μm in average, with significant differences between the glass solder configurations. Furthermore, one configuration was also tested after additional etching which did not lead to significant increase of surface roughness ($19.37 \pm 1.04$ μm) or adhesive strength ($57.2 \pm 5.8$ MPa). In conclusion, coating with glass solder matrix seems to be a promising surface modification technique that may enable direct insertion of ceramic implants in dental and orthopaedic surgery.

**Keywords:** Surface modification; glass solder matrix; ceramic implant; adhesive strength; roughness

## 1. INTRODUCTION

Ceramics are frequently used materials in the field of total joint replacement [1]. Especially for bearing surfaces, several mixtures of alumina and zirconia materials have been established due to their excellent tribological properties [2]. Furthermore, the high biocompatibility leads to acceptance by the human body [3]. Therefore, only few and low inflammatory effects by ceramics have been reported [4]. However, this property also involves a severe disadvantage. Due to the minimized interaction with biological tissue, ceramic surfaces do not connect with bone cells properly but rather get encapsulated in fibrous tissue [5,6]. However, present clinical data for zirconia implants is not sufficient to recommend ceramics implants for routine clinical use [7].

Several approaches of surface modifications and coatings for dental and orthopaedic implants have been reported. While some studies focused on covering titanium base bodies with different peptides in order to stimulate bone formation [8,9,10,11], others investigated the effect of calcium phosphate (CaP) or titanium plasma-spray (TPS) coatings [12,13,14,15,16] on implant osseointegration. A further common technique is the treatment of titanium samples by blasting the surface with diverse grits and acid etching, independently or in combination [12,13,14,17,18]. Approaches to coat titanium with colloidal zircon oxide [19] and hardystonite [16] have also been performed.

Surface modifications of zirconia implants become more important as the suitability of ceramic materials in particular for dental applications come to larger awareness. Such approaches mainly include topographical adaptations via sandblasting and/or acid etching [14,20,21,22] or sintering with pore formers [23,24]. However, coating with hydroxyapatite (HA) [23] and calcium liberating titanium oxide ($TiO_2$) [25] have also been examined.

Coatings with bioactive glass have already been reported, yet, only on titanium dental implants [15]. The present study investigates modifications of the ceramic surface by means of a glass solder matrix that may allow sufficient osseointegration of ceramic implants in the bone stock or additional mechanically stable coating of the ceramic implants with bioactive or structured layers.

## 2. MATERIALS AND METHODS

### 2.1. Test Specimens

The ceramic specimens used for present investigation were manufactured by Metoxit AG (Thayngen, Switzerland) according to DIN EN 60267. The TZP-A ceramic (tetragonal zirconia polycrystal with alumina) consists of $ZrO_2$, $Y_2O_3$ and $Al_2O_3$ with contents of 95%, 5% and 0.25%, respectively. For mechanical testing discs with 10 mm in diameter and 5 mm in height were fabricated.

## 2.2. Glass Matrix—Mixture, Application and Modification

The examined coatings are glasses of silica based materials taken from the DCMhotbond® series [26]. They can be used for surface conditioning of mixed ceramics or pure zirconia and mainly contain $SiO_2$ (60%–70%), $Al_2O_3$ (4%–10%), $K_2O$ (6%–10%), $Na_2O$ (6%–10%). Different configurations with varying contents of each component were applied to the TZP-A discs: HT1, LT1 and LT2. While the first is burned at high temperatures, the two other ones are processed at lower temperatures. The layer thickness was exemplarily determined on partly coated specimens using a high resolution (5 µm) caliper gauge (1101-150, INSIZE Co., Ltd., Suzhou New District, China). The corresponding properties, *i.e.*, curing temperature, layer thickness as well as the grit size of the powdery base material are summarized in Table 1.

**Table 1.** Properties and parameters of the investigated glass solder coatings.

| Glass solder | Grit size [µm] | Curing temperature [°C] | Layer thickness [µm] |
|---|---|---|---|
| HT1 | 12.6 | 1035 | 30 |
| LT1 | 24 | 850 | 50 |
| LT2 | 6 | 800 | 20 |

Prior to coating the ceramic specimens are preconditioned via sandblasting (see Table 2) and subsequent evaporation for providing a greaseless surface.

The powdery base material of the glass solder is mixed with an alcoholic fluid. The resulting emulsion of milky appearance is slowly and evenly applied to the ceramic discs using an airbrush operated at 1–1.5 bar and a distance of 10 cm until the surface is entirely covered. The sprayed object is placed on a firing tray and put into an oven (Vario 200, Zubler Geraetebau GmbH, Ulm, Germany) where curing is processed. The temperature rises at a rate of 5–30 K/min up to constant values of 800 up to 1035 °C which are held for 1–3 min, depending on the actual glass solder configuration.

Afterwards, the extremely plain and even glass matrix surface is roughened via sandblasting (see Table 2) and cleaned in an ultrasonic bath with distilled water. For further roughening the specimens coated with HT1 are also treated additionally by resting for 20 min in an acid mixture containing about 7% of 41% hydrofluoric acid and about 10% of 96% sulfuric acid. Then, the samples are neutralized and finally cleaned in an ultrasonic bath with calcium hydroxide solution (lime milk) and distilled water, respectively.

The discs for the mechanical investigations were coated on both end faces.

**Table 2.** Parameters for sandblasting with corundum (Al$_2$O$_3$, grit size: 110 µm).

| Surface type | Jet pressure [bar] | Angle to surface [°] | Distance to surface [cm] |
|---|---|---|---|
| ceramic base body | 2 | 60–80 | 2–3 |
| glass ceramic coating | 1 | 60–80 | 2–3 |

## 2.3. Roughness

Prior to determining the adhesive strength of the coating, the roughness of the surfaces was recorded. For this purpose a profilometer (Hommel-Etamic T1000, Jenoptik AG, Jena, Germany) was used. The surfaces of the coated ceramic discs were tested performing line scans with a tactile length of 8 mm in three different orientations (0°, 60° and 120°). The parameters "mean roughness index" R$_a$ and "average surface roughness" R$_z$ determined for each of the orientations were averaged giving the value for each specimen represented. Coated/sandblasted (HT1, LT1 and LT2) and coated/sandblasted/etched (HT1) specimens as well as untreated and sandblasted TZP-A specimens and sandblasted titanium specimens were investigated.

An exemplary image of the surface topography after coating with glass solder was created with a scanning electron microscope DSM 960A (Carl Zeiss AG, Oberkochen, Germany).

## 2.4. Adhesive Strength

To evaluate the adhesive strength of the coatings on the ceramic body the TZP-A discs were connected to titanium (Ti6Al4V) cylinders with a diameter of 10 mm (see Figure 1). The cylinders were blasted with Al$_2$O$_3$ (EK 80) on the end faces. The surfaces were connected using a bonding agent (HTK Ultra Bond 100, HTK Hamburg GmbH, Hamburg, Germany) that cured for 50 min at 180 °C under mechanical pressure. The pull-off test was performed with a universal testing machine (Z050, Zwick GmbH & Co. KG, Ulm, Germany) at a crosshead speed of 5 mm/min. Maximal force was measured and related to the surface area giving the adhesive strength for HT1 (sandblasted and sandblasted/etched), LT1 and LT2 specimens.

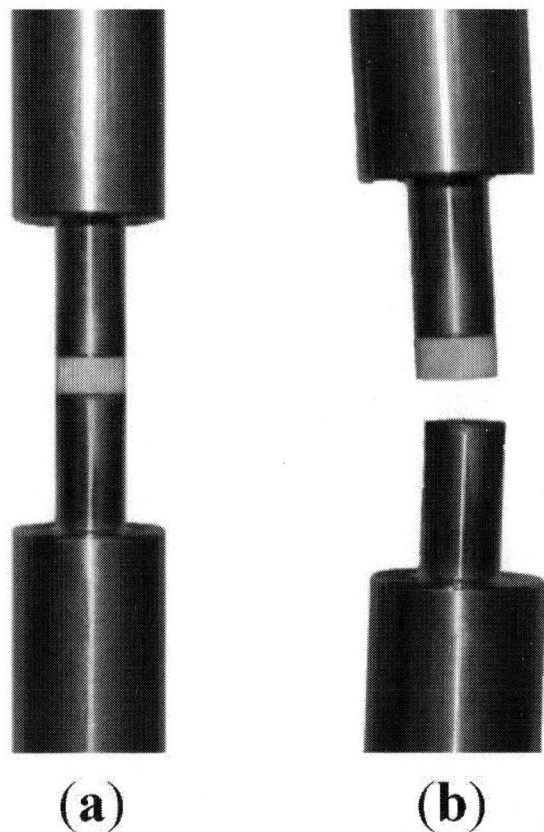

**(a)**              **(b)**

**Figure 1. (a)** Setup for adhesive strength test sample prior to testing; **(b)** Sample after testing.

## 2.5. X-ray Fluorescence Analysis

In order to make a statement on the qualitative adhesive strength, the titanium end faces needed to be inspected for residues of the surface coating. Therefore, X-ray fluorescence analysis was performed with a Niton® XL3t XRF analyzer (Thermo Fisher Scientific Inc., Munich, Germany). A circular area with a diameter of 8 mm was examined in a single measurement for each specimen. The values given by the analyzer were percentages [%] of the total amount of chemical elements detected on the investigated surface. Several elements that were contained in the coating material at higher ratios had to be ruled out: silicon Si (part of bonding agent), aluminum Al (part of titanium cylinders) and sodium Na (not detectable). In the end, the presence of potassium K on the titanium end face was found to be a suitable decisive criterion as it was contained neither in the titanium cylinders nor in the bonding agent nor in the ceramic base body itself but only in the coating material.

## 2.6. Statistical Analysis

Statistical analysis of results obtained from mechanical testing was performed using SPSS Statistics (v20, IBM Corp., Armonk, NY, USA). A one-way ANOVA and a post-hoc Bonferroni test with a significance level of $p = 0.05$ were conducted.

# 3. RESULTS AND DISCUSSION

## 3.1. Roughness

As a reference eight untreated and eight sandblasted TZP-A discs as well as the sandblasted (sb) end faces of eight titanium cylinders were investigated. Table 3 summarizes the corresponding roughness parameters. For untreated TZP-A a mean roughness index of $R_a = 0.20 \pm 0.03$ µm and an average surface roughness of $R_z = 1.57 \pm 0.16$ µm were determined, proving the smooth character of untreated ceramic surfaces. Sandblasting of the TZP-A specimens lead to roughness values of $R_a = 0.65 \pm 0.08$ µm and $R_z = 4.28 \pm 0.61$ µm. For titanium, the values were $R_a = 0.78 \pm 0.11$ µm and $R_z = 5.45 \pm 0.61$ µm, enabling osseointegration to a reasonable extent as shown by other authors before [12,13,18].

**Table 3.** Roughness parameters for reference samples (TZP-A and titanium).

| Sample type | No. of samples | Ø $R_a$ [µm] | Ø $R_z$ [µm] |
|---|---|---|---|
| TZP-A (untreated) | 8 | $0.20 \pm 0.02$ | $1.55 \pm 0.12$ |
| TZP-A (sb) | 8 | $0.65 \pm 0.08$ | $4.28 \pm 0.61$ |
| titanium (sb) | 8 | $0.78 \pm 0.11$ | $5.45 \pm 0.61$ |

Ceramic specimens coated with HT1 were investigated after sandblasting (sb) and after additional etching (sb/et). The roughness parameters were not influenced connotatively due to the process of etching as shown in Table 4. Neither the differences for $R_a$ nor for $R_z$ were significant ($p > 0.05$). This might indicate a certain resistance of the glass solder matrix against chemical treatment. Another explanation would be a balance between the roughening effect and the blunting of the topography generated by sandblasting.

**Table 4.** Roughness parameters for modified TZP-A ceramic surfaces.

| Sample type | No. of samples | Ø $R_a$ [µm] | Ø $R_z$ [µm] |
|---|---|---|---|
| HT1 (sb) | 6 | $3.61 \pm 0.23$ | $20.44 \pm 1.23$ |
| HT1 (sb/et) | 6 | $3.31 \pm 0.19$ | $19.37 \pm 1.04$ |
| LT1 (sb) | 6 | $2.90 \pm 0.48$ | $17.29 \pm 2.80$ |
| LT2 (sb) | 6 | $2.83 \pm 0.19$ | $16.99 \pm 1.35$ |

As described in Section 2.2, etching needs to be followed by neutralization and cleaning. Every work step during the manufacture of an implant bears a risk of failures and mistakes, and especially acid residues on implant surfaces may severely affect the biocompatibility. The etching treatment did not show a positive effect; therefore, the other surface modifications (LT1 and LT2) were only tested in a sandblasted condition. The corresponding roughness parameters are shown in Table 4. For the surface modifications LT1 and LT2, similar roughness values without significant differences were revealed. However, the roughness parameters of HT1 were significantly higher than those of LT1 and LT2. Generally, the roughness values of all glass solder coatings were significantly higher than those of untreated (factor 11 to 13 for $R_z$) and sandblasted TZP-A specimens (factor 4.0 to 4.8 for $R_z$) and those of sandblasted titanium specimens (factor 3.1 to 3.7 for $R_z$).

Figure 2 shows an exemplary scanning electron microscope (SEM) image of the surface topography of a TZP-A specimen coated with HT1 after sandblasting. The translucent appearance of each coating with glass solder matrix is illustrated inFigure 3.

**Figure 2.** SEM image (magnification: 200X) of a TZP-A specimen coated with glass solder matrix HT1.

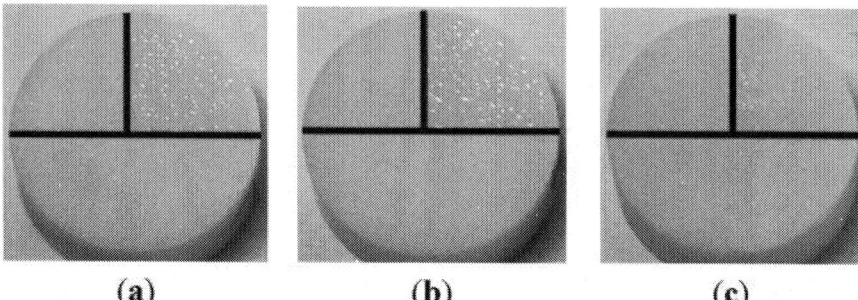

**(a)**                    **(b)**                    **(c)**

**Figure 3.** TZP-A specimens partly coated with glass solder: (**a**) HT1; (**b**) LT1; and (**c**) LT2 with each bottom half being uncoated, the right upper quarter being coated with raw glass solder and the left upper quarter being coated and sandblasted.

HT1 seems slightly more favorable than the others because it showed the highest average surface roughness as well as the lowest standard deviation with respect to $R_z$.

## 3.2. Adhesive Strength and X-ray Fluorescence Analysis

Adhesive strength tests were performed on the surface modifications HT1, LT1 and LT2 which were applied to the TZP-A discs and roughened via sandblasting thereafter. Again, the modification HT1 was also tested in an acid etched condition to investigate a possible influence of etching. The corresponding ultimate forces and stresses as well as the percentages of potassium on the titanium end faces are summarized in Table 5. Potassium and thus coating was released in every case. This is in accordance with the detected adhesive strength values, which are all lower than the tensile strength of the bonding agent (~100 MPa). However, the minimum adhesive strength as demanded by ASTM standard F-1147 (22 MPa) was reached in every case. No significant differences in adhesive strength could be observed ($p > 0.05$). Strength values obtained for HT1 with additional etching (57.2 ± 5.8 MPa) were slightly lower than those of HT1 in sandblasted condition (72.4 ± 11.8 MPa), which were the highest of all variations investigated. LT1 and LT2 showed a steady and in addition a similar behavior with LT1 having a slightly higher strength (71.3 ± 2.1 MPa). On the other hand, LT1 seems to be more brittle as quite large amounts of coating were released from the TZP-A base body (2.3% ± 1.3% *vs.* 0.6% ± 0.2%). However, the highest potassium rates were determined for both HT1 modifications (5.5% ± 2.9% (sb) and 3.9% ± 3.4% (sb/et)). Yet, none of the differences were significant ($p > 0.05$).

**Table 5.** Summary of results from adhesive strength testing and X-ray fluorescence analysis.

| Sample type | No. of samples | Ultimate force [N] | Adhesive strength [MPa] | K on Ti face [%] |
|---|---|---|---|---|
| HT1 (sb) | 3 | 5687 ± 928 | 72.4 ± 11.8 | 5.5 ± 2.9 |
| HT1 (sb/et) | 3 | 4491 ± 453 | 57.2 ± 5.8 | 3.9 ± 3.4 |
| LT1 (sb) | 3 | 5601 ± 167 | 71.3 ± 2.1 | 2.3 ± 1.3 |
| LT2 (sb) | 3 | 5284 ± 424 | 67.3 ± 5.4 | 0.6 ± 0.2 |

Highest values in surface roughness and adhesive strength were found for HT1. In consideration of the fact that all investigated coating configurations showed similar mechanical properties, only tendencies for a most suitable glass solder can be derived.

Despite the promising results the present study has still some limitations. First of all, the processes of applying the coating powder to the ceramic surface as well as all sandblasting operations were performed manually. In order to produce glass solder layers of consistent thickness and quality, the process should be established in an automated production chain. Furthermore, some issues still have to be examined, comprising the optimization of the application process and mechanical and cellbiological tests of the specimens. For instance, four-point-bending on coated ceramic rods has to be performed to determine a potential influence of the surface modification on the bending strength of ceramic base material. Also, cell proliferation on the coated surfaces has to be analyzed. Moreover, *in vivo* experiments investigating the integration of coated ceramic implants into the bone stock are required before clinical application of the new surface coating.

## 4. CONCLUSIONS

In the present study the mechanical properties of coatings for ceramic implants revealed promising results testing different glass solder matrix variations. Due to the investigated surface modifications roughness values could be achieved that were significantly higher than those of sandblasted TZP-A and sandblasted titanium which constitutes a kind of gold standard. Furthermore, it was found that all configurations possessed fairly sufficient adhesive strength indicating that the coating can resist the mechanical loads of dental and orthopaedic implants. Another finding was the lack of impact of additional etching on the surface topography and the adhesive strength. Therefore, this process was omitted. The three investigated configurations showed similar properties with the tendency of one variation being most suitable with respect to mechanical testing. In general, the coating with glass solder matrix constitutes an auspicious surface modification technique for enabling direct insertion of ceramic implants in dental and orthopaedic surgery.

## ACKNOWLEDGMENTS

The content of the proposed article was developed within the framework of a project funded by the German Ministry for Economy (BMWi) via Zentrales

Innovationsprogramm Mittelstand (ZIM). Furthermore, we kindly thank DOT GmbH (Rostock, Germany) for executing the X-ray fluorescence analyses and for conditioning the titanium cylinders for adhesive testing. We also thank Dental Creativ Management GmbH (Rostock, Germany) for supplying the glass solders, Metoxit AG (Thayngen, Switzerland) for supplying the ceramic specimens and the EMZ, University Medicine Rostock for enabling SEM analysis.

## CONFLICTS OF INTEREST

The authors E. Mick, J. Markhoff, A. Jonitz and R. Bader declare no conflict of interest. A. Mitrovic is managing partner of ZM Praezisionsdentaltechnik GmbH.

## REFERENCES

1.  Zietz, C.; Kluess, D.; Bergschmidt, P.; Haenle, M.; Mittelmeier, W.; Bader, R. Tribological aspects of ceramics in total hip and knee arthroplasty. *Semin. Arthroplast.* **2011**, *22*, 258–263.

2.  Milosev, I.; Kovac, S.; Trebse, R.; Levasic, V.; Pisot, V. Comparison of ten-year survivorship of hip prostheses with use of conventional polyethylene, metal-on-metal, or ceramic-on-ceramic bearings. *J. Bone Joint Surg. Am.* **2012**, *94*, 1756–1763.

3.  Bader, R.; Bergschmidt, P.; Fritsche, A.; Ansorge, S.; Thomas, P.; Mittelmeier, W. Alternative materials and solutions in total knee arthroplasty for patients with metal allergy. *Orthopade* **2008**, *37*, 136–142.

4.  Gallo, J.; Goodman, S.B.; Lostak, J.; Janout, M. Advantages and disadvantages of ceramic on ceramic total hip arthroplasty: A review. *Biomed. Pap. Med. Fac. Univ. Palacky Olomouc Czech. Repub.* **2012**, *156*, 204–212.

5.  Willmann, G. Keramische pfannen fuer hueftendoprothesen Teil 3: Zum problem der osteointegration monolithischer pfannen. *Biomed. Tech.* **1997**, *42*, 256–263.

6.  Anderson, J.M. Biological responses to materials. *Annu. Rev. Mater. Res.* **2001**, *31*, 81–110.

7.  Silva, N.; Sailer, I.; Zhang, Y.; Coelho, P.G.; Guess, P.C.; Zembic, A.; Kohal, R.J. Review: Performance of zirconia for dental healthcare. *Materials* **2010**, *3*, 863–896.

8.  Ferris, D.M.; Moodie, G.D.; Dimond, P.M.; Gioranni, C.W.; Ehrlich, M.G.; Valentini, R.F. RGD-Coated titanium implants stimulate increased bone formation *in vivo*. *Biomaterials* **1999**, *20*, 2323–2331.

9. Bernhardt, R.; van den Dolder, J.; Bierbaum, S.; Beutner, R.; Scharnweber, D.; Jansen, J.; Beckmann, F.; Worch, H. Osteoconductive modifications of Ti-implants in a goat defect model: Characterization of bone growth with SR μCT and histology. *Biomaterials* **2005**, *26*, 3009–3019. Elmengaard, B.; Bechtold, J.E.; Soballe, K. *In vivo* effects of RGD-coated titanium implants inserted in two bone-gap models. *J. Biomed. Mater. Res.* **2005**, *75*, 249–255.

10. Reyes, C.D.; Petrie, T.A.; Burns, K.L.; Schwartz, Z.; Garcia, A.J. Biomolecular surface coating to enhance orthopaedic tissue healing and integration. *Biomaterials* **2007**, *28*, 3228–3235.

11. Buser, D.; Schenk, R.K.; Steinemann, S.; Fiorellini, J.P.; Fox, C.H.; Stich, H. Influence of surface characteristics on bone integration of titanium implants. A histometric study in miniature pigs. *J. Biomed. Mater. Res.* **1991**, *25*, 889–902.

12. Martin, J.Y.; Schwartz, Z.; Hummert, T.W.; Schraub, D.M.; Simpson, J.; Lankford, J.; Dean, D.D.; Cochran, D.L.; Boyan, B.D. Effect of titanium surface roughness on proliferation, differentiation, and protein synthesis of human osteoblast-like cells (MG63). *J. Biomed. Mater. Res.* **1995**, *29*, 389–401.

13. Ferguson, S.J.; Langhoff, J.D.; Voelter, K.; von Rechenberg, B.; Scharnweber, D.; Bierbaum, S.; Schnabelrauch, M.; Kautz, A.R.; Frauchiger, V.M.; Mueller, T.L.; *et al.* Biomechanical comparison of different surface modifications for dental implants. *Int. J. Oral Maxillofac. Implant.* **2008**, *23*, 1037–1046.

14. Mistry, S.; Kundu, D.; Datta, S.; Basu, D. Comparison of bioactive glass coated and hydroxyapatite coated titanium dental implants in the human jaw bone. *Aust. Dent. J.* **2011**, *56*, 68–75.

15. Zhang, W.; Wang, G.; Liu, Y.; Zhao, X.; Zou, D.; Zhu, C.; Jin, Y.; Huang, Q.; Sun, J.; Liu, X.; *et al.* The synergistic effect of hierarchical micro/nano-topography and bioactive ions for enhanced osseointegration. *Biomaterials* **2013**, *34*, 3184–3195.

16. Coelho, P.G.; Bonfante, E.A.; Pessoa, R.S.; Marin, C.; Granato, R.; Giro, G.; Witek, L.; Suzuki, M. Characterization of five different implant surfaces and their effect on osseointegration: A study in dogs. *J. Periodontol.* **2011**, *82*, 742–750.

17. Herrero-Climent, M.; Lazaro, P.; Vicente Rios, J.; Lluch, S.; Marques, M.; Guillem-Marti, J.; Gil, F.J. Influence of acid-etching after grit-blasted on osseointegration of titanium dental implants: *In vitro* and *in vivo* studies. *J. Mater. Sci. Mater. Med.* **2013**, *24*, 2047–2055.

18. Sollazzo, V.; Pezzetti, F.; Scarano, A.; Piattelli, A.; Bignozzi, C.A.; Massari, L.; Brunelli, G.; Carinci, F. Zirconium oxide coating improves implant osseointegration *in vivo*. *Dent. Mater.* **2008**, *24*, 357–361.

19. Noro, A.; Kaneko, M.; Murata, I.; Yoshinari, M. Influence of surface topography and surface physicochemistry on wettability of zirconia (tetragonal zirconia polycrystal). *J. Biomed. Mater. Res. Appl. Biomater.* **2013**, *101*, 355–363.

20. Gahlert, M.; Roehling, S.; Wieland, M.; Sprecher, C.M.; Kniha, H.; Milz, S. Osseointegration of zirconia and titanium dental implants: A histological and histomorphometrical study in the maxilla of pigs. *Clin. Oral Implant. Res.* **2009**, *20*, 1247–1253.

21. Gahlert, M.; Roehling, S.; Wieland, M.; Eichhorn, S.; Kuechenhoff, H.; Kniha, H. A comparison study of the osseointegration of zirconia and titanium dental implants. A biomechanical evaluation in the maxilla of pigs. *Clin. Implant. Dent. Relat. Res.* **2010**, *12*, 297–305.

22. Rocchietta, I.; Fontana, F.; Addis, A.; Schupbach, P.; Simion, M. Surface-modified zirconia implants: Tissue response in rabbits. *Clin. Oral Implant. Res.* **2009**, *20*, 844–850.

23. Sennerby, L.; Dasmah, A.; Larsson, B.; Iverhed, M. Bone tissue responses to surface-modified zirconia implants: A histomorphometric and removal torque study in the rabbit. *Clin. Implant. Dent. Relat. Res.* **2005**, *7*, S13–S20.

24. Koch, F.P.; Weng, D.; Kraemer, S.; Biesterfeld, S.; Jahn-Eimermacher, A.; Wagner, W. Osseointegration of one-piece zirconia implants compared with a titanium implant of identical design: A histomorphometric study in the dog. *Clin. Oral Implant. Res.* **2010**, *21*, 350–356.

25. Hopp, M.; Zothner, A. Verfahren zur Konditionierung der Oberflächen von Dentalkomponenten und Verwendung des Verfahrens [in German]. Patent DE102009051655B3, 30 December 2010.

# Index